International Conference on Geosciences and Geotechnology (ICOGEST 2024)

IOP Conference Series: Earth and Environmental Science
Volume 1525

Makassar, Indonesia
6-7 November 2024

ISBN: 979-8-3313-2475-9
ISSN: 1755-1307

Printed from e-media with permission by:

Curran Associates, Inc.
57 Morehouse Lane
Red Hook, NY 12571

Some format issues inherent in the e-media version may also appear in this print version.

Institute of Physics
Dirac House, Temple Back
Bristol BS1 6BE UK

Phone: 44 1 17 929 7481
Fax: 44 1 17 920 0979

techtracking@iop.org

Additional copies of this publication are available from:

Curran Associates, Inc.
57 Morehouse Lane
Red Hook, NY 12571 USA
Phone: 845-758-0400
Fax: 845-758-2633
Email: curran@proceedings.com
Web: www.proceedings.com

TABLE OF CONTENTS

IOP Conf. Series: Earth and Environmental Science **1525** (2025) 011001 doi:10.1088/1755-1315/1525/1/011001

International Conference on Geosciences and Technology (ICOGEST) 2024: Global Collaboration in Earth Sciences for Natural Hazard Mitigation and Near-Surface Exploration

Muhammad Taufiq Rafie[1*], Andi Muhamad Pramatadie[1]

[1] Solid Geophysics Laboratory, Department of Geophysics, Faculty of Mathematics and Natural Sciences, Hasanuddin University, Makassar, Indonesia

*E-mail: taufiqrafie@unhas.ac.id

The International Conference on Geosciences and Technology (ICOGEST) 2024 is a newly established international forum organized by the Department of Geophysics, Faculty of Mathematics and Natural Sciences, and the Department of Geological Engineering, Faculty of Engineering, Hasanuddin University, Indonesia. This conference took place on 6-7 November 2024 at Unhas Hotel & Convention, Makassar, Indonesia.

In response to the increasingly complex global challenges–including natural disasters such as earthquakes, tsunamis, volcanic eruptions, and the growing impacts of climate change-ICOGEST was initiated with a vision to strengthen the role of earth sciences in both natural hazard mitigation and responsible exploration of natural resources.

The 2024 conference adopted the theme: "**Global Collaboration in Earth Sciences for Natural Hazard Mitigation and Near-Surface Exploration**". This theme emphasizes the necessity of international, interdisciplinary cooperation to improve predictive capabilities, scientific understanding, and response strategies to natural hazard, while also optimizing the sustainable use of near-surface geological resources. In this spirit, ICOGEST 2024 served as a collaborative platform for academics, researchers, and practitioners from diverse institutions, backgrounds, and countries to exchange scientific knowledge, share experiences, present innovative research findings, and build long-term collaborations.

The conference featured seven keynote and invited speakers from Indonesia, Singapore, Japan, Australia, and Romania, including the notable contributions from institutions such as the Earth Observatory of Singapore, Chiba University, and PT Leica Geosystem Indonesia. The scientific program included 45 oral presentations, grouped into four main topics: Atmospheric Science, Solid Earth Sciences, Ocean Sciences, and Interdisciplinary Earth Science and Technology.

ICOGEST 2024 was attended by more than 180 participants among global researchers, academics, practitioners, and students who contribute and showcase their work. The two-day program consisted of plenary session, which were held simultaneously in three rooms (Innovate A, Innovate B, Innovate C). These sessions covered a wide range of topics, including tectonic deformation, geological investigation, seismic and engineering hazard, geophysical exploration, hydrogeological studies, mining, and remote sensing applications.

All submitted papers were reviewed through a double-blind peer review process coordinated by the conference editors using the Morressier submission system. Each manuscript was evaluated by experts to ensure academic rigor, clarity, and relevance. Out of 63 submission, 50

IOP Conf. Series: Earth and Environmental Science **1525** (2025) 011001 doi:10.1088/1755-1315/1525/1/011001

papers are ready to be submitted to the IOP Conference Series: Earth and Environmental Science. The organizing committee would like to thank all authors, reviewers, session chairs, and participants for their valuable contributions. We extend our deepest appreciation to our keynote speakers for sharing their insights, and to PT Leica Geosystem Indonesia for their support and partnership. We also express our gratitude to IOP Publishing for facilitating the dissemination of this work to the global scientific community. We hope that the proceedings of ICOGEST 2024 will provide valuable insights and courage further collaboration in building a more resilient and sustainable future.

The Organizing Committee

Head of Geophysics Department	Dr. Eng. Muh. Alimuddin Hamzah Assagaf, M.Eng (UNHAS) Dr. Sakka, M.Si. (UNHAS)
Director	Dr. Eng. Amiruddin, S.Si., M.Si. (UNHAS) Prof. Dr. Eng. Ir. Muhammad Isran Ramli (UNHAS)
Steering and Scientific Committees	Prof. Dr. Halmar Halide, M.Sc. (UNHAS) Prof. Dr. Ir. Muh. Altin Massinai, M.T.Surv., IPM (UNHAS) Prof. Dr. Dahlang Tahir, S.Si., M.Si. (UNHAS) Dr. Sakka, M.Si. (UNHAS) Prof. Dr. A. M. Imran Umar (UNHAS) Prof. Dr. Adi Tonggiroh (UNHAS) Ilham Alimuddin, S.T., M.Gis, Ph.D. (UNHAS) Syamsuddin, S.Si., M.T. (UNHAS) Sabrianto Aswad, S.Si., M.T. (UNHAS)
Chairman	Dr. Eng. Muh. Alimuddin Hamzah Assagaf, M.Eng (UNHAS)
Vice Chairman	Dr. Eng. Hendra Pachri, S.T., M.Eng. (UNHAS)
General Secretary	Andi Muhamad Pramatadie, S.T., M.Eng., Ph.D. (UNHAS)
Treasurer	Syafruddim, S.T., M.Eng. (UNHAS)
Official Team	Saaduddin, S.Pd., M.Sc. (UNHAS) Dr. Eng. Sulhuzair, S.T., M.T. (UNHAS) Asti S.T., M.T. (UNHAS) Arifuddin Jamil, S.T., M.Eng. (UNHAS) Andika, S.Si., M.Si. (UNHAS) Bahrul Hidayat, S.T., M.T. (UNHAS) Agustinus Ribal, S.Si., M.Sc., Ph.D. (UNHAS) Dr. Muhammad Taufiq Rafie, S.Si., M.T. (UNHAS) Erfan Syamsuddin, S.Si., M.Si. (UNHAS) Dr. Ir. Ulva Ria Irfan, M.T. (UNHAS) Dr. Eng. Meutia Farida (UNHAS) Aswar Syafnur, S.Si., M.Eng. (UNHAS)
Editors	Dr. Muhammad Taufiq Rafie, S.Si., M.T. (UNHAS) Andi Muhamad Pramatadie, S.T., M.Eng., Ph.D. (UNHAS)

IOP Conf. Series: Earth and Environmental Science **1525** (2025) 011001 doi:10.1088/1755-1315/1525/1/011001

Documentation

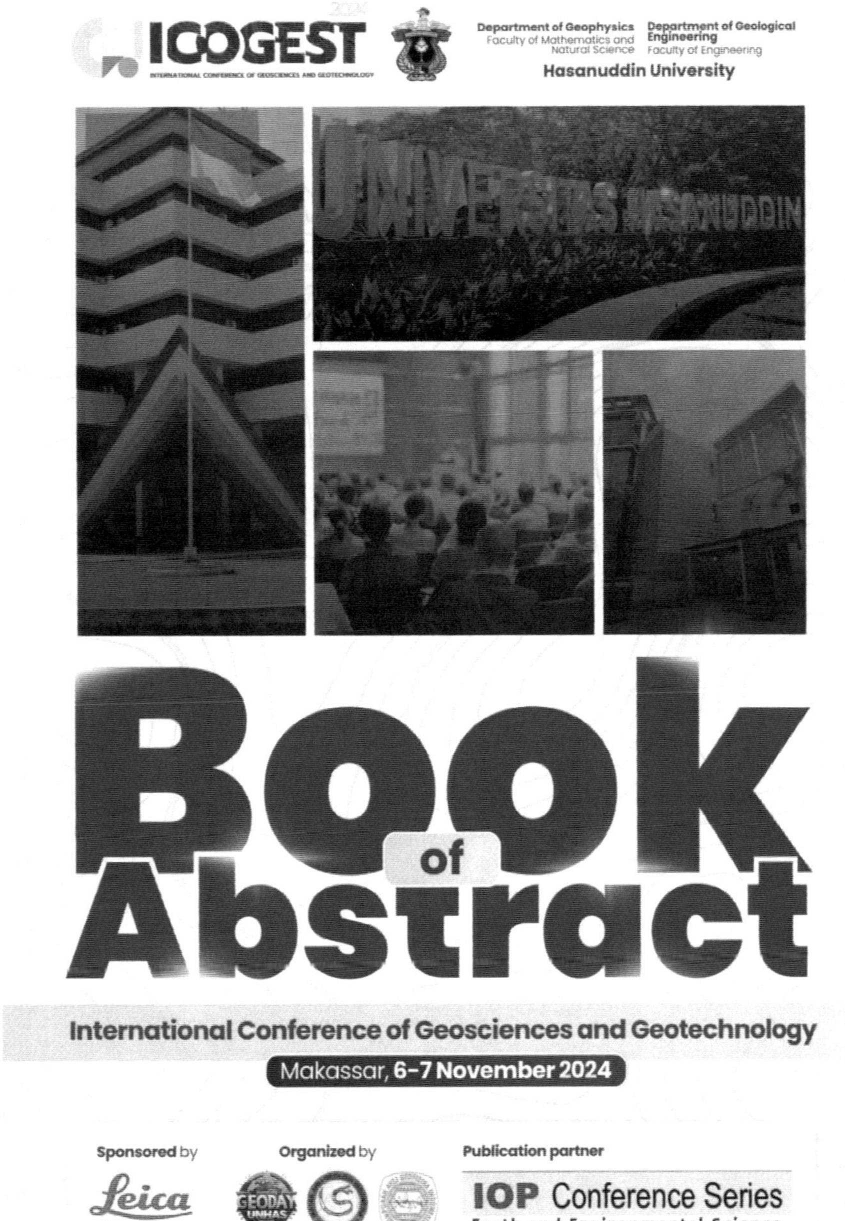

IOP Conf. Series: Earth and Environmental Science **1525** (2025) 011001 doi:10.1088/1755-1315/1525/1/011001

Department of Geophysics, Faculty of Mathematics and Natural Science
Department of Geological Engineering, Faculty of Engineering
Hasanuddin University

Foreward

Dear ICOGEST 2024 Participants,

On behalf of the organizing committee, it is a great pleasure to welcome you to the International Conference on Geosciences and Technology (ICOGEST) 2024. We are honored to host this gathering of esteemed researchers, scientists, and professionals from across the globe, united by a common goal: advancing collaboration in earth sciences to tackle the pressing challenges of natural hazard mitigation and near-surface exploration.

The theme of this year's conference, "Global Collaboration in Earth Sciences for Natural Hazard Mitigation and Near-Surface Exploration," underscores the importance of international cooperation in addressing the complex issues that transcend national boundaries. From mitigating natural disasters to sustainably exploring our planet's resources, we are reminded that only through shared knowledge and joint efforts can we achieve meaningful progress.

We hope this conference provides you with a fruitful platform for exchanging ideas, establishing new partnerships, and fostering innovative research that will benefit communities worldwide. Thank you for joining us in this endeavor. May your time at ICOGEST 2024 be inspiring, enlightening, and productive.

Warm regards,

Prof. Dr. Eng. Amiruddin, S. Si., M.Si.
Dean, Faculty of Mathematics and Natural Sciences
Hasanuddin University

IOP Conf. Series: Earth and Environmental Science **1525** (2025) 011001 doi:10.1088/1755-1315/1525/1/011001

ICOGEST 2024
INTERNATIONAL CONFERENCE OF GEOSCIENCES AND GEOTECHNOLOGY

Department of Geophysics
Faculty of Mathematics and Natural Science
Department of Geological Engineering
Faculty of Engineering
Hasanuddin University

List of Plenary Speakers

"Multi-Proxy Approaches Toward A Long View of Past Coastal Disasters"

Adam Switzer
Nanyang TU EOS, Singapore

"A Seismic Hazard Analysis Framework for The Sunda Arc OffshoreIndonesia Using Geodynamic Simulations"

Wahyu Triyoso
Bandung Institute of Technology, Indonesia

"Use of MASW Surveys in Australia"

Koya Suto
Terra Australis geophysica, Australia

IOP Conf. Series: Earth and Environmental Science **1525** (2025) 011001 doi:10.1088/1755-1315/1525/1/011001

ICOGEST 2024
INTERNATIONAL CONFERENCE OF GEOSCIENCES AND GEOTECHNOLOGY

Department of Geophysics, Faculty of Mathematics and Natural Science
Department of Geological Engineering, Faculty of Engineering
Hasanuddin University

Gilang Widyawisaksana
Leica Geosystems, Indonesia

"Innovative Geospatial Technology and AI for Near-Surface Exploration and Risk Reduction in Indonesia"

Adi Maulana
Hasanuddin University, Indonesia

"Geoscience Data Integration Model for Limestone Exploration: A New Approach in Metal Industry Downstream"

Victor Mocanu
University of Bucharest, Romania

"Ridge Subduction and Mantle Flow. South America Case Study"

Josaphat Tetuko S.S.
Chiba University, Japan
Universitas Sebelas Maret, Indonesia

"The Use of SAR Technology for Disaster Mapping and Management"

IOP Conf. Series: Earth and Environmental Science **1525** (2025) 011001 doi:10.1088/1755-1315/1525/1/011001

ICOGEST 2024 INTERNATIONAL CONFERENCE OF GEOSCIENCES AND GEOTECHNOLOGY

Department of Geophysics, Faculty of Mathematics and Natural Science
Department of Geological Engineering, Faculty of Engineering
Hasanuddin University

Schedule of International Conference on Geoscience and Geotechnology (ICOGEST) 2024 Hasanuddin University

Wednesday, November 6, 2024		
Time	**Agenda**	**PIC**
07:30 - 08:00	Registration	The ICOGEST Committee
08:00 - 08:05	Opening National anthem of Indonesia Raya	
08:05 - 08:15	Opening Traditional Dance Performanc	
08:15 - 08:25	Pray	The ICOGEST Committee
08:25 - 08:35	Report from the Chairperson of the ICoGEST	Dr. Muh. Alimuddin Hamzah & Dr. Hendra Pachri
08:35 - 08:45	Speech from the UNHAS Rector and opening the event	Prof. Dr. Jamaluddin Jompa, M.Si.
08:45 – 08:55	MoU Signing Ceremony: Leica and Engineering and MIPA Faculty	
08:55 – 09:00	Photo Session and Closing	
09:00 - 09:15	Coffee Break	
09:15 - 10:00	Dr. Wahyu Triyoso, M.Si.	Moderator: Dr. Muh. Taufiq Rafie
10:00 - 10:45	Prof. Adi Maulana	Moderator: Dr. Sc. Muhammad Sulhuzair Burhanuddin, S.T. M.T
10:45 - 11:30	Leica: Gilang Widya Wisaksana, S.T., MBA	Moderator: Arifuddin Jamil, S.T., M.Eng.
11:30 - 13:00	Lunch break and Pray	
13:00 - 15:00	Parallel Session *Innovate, Inspiring, and Vision Room*	CESTCO *UHC Ballroom*
15:00 - 15:30	Break (Snack)	Break
15:30 - 17:00	Parallel Session *Innovate, Inspiring, and Vision Room*	CESTCO *UHC Ballroom*

IOP Conf. Series: Earth and Environmental Science **1525** (2025) 011001 doi:10.1088/1755-1315/1525/1/011001

ICOGEST 2024
INTERNATIONAL CONFERENCE OF GEOSCIENCES AND GEOTECHNOLOGY

Department of Geophysics
Faculty of Mathematics and Natural Science
Department of Geological Engineering
Faculty of Engineering
Hasanuddin University

Thursday, November 7, 2024

Time	Agenda	PIC
07:30 - 08:00	Registration	The ICOGEST Committee
08:30 - 09:15	Prof. Adam Switzer	Moderator: Dr. Alimuddin Hamzah
09:15 - 10:00	Prof. Victor Mocanu	Moderator: Bahrul Hidayat, S.T., M.T
10:00 - 10:30	Break	
10:30 - 11:15	Koya Suto, M.E.	Moderator: A. M. Pramatadie, Ph.D
11:15 - 12:00	Prof. Josaphat Tetuko S.S.	Moderator: Ilham Alimuddin, Ph.D.
12:00 - 13:00	Lunch break and Pray	
13:00 - 14:00	Closing Ceremony The ICOGEST Committee	

IOP Conf. Series: Earth and Environmental Science **1525** (2025) 011001 doi:10.1088/1755-1315/1525/1/011001

ICOGEST INTERNATIONAL CONFERENCE OF GEOSCIENCES AND GEOTECHNOLOGY

Department of Geophysics, Faculty of Mathematics and Natural Science
Department of Geological Engineering, Faculty of Engineering
Hasanuddin University

Schedule of Parallel Session International Conference on Geoscience and Geotechnology (ICOGEST) 2024

Time		Innovate Room A	Innovate Room B	Innovate Room C
Moderator		Sabrianto Aswad, S.Si., M.T.	Aswar Syafnur, S.Si., M.Eng.	Syafruddim, ST.M.Eng
Facilitators		Muhammad Ridwan Indira Nazwa Nandita	Afrisal idris Yuli	Ariel Ar-Rahman Zalsa Nabila HR
13.00	13.12	SES 01	IEST 01	IEST 11
13.12	13.24	SES 02	IEST 02	IEST 12
13.24	13.36	SES 03	IEST 03	IEST 13
13.36	13.48	SES 04	IEST 04	IEST 14
13.48	14.00	SES 05	IEST 05	IEST 15
14.00	14.12	SES 06	IEST 06	IEST 16
14.12	14.24	SES 07	IEST 07	IEST 17
14.24	14.36	SES 08	IEST 08	IEST 18
14.36	14.48	SES 09	IEST 09	IEST 19
14.48	15.00	SES 10	IEST 10	IEST 20
15.00	15.30	Coffee Break		
Time		Innovate Room A	Innovate Room B	Innovate Room C
Moderator		Asti, ST,M.Eng	Sahabuddin, ST,M.Eng	Saaduddin, M.Sc.
Facilitators		Muh. Anil Fauzan Akbar Ihya Haeratun Hisani	Alif Amanullah Nurul Hikma	Glenn Rex Eugenio Jenifer Pither Pali
15.30	15.42	SES 11	IEST 21	OS 01
15.42	15.54	SES 12	IEST 22	OS 02
15.54	16.06	SES 13	IEST 23	OS 03
16.06	16.18	SES 14	IEST 24	OS 04
16.18	16.30	SES 15	AS 01	OS 05
16.30	16.42	SES 16	AS 02	OS 06
16.42	16.54			
16.54	17.06			

ICOGEST 2024
INTERNATIONAL CONFERENCE OF GEOSCIENCES AND GEOTECHNOLOGY

Department of Geophysics
Faculty of Mathematics and Natural Science

Department of Geological Engineering
Faculty of Engineering

Hasanuddin University

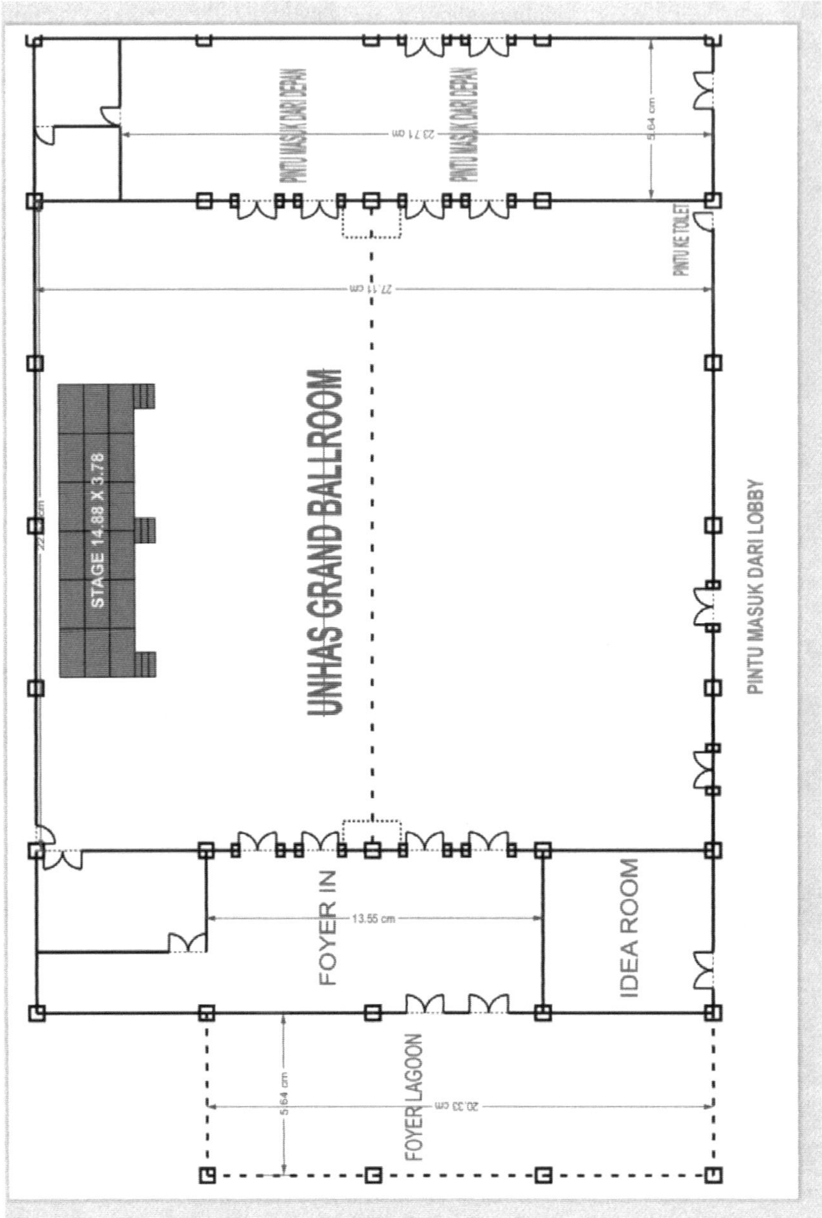

IOP Conf. Series: Earth and Environmental Science **1525** (2025) 011001 doi:10.1088/1755-1315/1525/1/011001

ICOGEST 2024 INTERNATIONAL CONFERENCE OF GEOSCIENCES AND GEOTECHNOLOGY

Department of Geophysics Faculty of Mathematics and Natural Science
Department of Geological Engineering Faculty of Engineering
Hasanuddin University

Speakers of Plenary Session

IOP Conf. Series: Earth and Environmental Science **1525** (2025) 011001 doi:10.1088/1755-1315/1525/1/011001

Department of Geophysics
Faculty of Mathematics and Natural Science
Department of Geological Engineering
Faculty of Engineering
Hasanuddin University

Presenters of Parallel Session

1. Solid Earth Science

Code	Presenter	Authors	Title	Page
SES 01	Dewi Pratiwi	Dewi Pratiwi, Muhammad Taufiq Rafie, Sabrianto Aswad	A Brief Note to Machine Learning for Earthquake Statistics: Unsuperviced Machine Learning Approach	21
SES 02	Haidir Jibran	Haidir Jibran, Muhammad Taufiq Rafie, Muhammad Altin Massinai	Seismotectonic Heterogeneity in Sulawesi Island: An Application of Unsupervised Machine Learning	22
SES 03	Ahmad Hidayah	Ahmad Hidayah, Bambang Harimei	Identification of Earthquake-Prone Zones in West Sulawesi Province for Disaster Mitigation Using the Probabilistic Seismic Hazard Analysis (PSHA) Method	23
SES 04	Edy Santoso, S.Si,M.Si	Edy Santoso, Sigit Pramono, M.Syamsu Rosid, Adhi Harmoko, Dwikorita Karnawati	Prediction of Ground Motion Utilizing Machine Learning Techniques in The Lombok Region	24
SES 06	Andi Zulkifli	Andi Zulkifli, Muhammad Altin Massinai, Sabrianto Aswad, Syamsuddin, Muh. Farid Wajedy	Determination of Resistivity Values and Thickness of The Subsurface Layer Using "Least Square" Inversion	25
SES 07	Abdul Mun'im Thariq	Abdul Mun'im Thariq, Hamsu Abdul Gani, Darmawang	Aquifer Mapping Using Geoelectric Method in Bulukumba Regency, South Sulawesi	26
SES 08	Mulyadi Azwar Sannang	Mulyadi Sannang, Hendra Pachri, Ilham Alimuddin	Open Pit Coal Mine Failure Model Based on Statistical Data	27
SES 09	Erfan Syamsuddin, M.Si	Adi Maulana, Erfan Syamsuddin, Ulva Ria Irvan, Alimuddin Hamzah Assagaf, Sri Widodo, Andi Muhammad Anshar	Subsurface Limestone Mapping Using Electrical Resistivity Tomography (ERT)	28
SES 10	Erfan Syamsuddin, M.Si	Adi Maulana, Ulva Ria Irfan, Andi Muhammad Anshar, Erfan Syamsuddin, Alimuddin Hamzah Assagaf, Sri Widodo	Utilizing X-Ray Fluorescence (XRF) for Evaluating Limestone Quality in Production Processes	29
SES 11	Almubdy Siraj Ramadhan	Almubdy Siraj Ramadhan, Muhammad Altin Massinai, Andi Muhammad Pramatadie	Investigation of Landslide Occurrence Using Microtremor Case Study of BuloBulo Village, Barru, Indonesia	30
SES 12	Harisma	Harisma, Muhammad Zamrun Firihu, Viska Inda Variani, Muliddin, Erzam S. Hasan	Preliminary Results of the Neogene History of Marine Succession of the Tondo Formation in the Buton Island, SE Sulawesi, Indonesia	31
SES 13	Dr. Rustan Efendi, M.Si	Rustan Efendi, Ilham Nuryadin, Abdullah, Sofian, Erfan Syamsuddin	Mapping of geological structures based on horizontal gradient of gravity anomaly in Palu Valley, Central Sulawesi, Indonesia	32

IOP Conf. Series: Earth and Environmental Science **1525** (2025) 011001 doi:10.1088/1755-1315/1525/1/011001

ICOGEST 2024 INTERNATIONAL CONFERENCE OF GEOSCIENCES AND GEOTECHNOLOGY

Department of Geophysics, Faculty of Mathematics and Natural Science; Department of Geological Engineering, Faculty of Engineering; Hasanuddin University

ICOGEST 2024 INTERNATIONAL CONFERENCE OF GEOSCIENCES AND GEOTECHNOLOGY
Department of Geophysics, Faculty of Mathematics and Natural Science; Department of Geological Engineering, Faculty of Engineering
Hasanuddin University

3. Atmospheric Science

4. Ocean Science

IOP Conf. Series: Earth and Environmental Science **1525** (2025) 011001 doi:10.1088/1755-1315/1525/1/011001

IOP Conf. Series: Earth and Environmental Science **1525** (2025) 011001 doi:10.1088/1755-1315/1525/1/011001

IOP Conf. Series: Earth and Environmental Science **1525** (2025) 011002 doi:10.1088/1755-1315/1525/1/011002

Peer Review Statement

All papers published in this volume have been reviewed through processes administered by the Editors. Reviews were conducted by expert referees to the professional and scientific standards expected of a proceedings journal published by IOP Publishing.

- **Type of peer review:** Double Anonymous
- **Conference submission management system:** Morressier
- **Number of submissions received:** 63
- **Number of submissions sent for review:** 54
- **Number of submissions accepted:** 50
- **Acceptance Rate (Submissions Accepted / Submissions Received × 100):** 79.4
- **Average number of reviews per paper:** 1.5
- **Total number of reviewers involved:** 12
- **Contact person for queries:**
  Name: Muhammad Taufiq Rafie
  Email: taufiqrafie@unhas.ac.id
  Affiliation: Universitas Hasanuddin - Universitas Hasanuddin Fakultas Matematika dan Ilmu Pengetahuan Alam

Decadal Rainfall Variability in the Drylands of East Nusa Tenggara

Tri Ayu Lestari[1*] and Andika Andika[2]

[1] Physics Education Study Program, FKIP, Universitas Nusa Cendana, Kupang, Indonesia
[2] Hydrometeorology Laboratory, Geophysics Department, FMIPA, Universitas Hasanuddin, Makassar, Indonesia

*E-mail: tri_lestari@staf.undana.ac.id

Abstract. The drylands of East Nusa Tenggara (NTT) experience a predominantly dry climate characterized by low annual rainfall, which presents significant challenges to agricultural productivity in the region. This work used the monthly Climate Hazards Group Infrared Precipitation with Stations (CHIRPS) dataset from 1991 to 2020 to analyze decadal rainfall variability in NTT's drylands. The analysis categorized the data into three ten-year periods (1991-2000, 2001-2010, 2011-2020) to identify patterns and trends over time. The results reveal notable fluctuations in rainfall, with marked changes observed over the past three decades. During the first decade (1991-2000), many areas experienced wet anomalies. In contrast, the second decade (2001-2010) was characterized by widespread dry anomalies across most of the region. Rainfall levels increased once more in the third decade (2011-2020). Additionally, the correlation between the El Niño-Southern Oscillation (ENSO) phenomenon and rainfall in this area is relatively low, suggesting that other factors may influence rainfall variability in NTT. Understanding the elements that contribute to rainfall variability in NTT's drylands is essential for developing effective mitigation strategies against potential droughts and for promoting sustainable natural resource management in the region. The findings of this study could have a significant impact on policy making and resource allocation, empowering stakeholders to make informed decisions.

Keywords: Decadal Analysis, Drylands, East Nusa Tenggara, Rainfall Variability.

1. Introduction

East Nusa Tenggara (NTT) is known as the dryland Islands, which has minimal or no waterlogging characteristics throughout the year, making it the province with the highest proportion of drylands (in Indonesian called Lahan Kering Kepulauan or LKK) in Indonesia. According to (1), NTT receives less than 2,000 mm of rainfall annually and endures a dry season over seven months, resulting in water availability being the primary constraint on land productivity. Moreover, the hilly topography, shallow and rocky soils, and inadequate utilization of water sources such as rivers and reservoirs contribute to the difficulties faced by NTT (2).

Understanding the characteristics of weather and climate is crucial for managing the drylands. Climate variability plays a key role in the sustainability of dry island areas over extended periods. Climate variability analysis can be conducted using a decadal approach, which compares rainfall data at ten-year intervals. This approach is essential because it can reveal long-term changes that may not be evident in annual or seasonal analyses. Previous studies have shown that decadal analysis is effective in identifying rainfall characteristics, which are crucial indicators in climatology studies. For example, (3,4) conducted decadal analyses in Nigeria, finding that each decade has a unique rainfall pattern. Additionally, (5) highlighted decadal rainfall variability in Senegal, focusing on the frequency and intensity of extreme rainfall. This study revealed that rainfall patterns are not only determined by total amounts but also by other characteristics. The (5) found a relationship between decadal variability and global climate phenomena affecting rainfall in tropical regions. Thus, decadal analysis provides a comprehensive picture of the complex climate dynamics that are important for supporting climate change adaptation and mitigation.

This study examines how climate change affects drylands, focusing on analyzing rainfall data from the last three decades, specifically from 1991 to 2020. By understanding the trends and patterns of rainfall changes during this period, this research can provide valuable insights for researchers, policymakers, and other stakeholders to formulate effective mitigation strategies against the impacts of climate change. Thus, the results of this study are expected to contribute to efforts in maintaining food security and sustainability of drylands in NTT, as well as to guide better natural resource management in the future.

2. Data & Method

We used data from the Climate Hazards Group Infrared Precipitation with Stations (CHIRPS), developed by the Famine Early Warning Systems Network (FEWS NET) (6,7). The CHIRPS dataset covers land areas with a spatial resolution ranging from 0.05 to 0.1 degrees. CHIRPS can be accessed at https://data.chc.ucsb.edu/products/CHIRPS-2.0/. This study used the monthly CHIRPS data from 1991 to 2020. CHIRPS data for monitoring drought over an extended period has been assessed as effective and reliable. Accuracy and validation processes conducted by various researchers, including (8–11), demonstrate that CHIRPS data can provide relevant information about rainfall and drought occurrences over a long period.

3. Results and Discussion

3.1 Climatological Analysis

The average annual rainfall in NTT from 1991 to 2020 is 1,444 mm, with monthly values shown in Figure 1. December records the highest rainfall at 267.41 mm per month, while August has the lowest at 11.16 mm per month. Rainfall is classified as low (0–100 mm per month), medium (101–300 mm per month), and high (>500 mm per month), with six months (March to October) categorized as low and the remaining months as medium; no months fall into the high category. The rainfall pattern in NTT exhibits a distinct wet and dry season occurring once a year, with the rainy season peaking in December, January, and February (DJF). This seasonal pattern aligns with the findings of (12), who identified a monsoonal rainfall type in NTT, characterized by one peak and one trough in its annual cycle.

IOP Conf. Series: Earth and Environmental Science **1525** (2025) 012001 doi:10.1088/1755-1315/1525/1/012001

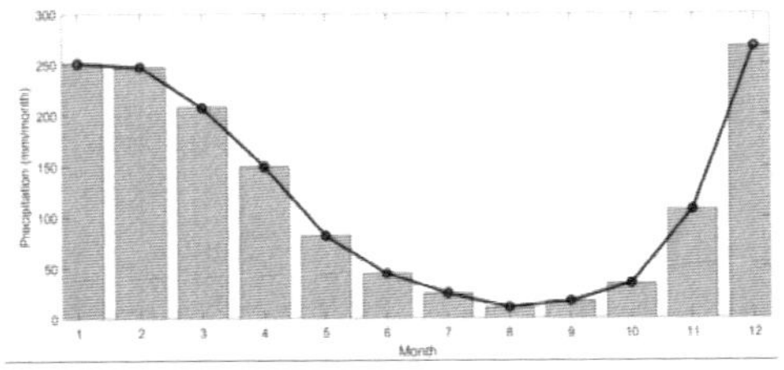

Figure 1. Monthly Rainfall Climatology (1991-2020) in NTT.

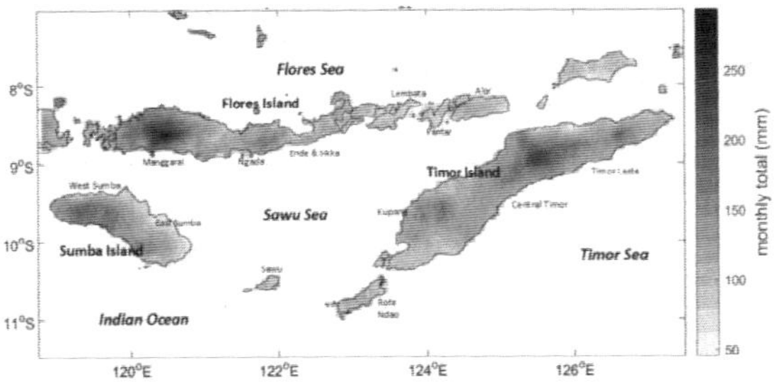

Figure 2. Monthly Climatologies of Rainfall Distribution (1991-2020).

Based on the climatological rainfall plot in NTT (Figure 2), most areas experience low to moderate rainfall, ranging from 50 to 250 mm per month. East Sumba, particularly along its coastline, records the lowest rainfall at approximately 50 mm per month. Research by (13) indicates that this area faces extreme drought, with over 60 consecutive days without rain, exacerbating water scarcity and food security challenges. In contrast, the central part of Manggarai on Flores Island receives the highest rainfall, reaching around 250 mm per month. (14) highlights the role of topography in shaping rainfall patterns across NTT, where mountainous regions such as Manggarai and central Flores act as orographic barriers that enhance localized precipitation.

Rainfall in NTT is largely influenced by orographic lifting, where moist air is forced to rise over elevated terrain, cooling adiabatically and condensing into clouds that generate precipitation (14). Higher elevations strengthen this effect, resulting in increased rainfall in mountainous areas. Conversely, coastal and lowland regions like East Sumba, Lembata, and Sawu receive less rainfall due to the absence of significant topographical barriers to lift moist air. Additionally, descending air on the leeward side of mountains warms adiabatically, reducing humidity and inhibiting cloud formation, further contributing to drier conditions. Thus, mountainous terrain plays a key role in rainfall distribution across NTT, with elevated regions receiving more rain and low-lying coastal areas remaining drier.

3.2 Decadal Analysis

3.2.1 Average Rainfall Decadal Anomalies

The decadal average values were derived from the monthly means for each period. The first decade spans from 1991 to 2000, the second from 2001 to 2010, and the third from 2011 to 2020. Rainfall anomalies (Figure 3) were computed using the absolute difference method, where the anomaly value was obtained by subtracting the 30-year climatological average (1991–2020) from the actual rainfall data. This method provides a direct interpretation of deviations in millimeters per month without standardization. The results reveal distinct anomaly patterns for each decade: positive anomalies indicate above-average rainfall (wet anomalies), while negative anomalies indicate below-average rainfall (dry anomalies).

In the first decade, wet and dry anomalies are distributed across the region. The most significant wet anomaly is observed on Sumba Island, with a value of approximately 10 mm/month, while the highest dry anomaly occurs from Lembata to Alor. The dominance of wet anomalies includes the western region of NTT, encompassing Sumba, almost the entire Flores

Island, Sawu Island, and a small part of Timor Island and Rote Ndao. Lower values of dry anomalies are also observed in several areas of Timor Island and slightly in Rote Ndao, but they are primarily concentrated from Lembata to Alor. Overall, most regions experience wet anomalies, with the highest values found on Sumba Island.

The second decade presents a pattern opposite to the first decade, where dry anomalies are more prevalent. Wet anomalies with relatively low values are only detected in a small area, specifically in the northern part of Timor Island and the Lembata to Alor region. Dry anomalies are highly dominant and widespread across Sumba Island, Flores Island, Sawu Island, and Rote Ndao, with values reaching approximately -10 mm/month. This decade demonstrates widespread extreme drought across NTT.

In the third decade, wet anomalies become more dominant again, with only a small portion of the region experiencing dry anomalies. This decade shows a trend toward wetter rainfall anomaly. Based on the results of the rainfall anomaly calculations over the three decadal periods, it is found that the three decades exhibit different spatial characteristics, with some regions showing very contrasting differences. While monsoonal winds and orographic factors influence the climatological average, the characteristics of rainfall based on the anomalies from these three decadal periods are not yet fully understood.

3.2.2 Comparison of Decadal Rainfall Distribution

The decadal comparison aims to understand long-term patterns of rainfall changes, which is an essential part of climate change analysis. By comparing rainfall conditions for each decade, we

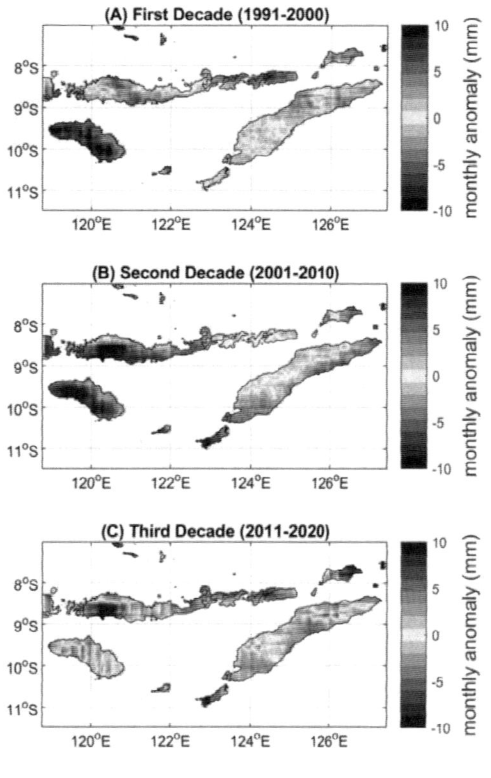

Figure 3. Average of Rainfall Anomalies for Each Decade.

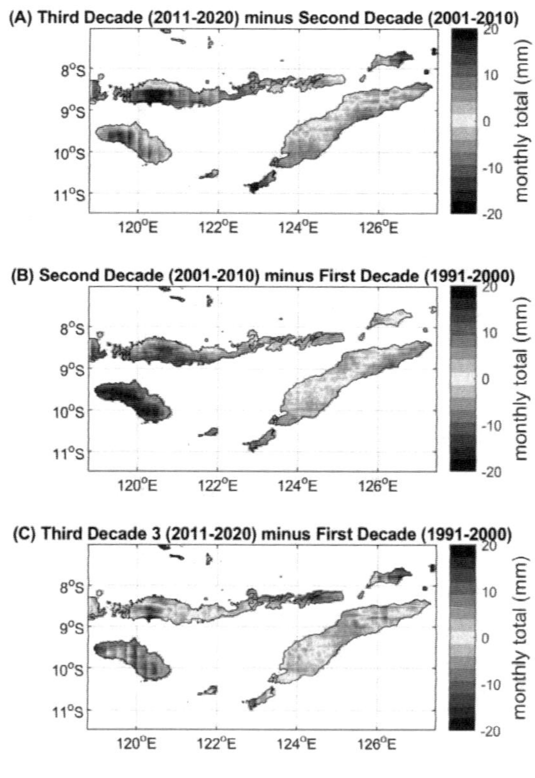

Figure 4. Comparison of Decadal Rainfall Distribution

IOP Conf. Series: Earth and Environmental Science **1525** (2025) 012001 doi:10.1088/1755-1315/1525/1/012001

can observe potential wet or dry trends that may occur periodically or persist over time. The decadal analysis reveals clearer fluctuations in rainfall over an extended time frame, allowing us to observe shifts between wet and dry conditions (Figure 4).

The comparison between the third decade (2011-2020) and the second decade (2001-2010) shows notable differences in rainfall. Positive values dominate the region, indicating that the third decade is wetter than the previous decade. The largest positive values are found in areas that previously experienced drought during the 2001-2010 decade, signaling a significant recovery in rainfall.

In contrast, the comparison between the second decade (2001-2010) and the first decade (1991-2000) reveals that most areas of NTT experienced a reduction in rainfall (negative values). This indicates that 2001-2010 was generally drier than 1991-2000.

The comparison between the third decade (2011-2020) and the first decade (1991-2000) reveals a combination of positive and negative values. This most recent decade exhibits variations between wet and dry conditions compared to the first decade. In some areas, such as Sumba Island and parts of northern Flores, negative values indicate that these regions are still trending drier than in the first decade (1991-2000). Conversely, other regions, especially in the north and central areas, such as Lembata, Alor, and eastern Timor Island, show positive values, indicating increased rainfall in certain locations compared to the first decade.

3.3 Discussion

These differing conditions across the three decades align with the findings of (15), who analyzed rainfall data from 1981 to 2021 using meteorological station data. (15) stated that rainfall over the long term generally shows an increase, although short-term fluctuations are not always predictable. He explained that the nature of rainfall in Indonesia's Maritime Continent, including NTT, is non-stationary, meaning that while rainfall may increase climatologically, there are still significant year-to-year fluctuations.

Overall, our analysis supports (15)'s conclusion that while there is a long-term increase in rainfall in NTT, significant fluctuations still occur. Previous studies, such as those conducted by (12,16) also analyzed rainfall patterns in Indonesia and showed that ENSO is one of the factors affecting rainfall anomalies. We conducted a temporal correlation analysis between rainfall in NTT and the Oceanic Niño Index (ONI) 3.4 to examine the impact of ENSO on temporal rainfall changes. The results showed a low correlation, <0.45 on spatial distribution (figure not shown), indicating no strong relationship between rainfall in NTT and ENSO variability. This finding suggests that, aside from ENSO, other factors may shape the characteristic rainfall patterns in each decade.

This low correlation suggests the possibility of other climate variability factors influencing rainfall in NTT. In (15), proposes additional potential factors, such as the Pacific Decadal Oscillation (PDO) and Indian Ocean Dipole (IOD), which can significantly impact rainfall patterns. For example, the PDO can affect sea surface temperatures, which contribute to rainfall patterns. At the same time, the IOD is related to the movement of water masses in the Indian Ocean, which also influences rainfall in the NTT region. Therefore, further analysis is necessary to understand these factors, which are crucial for developing strategies to mitigate the frequent droughts experienced in NTT.

IOP Conf. Series: Earth and Environmental Science **1525** (2025) 012001 doi:10.1088/1755-1315/1525/1/012001

4. Conclusion

This study demonstrates that the decadal rainfall variability in the drylands of NTT Tenggara has a fluctuating and significant pattern from 1991 to 2020. The results indicate that the period from 1991 to 2000 was marked by significant wet anomalies, followed by dry anomalies from 2001 to 2010, and an increase in rainfall from 2011 to 2020. These findings also show that rainfall patterns in NTT are non-stationary, where significant fluctuations can affect long-term trends. The time frame selection in this analysis greatly influences the final interpretation results.

A deeper analysis reveals that the ENSO has a low correlation with rainfall variability in this region. This suggests that other factors may be more dominant in influencing changes in rainfall patterns. Therefore, further research is needed to identify and understand the factors that play a crucial role in rainfall changes in NTT, particularly regarding drylands.

References

1. Dariah A, Subiksa IG., Sutono. Sistem Pengelolaan Tanah Pada Lahan Kering Beriklim Kering. Jakarta: IAARD Press, Badan Penelitian dan Pengembangan Pertanian, Kementerian Pertanian; 2013. 63 hal.
2. Mulyani A, Nursyamsi D, Irsal L. Acceleration of Agricultural Development in Dryland with Dry Climate in Nusa Tenggara. Pengemb Inov Pertan. 2014;6(1):187–98.
3. Okoro UK. Decadal Rainfall Trends and Variability Across Nigeria. Int J Environ Clim Chang. 2023;13(11):2654–65.
4. Ofordu CS, Onilude QA, Adedoyin ED, Mba NC, Adeoti OO, Osundun SO, et al. Decadal Assessment and Distribution of Rainfall Anomaly Index (1991 – 2020) for Benin City, Edo State, Nigeria. J Appl Sci Environ Manag. 2022;26(10):1629–35.
5. Badji A, Mohino E, Diakhaté M, Mignot J, Gaye AT. Decadal Variability of Rainfall in Senegal: Beyond the Total Seasonal Amount. J Clim. 2022;35(16):5339–58.
6. Pedreros D, Funk C, Peterson P, Landsfeld M, Verdin A, Michaelsen J, et al. The FEWS NET's Rainfall Enhancement Process. 2012;2012.
7. Funk C, Nicholson SE, Landsfeld M, Klotter D, Peterson P, Harrison L. The Centennial Trends Greater Horn of Africa precipitation dataset. Sci Data. 2015;2:1–15.
8. Saragih NF, Sitepu S, Simanungkalit GT, Sinambela M, Rajagukguk E, Larosa FG, et al. Validation of CHIRPS Estimation Rainfall Data Using Numerical Accuracy Test with Precipitation Observation Data. IOP Conf Ser Earth Environ Sci. 2022;1083(1).
9. Auliyani D, Wahyuningrum N. Rainfall variability based on the Climate Hazards Group InfraRed Precipitation with Station Data (CHIRPS) in Lesti watershed, Java Island, Indonesia. IOP Conf Ser Earth Environ Sci. 2021;874(1).
10. Hsu J, Huang WR, Liu PY, Li X. Validation of chirps precipitation estimates over Taiwan at multiple timescales. Remote Sens. 2021;13(2):1–18.
11. Narulita I, Fajary FR, Mulyono A, Kusratmoko E, Djuwansah MR. Application of Climate Hazards Group InfraRed Precipitation with Station (CHIRPS) satellite data for drought mitigation in Bintan island, Indonesia. IOP Conf Ser Earth Environ Sci. 2021;789(1).
12. Aldrian E, Dwi Susanto R. Identification of three dominant rainfall regions within Indonesia and their relationship to sea surface temperature. Int J Climatol. 2003;23(12):1435–52.
13. Karuniasa M, Pambudi PA. The analysis of the El Niño phenomenon in the East Nusa Tenggara Province, Indonesia. J Water L Dev. 2022;52:180–5.
14. Afghani FA, Wibowo OT, Mashuri I, Agha H. Analysis of Monthly Rainfall Characteristics in Nusa Tenggara Timur and its Spatial and Temporal Shifts. 2024;5(1):76–83.
15. Sudarman GG, Hadi TW, Ningsih NS, Sopaheluwakan A, Syahputra MR, Marpaung F. Nonstationary Changes in Annual Rainfall over Indonesia's Maritime Continent. Adv Meteorol. 2024;2024.
16. Hamada JI, Yamanaka MD, Matsumoto J, Fukao S, Winarso PA, Sribimawati T. Spatial and temporal variations of the rainy season over Indonesia and their link to ENSO. J Meteorol Soc Japan. 2002;80(2):285–310.

Heterogeneity of Tectonic Regimes in Sulawesi Island: Applying Unsupervised Machine Learning to Focal Mechanism Data

Haidir Jibran[1] and Muhammad Taufiq Rafie[2*]

[1]Study Program of Geophysics, Geophysics Department, Faculty of Mathematics and Natural Sciences, Hasanuddin University, 90245, Indonesia
[2]Solid Geophysics Laboratory, Faculty of Mathematics and Natural Sciences, Hasanuddin University, 90245, Indonesia

*E-mail: taufiqrafie@unhas.ac.id

Abstract. Sulawesi Island was formed in the tripel junction area between the Sunda plate, Australian plate, and Philippine Sea plate. The collision between the three plates resulted in Sulawesi island having a variety of fault mechanisms. To better understand the tectonic processes in Sulawesi, we clustered earthquake events in the region, which were then analyzed based on their focal mechanisms. The data we used is the distribution of focal mechanisms around the island of Sulawesi for 1390 events from 1967 to 2022 that we obtained from a combination of the ISC GEM, Global CMT, and BMKG earthquake catalogs. We applied one of the unsupervised machine learning technique that is Density-Based Spatial Clustering of Application with Noise (DBSCAN) to cluster these earthquake events. This method clusters earthquake events based on the density of the events using two parameters: the minimum number of events that occur in cluster (minPts), and distance radius for neighbouring points (ε). Futhermore, we analysed the density of the focal mechanism in each cluster to comprehend its fault mechanism heterogeneity. We obtained ten earthquake clusters, using the parameters minPts=15 and ε=0.40, where these parameters explain the reliability of the tectonic setting in Sulawesi. Based on the clusterization, we infer that sevent clusters have compressional regimes located in the northern arm of Sulawesi, the Molucca Sea Double Subduction area, the West Sangihe trench, the North-Vergent thrust, and the Nusa Tenggara Back Arc thrust. Additionally, we found that the clusters associated with the Palu-Koro Fault and Matano Fault exhibit characteristics of a transform regime, while the cluster located between the Lalanga and Tongian ridges is characterized by an extensional setting.

Keywords: Celebes, DBSCAN, Tectonic Regime

1. Introduction

Sulawesi Island has tectonic activity that reflects the complexity of its tectonic formation [1]. The "Latest Miocene Collision" that formed the island of Sulawesi triggered significant deformation of the region, including initiating rotational movement of the island's microblocks [2]. As a result of the movement of these microblocks, and the subduction of the Celebes Sea along the North Sulawesi trench with the Central Sulawesi Fault System [3] the island's seismicity activity has a high degree of heterogeneity.

As a step towards understanding the tectonic processes in a region, a study of the kinematics and stress fields associated with the seismicity of the region is necessary [4]. However, one of the basic assumptions in conducting these studies is the homogeneity of the region to be studied. To accommodate this prerequisite, we clustered earthquake events that occurred in the area around the island of Sulawesi based on their similar seismicity. Through these clusters, we resolved what they imply to the heterogeneous tectonic regimes in Sulawesi Island.

2. Data and Methodology

In this study, we used focal mechanism catalogues occurring from April 1967 to september 2022, with 4.0 < M_w < 7.0 and focal depth ≤ 50km, compiled by Global Centroid Moment Tensor (CMT) [5,6], International Seismological Centre – Global Earthquake Model (ISC-GEM) [7,8,9], and BMKG [10]. From these compilation, we obtained 1390 focal data (Figure 1) which will be used for the clustering process.

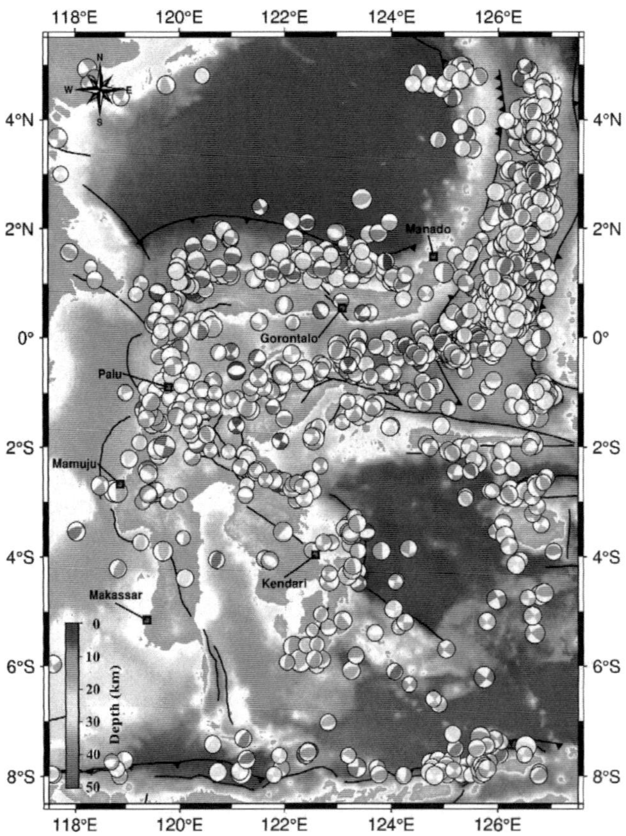

Figure 1. Map showing the distribution of focal mechanisms data, with 4.0 < Mw < 7.0 and depth ≤ 50 km. The surface fault traces around Sulawesi Island were obtained from PUSGEN (2017) [11]. The red box indicates the major city with highly dense population.

IOP Conf. Series: Earth and Environmental Science **1525** (2025) 012002 doi:10.1088/1755-1315/1525/1/012002

We first clustered the data using Density-Based Spatial Clustering of Application with Noise (DBSCAN) [12], an unsupervised mechine learning method, to identify clusters where the focal data originated from uniform principal mechanisms, corresponding to relatively homogeneous areas. DBSCAN forms clusters based on data density, using two parameters: the minimum number of points required for a cluster (minPts) and the distance radius for neighboring points (ε) [13]. This method is widely used in areas with a pronounced spatially clustered seismic activity [13,14].

To optimize the clustering parameters, we experimented with various combinations of minPts and ε. We used a minPts value of 20, because this value is utilized as the minimum number of events for testing the iterative joint inversion algorithm [15]. Additionally, we explored minPts values of 15 and 10 to assess the variation in cluster characteristics according to the tectonic setting. Each minPts value was paired with ε values ranging from 0.05 to 0.50, in increments of 0.05 (Table 1). We then applied ternary diagrams and plotted the P-/T- axis distributions for each cluster generated by different parameter pairs to determine which parameter combinations best represented the distinctive characteristics of each tectonic setting.

Table 1. Number of clusters formed for each pair of parameters

Parameters	Number of Cluster	Parameters	Number of Cluster	Parameters	Number of Cluster
minPts10ε0.05	3	minPts15ε0.05	0	minPts20ε0.05	0
minPts10ε0.10	9	minPts15ε0.10	6	minPts20ε0.10	2
minPts10ε0.15	8	minPts15ε0.15	5	minPts20ε0.15	5
minPts10ε0.20	12	minPts15ε0.20	6	minPts20ε0.20	6
minPts10ε0.25	15	minPts15ε0.25	7	minPts20ε0.25	6
minPts10ε0.30	16	minPts15ε0.30	9	minPts20ε0.30	6
minPts10ε0.35	11	minPts15ε0.35	10	minPts20ε0.35	6
minPts10ε0.40	11	minPts15ε0.40	10	minPts20ε0.40	5
minPts10ε0.45	11	minPts15ε0.45	11	minPts20ε0.45	7
minPts10ε0.50	8	minPts15ε0.50	10	minPts20ε0.50	7

Table 2. Distribution of data in each cluster generated with parameters minPts=15 and ε=0.40

Cluster ID	Number of Data	Cluster ID	Number of Data
1	49	6	19
2	81	7	18
3	62	8	742
4	16	9	30
5	24	10	33

IOP Conf. Series: Earth and Environmental Science **1525** (2025) 012002 doi:10.1088/1755-1315/1525/1/012002

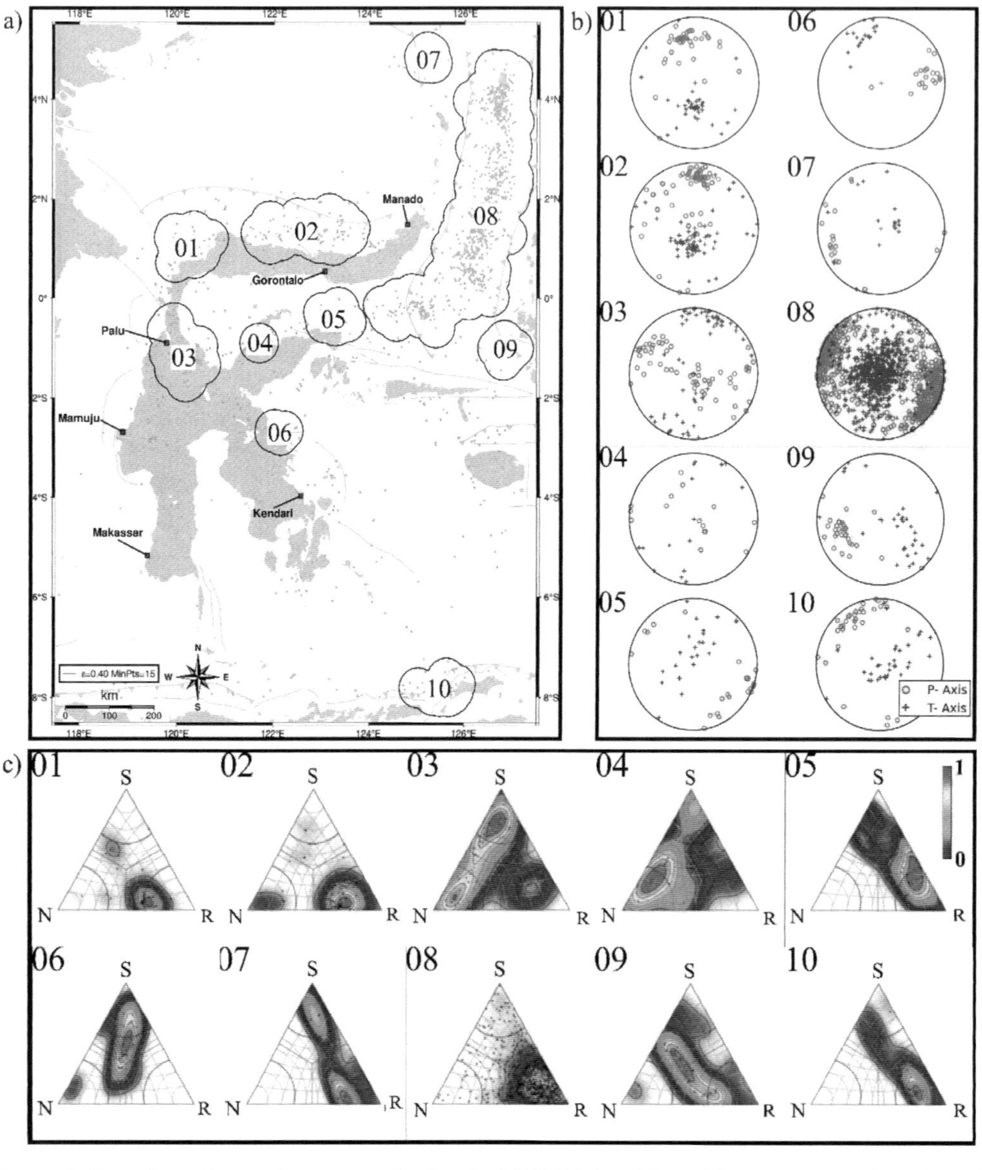

Figure 2. Best clustering results generated using the DBSCAN algorithm with a parameter of minPts=15 and ε=0.40. a) The map shows boundaries of each cluster, with red boxes indicating large cities with high population density. b) The distribution of the P-/T- axes within each cluster is illustrated, where red circles indicate the P-axis direction and blue plus represent the T-axis direction. c) A ternary diagram shows the distribution of event mechanisms in each cluster: S represents strike-slip, N indicates normal, and R denotes reverse faulting.

3. Result and Discussion

In this study, the optimal parameters for the DBSCAN algorithm were determined to be minPts=15 and ε=0.40. These parameters were chosen based on the distribution patterns of focal mechanism types within each cluster, as represented in ternary diagrams and the P-/T- axis distribution. Using these settings, the algorithm successfully clustered the seismic events into 10 clusters (Figure 2), each containing a different number of events (Table 2). Of these clusters, six were located on the island of Sulawesi, while the remaining four were distributed around the island. In addition to the clustered earthquake events, we identified 315 earthquake events categorized as noise, which are evenly distributed across the study area. These events were excluded from the stress inversion analysis due to their insufficient density based on the defined parameters. This exclusion ensures that the data used for stress inversion consist solely of earthquakes generated from a uniform stress orientation, thereby preventing the inclusion of events influenced by different stress regimes. By doing so, potential uncertainties in the analysis are minimized.

Clusters 01 and 02 consist of 49 and 81 earthquake events, respectively, distributed across the northern part of the northern arm of Sulawesi. Both clusters exhibit a compressional regime, characterized by a predominance of reverse fault mechanisms, resulting from the subduction activity of the Sulawesi Sea in the region. Cluster 03, which contains 62 events, displays a more diverse range of focal mechanisms. The ternary diagram reveals two dominant earthquake types within this cluster: strike-slip and normal. The primary fault in this cluster is the Palu-Koro Fault, which exhibits a transform regime characterized by left-lateral movement, resulting in events with strike-slip mechanisms. Additionally, normal mechanisms are generated by smaller faults in the area, including the Palolo Fault and Bada Fault [11]. Clusters 04 and 05, containing 16 and 24 events respectively, exhibit complex seismic characteristics due to their location in regions where the latest Miocene collision occurred [1]. According to the ternary diagram, Cluster 04 is dominated by normal faulting, indicating an extensional setting influenced by the behavior of the Lalanga ridge and Tongian ridge [1]. In contrast with Cluster 04, Cluster 05 is dominated by reverse fault mechanisms, which are attributed to its position within the North-Vergent thrust zone [1]. Cluster 06 consists of earthquakes occurring around the Matano Fault. The ternary diagram for this cluster indicates a predominance of strike-slip mechanisms, resulting from the shear faulting activity along the Matano Fault. The remaining four clusters, located outside the mainland of Sulawesi Island, are situated in subduction zones and therefore experience compressional regimes. Cluster 07 is positioned in the West Sangihe trench, Clusters 08 and 09 are found within the Molucca Sea Double Subduction, and Cluster 10 is located in Nusa Tenggara Back Arc thrust. [11].

4. Conclution

Based on the results of this research, the application of unsupervised machine learning using the DBSCAN algorithm can be used effectively in grouping earthquake events in the Sulawesi region according to the similarities in their generating mechanisms. Utilizing the parameters minPts=15 and ε=0.40 as the optimized parameter, we identified 10 distinct clusters. Notably, the clusters associated with the Palu-Koro Fault and Matano Fault exhibit a transform regime, while areas between the Lalanga and Tongian ridges are characterized by extensional setting. The remaining seven clusters display compressional regimes, predominantly formed in subduction zones, including those in the northern arm of Sulawesi, the Molucca Sea Double Subduction area, the West Sangihe trench, the North-Vergent thrust, and the Nusa Tenggara Back Arc thrust.

References

[1] Serhalawan Y and Chen P F 2024 Seismotectonics of Sulawesi, Indonesia *Tectonophysics* **883** 230366 https://doi.org/10.1016/j.tecto.2024.230366.

[2] Socquet A, Simons W, Vigny C, McCaffrey R, Subarya C, Sarsito D, Ambrosius B and Spakman W 2006 Microblock rotations and fault coupling in SE Asia triple junction (Sulawesi, Indonesia) from GPS and earthquake slip vector data *Journal of Geophysical Research: Solid Earth* **111** B8 https://doi.org/10.1029/2005JB003963.

[3] Walpersdorf A, Rangin C and Vigny C 1998 GPS compared to long-term geologic motion of the north arm of Sulawesi *Earth and Planetary Science Letters* **159** (1-2) 47-55 https://doi.org/10.1016/S0012-821X(98)00056-9.

[4] Scarfì L, Langer H, Messina A and Musumeci C 2021 Tectonic regimes inferred from clustering of focal mechanisms and their distribution in space: application to the Central Mediterranean Area *Journal of Geophysical Research: Solid Earth* **126** (1) 2020JB020519 https://doi.org/10.1029/2020JB020519.

[5] Dziewonski A M, Chou T A and Woodhouse J H 1981 Determination of earthquake source parameters from waveform data for studies of global and regional seismicity *Journal of Geophysical Research: Solid Earth* **86** (B4) 2825-2852 https://doi.org/10.1029/JB086iB04p02825.

[6] Ekström G, Nettles M and Dziewoński A M 2012 The global CMT project 2004–2010: Centroid-moment tensors for 13,017 earthquakes *Physics of the Earth and Planetary Interiors* **200** 1-9 https://doi.org/10.1016/j.pepi.2012.04.002.

[7] Storchak D A, Di Giacomo D, Bondár I, Engdahl E R, Harris J, Lee W H, Villaseñor A and Bormann P 2013 Public release of the ISC–GEM global instrumental earthquake catalogue (1900–2009) *Seismological Research Letters* **84** (5) 810-815 https://doi.org/10.1785/0220130034.

[8] Storchak D A, Di Giacomo D, Engdahl E R, Harris J, Bondár I, Lee W H, Bormann P and Villaseñor A 2015 The ISC-GEM global instrumental earthquake catalogue (1900–2009): introduction *Physics of the Earth and Planetary Interiors* **239** 48-63 https://doi.org/10.1016/j.pepi.2014.06.009.

[9] Di Giacomo D, Engdahl E R and Storchak DA 2018 The ISC-GEM earthquake catalogue (1904–2014): status after the extension project *Earth System Science Data* **10** (4) 1877-1899 https://doi.org/10.5194/essd-10-1877-2018.

[10] Massinai M F I, Lantu, Aswad S and Massinai M A 2015 Tectonics earthquake distribution pattern analysis based focal mechanisms (Case study Sulawesi Island, 1993–2012) In *AIP Conference Proceedings* **1658** 1 https://doi.org/10.1063/1.4915021.

[11] National Center for Earthquake Studies of Indonesia (PUSGEN) 2017 *Indonesia's Earthquake Source and Hazard Map* (in Bahasa Indonesia) Indonesia Research and Development Agency of Ministry of Public Work and Housing ISBN 978-602-5489-01-3.

[12] Ester M, Kriegel H P, Sander J and Xu X 1996 A density-based algorithm for discovering clusters in large spatial databases with noise In *kdd* **96** 34 226-231.

[13] Wollin C, Bohnhoff M, Vavryčuk V and Martínez-Garzón P 2019 Stress inversion of regional seismicity in the Sea of Marmara Region, Turkey *Pure and Applied Geophysics* **176** 3 1269-1291 https://doi.org/10.1007/s00024-018-1971-1.

[14] Schoenball M and Ellsworth W L 2017 A systematic assessment of the spatiotemporal evolution of fault activation through induced seismicity in Oklahoma and southern Kansas *Journal of Geophysical Research: Solid Earth* **122** 12 10-189 https://doi.org/10.1002/2017JB014850.

[15] Vavryčuk V 2014 Iterative joint inversion for stress and fault orientations from focal mechanisms *Geophysical Journal International* **199** 1 69-77 https://doi.org/10.1093/gji/ggu224.

International Conference on Geosciences and Geotechnology (ICOGEST 2024) IOP Publishing
IOP Conf. Series: Earth and Environmental Science **1525** (2025) 012003 doi:10.1088/1755-1315/1525/1/012003

Mapping of geological structures based on horizontal gradient of gravity anomaly in Palu Valley, Central Sulawesi, Indonesia

Rustan Efendi*[1], Ilham Nuryadin[1], Abdullah[1], Sofian[2]

[1] Geophysics Engineering Department Tadulako University Palu Indonesia
[2] Meteorology, Climatology and Geophysics Agency of Palu Indonesia

* E-mail: rstefendi67@gmail.com

Abstract. Palu is one of the cities that is a residential center in carrying out socio-political and economic activities located in the Palu valley. The existence of the Palu-Koro fault makes this city very vulnerable to earthquake shocks. Earthquake activity causes strong rock deformation that allows the formation of local faults in the Palu Valley. One of the mitigation steps that can be taken is to map the geological structure in this city. This study aims to map the geological structure using horizontal gradients on gravity anomaly data. Gravity measurements were carried out in Palu City and its surroundings with a research area of approximately 11 km x 6 km. Furthermore, data processing and analysis were carried out using horizontal gradient techniques on residual gravity anomaly data. The study produced several local faults that were mapped and spread across the western and eastern regions of the Palu Valley in varying directions. In the western region of the Palu Valley, a fault was identified that extends from north to south. This fault is estimated to be the Palu-Koro fault plane formed by an earthquake.

Keyword : Gravity Anomaly, Horizontal Gradient, Fault, Palu Valley

1. Introduction

Palu Valley is one of the areas in Central Sulawesi where various community activities take place. This area is occupied by the center of government, economy, settlements and socio-culture for the people of Palu City and Sigi Regency. This area also has complex geological conditions as indicated by the existence of the Palu Koro fault that crosses western part of this Valley. Existence of the Palu Koro fault crossing the Palu Valley makes this area very prone to earthquakes. One of the steps to mitigate earthquake disasters is to map the geological structure in this city. Geological structures in the form of faults have been identified in the Palu valley which are suspected to be local faults and the Palu Koro fault using Euler Deconvolution and geomagnetic data modeling by [1], [2]. The results of research using gradient techniques on gravity anomaly data have identified fracture zones located on the left and right sides of Palu Koro fault[3].

The earthquake of September 28, 2018 has caused strong deformation of rocks, allowing the formation of local faults in the Palu valley. Thus, it is necessary to re-map the geological conditions of the Palu valley after the earthquake of September 28, 2018. One technique that can be used to identify geological structures is the horizontal gradient on gravity anomaly data. Advantage of gradient technique is able to reduce ambiguity so that it can identify geological structures clearly and accurately as has been done by[4], [5]. This study applies the horizontal gradient technique to gravity anomaly data to map geological structures after the earthquake of September 28, 2018.

2. Geological Setting

The geological conditions of the study area and its surroundings consist of alluvium and coastal deposits consisting of coral limestone and gravel, sand, mud. The youngest unit is alluvium which is formed as a result of the destruction of the hilly area that borders it. Other formations found in this area are Molasses Celebes Sarasin which includes mudstone, limestone and conglomerate. The Tinombo Formation includes volcanic rocks, limestone, flint, shale and sandstone. As well as a complex of metamorphic rocks including genes, marble, amphibolite schist and mica schist [6], [7] as shown in figure 1.

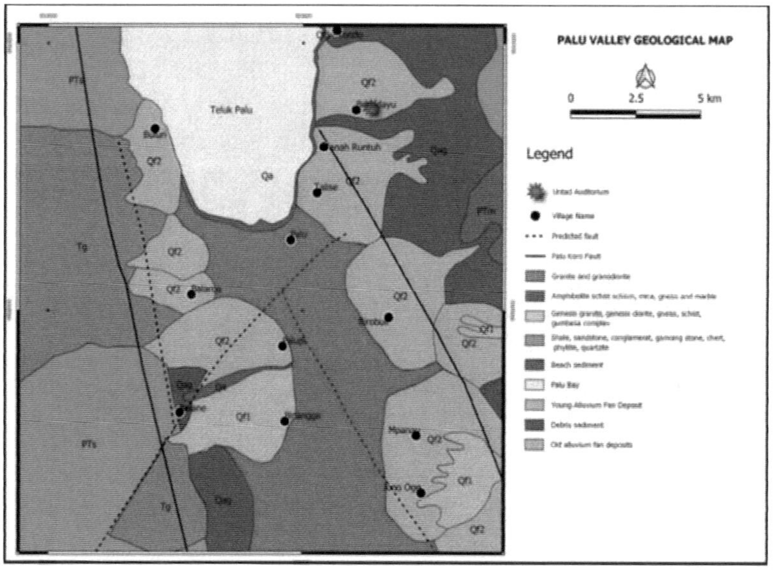

Figure 1. Geological map of the research location[7]

3. Methodology

3.1. Gravity Data

The gravity data used in this study are the results of measurements using the Burris Gravity Meter in July to August 2023. The measurement area is approximately 11 km x 6 located in Palu City with a total of approximately 840 measurement stations. The measurement results are processed by making several corrections to obtain the Complete Bouguer Anomaly as shown in Figure 2a. Furthermore, the Butterworth filter is used to obtain the residual anomaly as shown in Figure 2b.

Gravity data used in this study are the results of measurements using the Burris Gravity Meter in July to August 2023. The measurement area is approximately 11 km x 6 located in Palu City with a total of approximately 840 measurement stations. The measurement results are processed by making several corrections to obtain the Complete Bouguer Anomaly as shown in Figure 2a. Furthermore, the Butterworth filter is used to obtain the residual anomaly as shown in Figure 2b

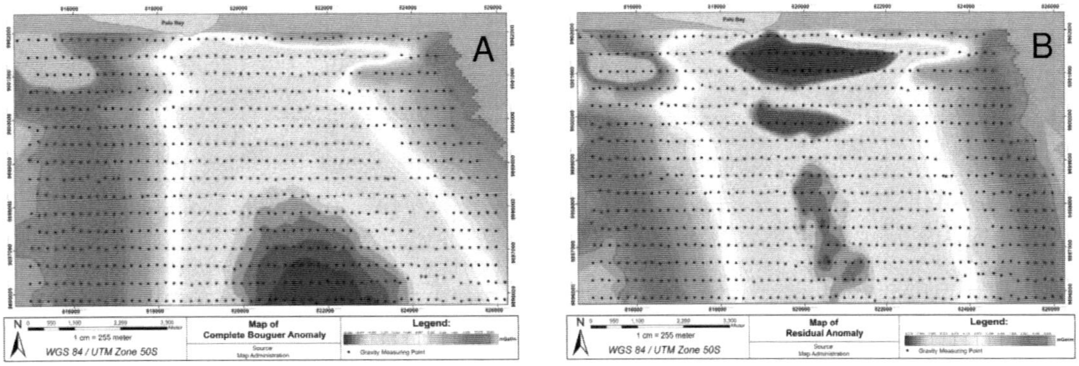

Figure 2. (a) Complete Bouguer Anomaly Map (b) Residual Gravity Anomaly Map

3.2. Horizontal Gradient Magnitude

In general, horizontal gradient analysis on gravity data describes the vertical and lateral location of the source below the surface. Therefore, this analysis provides a comprehensive picture of the fault pattern in the study area. This is very helpful in understanding the structural and tectonic framework in a valley [8]. Mathematically, Horizontal Gradient Magnitude is expressed as follows [4]:

$$HGM = \sqrt{\left(\frac{\partial F}{\partial x}\right)^2 + \left(\frac{\partial F}{\partial y}\right)^2}$$

Where $\partial F/\partial x$ and $\partial F/\partial y$ are the horizontal gradients of F in x and y.

4. Results and Discussion

Geological structure mapping is based on horizontal gradient magnitude (HGM) on the residual anomaly. The distribution of HGM magnitude in the study area is shown in Figure 3a where the maximum value is colored red and the low value is colored blue. In Figure 3a, the maximum value of HGM magnitude is spread in the west and east of the Palu valley with an irregular straightness direction. In the western part, the maximum value of HGM is denser than in the central and eastern parts of the Palu valley. Furthermore, the maximum value of HGM is used to map the traces of geological structures in the study area. Based on the maximum value of HGM, several geological structures can be mapped including faults with irregular directions as shown in Figures 3a and 3b.

The identified faults are interpreted as regional faults and local faults. Regional faults are interpreted based on the tendency of continuity of HGM values, while the maximum HGM value which tends to be discontinuous is interpreted as a local fault. Regional faults are identified in the western part and are thought to be the Palu Koro fault plane which extends from north to south, while regional faults identified in the eastern part which are thought to be opposite of Palu Koro fault are named Janedo fault.

The results of this identification are validated with the results of research conducted by [9] and geological maps from [6], [7]. conducted direct observations at 16 locations that experienced shifts and cracks. [9] conducted direct observations at 16 locations that experienced shifts and cracks caused by the earthquake of September 28, 2018. The results of the study by [9] are in accordance with the results of the identification of faults based on the maximum HGM value. Local faults were identified on the left and right sides of the Palu Koro fault and the Janedo fault. The pattern of local faults shows an appearance that is not connected to each other and forms a destruction zone. This destruction zone also includes the Balaroa and Petobo liquefaction areas of Palu City as shown in Figure 3b.

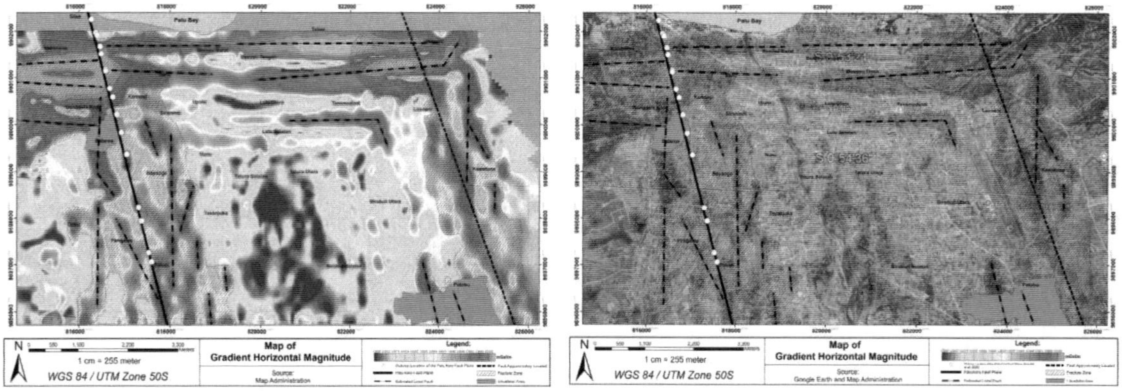

Figure 3. (a) Horizontal Magnitude Map of Gravity Anomaly. (b) Horizontal Magnitude Map of Gravity Anomaly Overlaid with Google Earth Imagery.

5. Conclusion

The results of the study show that HGM on gravity data can map geological structures in the form of regional faults and local faults formed in the Palu valley. The regional fault suspected to be Palu Koro fault plane is located in the west and the fault estimated to be located in the east is called Janedo fault. The results of the study can also map local faults located on the left and right sides of the Palu Koro fault and the Janedo fault. Location of these local faults forms a destruction zone around the Palu Koro fault and the Janedo fault.

References

[1] R. Efendi, E. Syamsuddin, Sandra, and Sabrianto, "Subsurface structure model using geomagnet data of Palu valley after the 2018 earthquake," *IOP Conf. Ser. Earth Environ. Sci.*, vol. 575, no. 1, 2020, doi: 10.1088/1755-1315/575/1/012236.

[2] R. Efendi, Sandra, K. Maulidin, and H. Leopatty, *Geomagnetic Data Interpretation Using 2D Modeling of*

IOP Conf. Series: Earth and Environmental Science **1525** (2025) 012003 doi:10.1088/1755-1315/1525/1/012003

Subsurface Rock Layer Structures in the Palu Valley Province, Central Sulawesi, vol. 1. Atlantis Press International BV, 2023. doi: 10.2991/978-94-6463-228-6_8.

[3] Setianingsih, R. Efendi, W. G. A. Kadir, D. Santoso, C. I. Abdullah, and S. Alawiyah, "Gravity Gradient Technique to Identify Fracture Zones in Palu Koro Strike-slip Fault," *Procedia Environ. Sci.*, vol. 17, pp. 248–255, 2013, doi: 10.1016/j.proenv.2013.02.035.

[4] L. T. Pham, "Mapping the structural configuration of the northern part of the Central Indian Ridge from satellite gravity data using derivatives of the horizontal gradient," *Adv. Sp. Res.*, vol. 74, no. 4, pp. 1648–1663, 2024, doi: https://doi.org/10.1016/j.asr.2024.05.054.

[5] L. T. Pham, S. P. Oliveira, C. V. A. Le, N. T. Bui, A. H. Vu, and D. A. Nguyen, "Gravity Data Enhancement Using the Exponential Transform of the Tilt Angle of the Horizontal Gradient," *Minerals*, vol. 13, no. 12, 2023, doi: 10.3390/min13121539.

[6] R. Sukamto, "Peta Geologi Lembar Palu, Sulawesi, Skala 1:250.000," *Pus. Penelit. dan Pengemb. Geol. Bandung*, 1973.

[7] Soehaimi dkk, "Geologi Lajur Sesar Aktif Palu Koro segmen Palu Sulawesi Tengah," *PPPG, Bandung*, 1999.

[8] S. O. Elkhateeb, "Analysis of the Bouguer Gravity Data of Northwestern Sinai Basin Using the Horizontal Gradient Magnitude," *J. Egypt. Geophys. Soc.*, vol. 2, no. 1, pp. 31–35, 2024, doi: 10.21608/jegs.2024.374754.

[9] A. I. Abdullah and Abdullah, "A field survey for the rupture continuity of Palu-Koro fault after Donggala earthquake on September 28th, 2018," *J. Phys. Conf. Ser.*, vol. 1434, no. 1, 2020, doi: 10.1088/1742-6596/1434/1/012009.

Identification of Earthquake-Prone Zones in West Sulawesi Province for Disaster Mitigation Using the Probabilistic Seismic Hazard Analysis (PSHA)

Ahmad Hidayah[1*], Bambang Harimei[1], Muhammad Fikri Hayqal[2]

[1] Geophysics, Hasanuddin University, Makassar, Indonesia
[2] Agency of Meteorology, Climatology and Geophysics, Gowa, Indonesia

*E-mail: hidayaha20h@student.unhas.ac.id

Abstract. The West Sulawesi province is one of the earthquake-prone areas in Indonesia. One of the causes of earthquakes in this region is the presence of thrust fault associated with folds, commonly known as the Mamuju-Majene Fault Thrust Belt (FTB). This thrust fault generally moves from north to south, crossing both the sea and the land of West Sulawesi. This study aims to determine the Peak Ground Acceleration (PGA) value using the Probabilistic Seismic Hazard Analysis (PSHA) method and to analyse earthquake-prone zones based on seismic hazard curves in West Sulawesi. The study utilizes earthquake data from the BMKG and USGS catalogs for the period 1973–2023. The research stages include identifying seismic source models, characterizing seismic sources, determining the logic tree and attenuation function, also conducting Probabilistic Seismic Hazard Analysis (PSHA) in the form of hazard maps and hazard curves. The result of PSHA processing using USGS PSHA 2007 software is in hazard value on bedrock for a probability of exceedance of 2% in 50 years (earthquake recurrence period of 2475 years) yielded a range of acceleration values of 0.05–0.9 g for PGA, 0.2–2.5 g for short periods (T = 0.2 seconds), and 0.05–0.7 g for long periods (T = 1 second). The PSHA analysis indicates that high PGA values predominantly occur in North Mamuju, Mamuju, Majene and Polewali Mandar regency. This is because these three districts are closest to fault sources, including the Palukoro Fault and the Makassar Strait Fault.

1. Introduction

West Sulawesi is one of the earthquake-prone areas in Indonesia [1]. One of the causes of earthquakes in this region is the presence of a thrust fault associated with fold, known as the Mamuju-Majene Fold Thrust Belt (FTB). This thrust fault generally stretches from north to south, crossing both the sea and land of West Sulawesi [2].

Disaster history records that the earthquake in West Sulawesi on January 15, 2021, was not the first to occur. West Sulawesi had experienced earthquakes with magnitudes of M ≥ 6 on the Richter scale in 1820, 1969, 1976, and 1984. These earthquakes caused many casualties and injuries, partly because they triggered tsunamis [3].

Therefore, special attention is needed for the West Sulawesi region, which is significantly affected by earthquake disasters, as earthquakes remain unpredictable in terms of timing, magnitude, and location. One effort to minimize the impact of earthquake disasters is to create an earthquake hazard map, reflected in a peak ground acceleration map. The method used in earthquake hazard mapping is Probabilistic Seismic Hazard Analysis (PSHA), which will be applied in this study. The advantage of the PSHA method is that it accounts for the influence of uncertainties in the analysis, such as earthquake size, location, and frequency, allowing these uncertainties to be identified, calculated, and integrated using a rational approach. This provides a more complete picture of earthquake occurrences and integrates the seismic hazards of a location from various seismic sources.

1.1 Data Collection

The earthquake data required for this research was collected from secondary data catalogs of BMKG and USGS for the period 1973 - 2023, covering the West Sulawesi region and its surroundings. The magnitude scale used is the moment magnitude (Mw), with a maximum depth of 300 km.

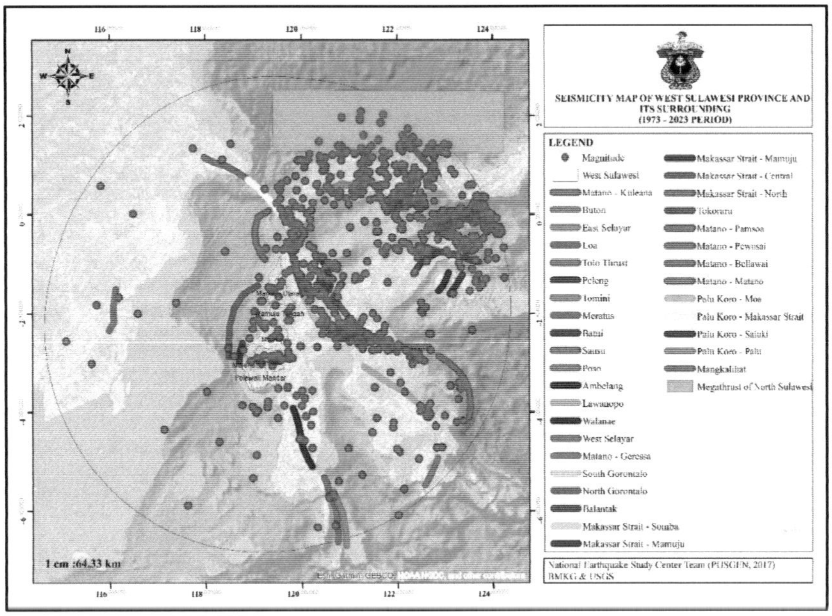

Figure 1. Seismicity Map of West Sulawesi and its surroundings

1.2 Magnitude Scale Conversion

The earthquake data obtained from BMKG and USGS from 1973 to 2023 consists of various types of magnitudes. The purpose of magnitude conversion is to standardize the earthquake magnitudes from different scales into the Mw scale. In this research, the magnitude conversion refers to the following table [4].

Table 1. Magnitude Conversion

Magnitude	Magnitude Conversion Relation
M_S	$M_W = 1.27(M_L - 1) - 0.016M_L^2)$
M_B	$M_W = 0.63M_s + 2.5$
M_b	$M_B = 1.5M_b - 2.2$
M_W	$M_W = 0.85M_b + 1.03$

1.3 Identification and Modelling of Seismic Sources

The seismic source zone is defined as an area that has the same seismic degree, where every point within the zone has an equal probability of future earthquake occurrences [5]. A seismic source model is necessary as a connection between earthquake occurrences and the calculation model used to determine the level of seismic hazard. In modeling, the boundaries of a seismic source zone used are as follows [6]:

1. The distance from the seismic source location to the examined site is limited to 500 km.
2. The seismic source depth is limited to 300 km.
3. The attenuation function used for each model is the one considered appropriate for the seismicity characteristics and seismic source model for the Indonesian region.

Seismic sources in the Probabilistic Seismic Hazard Analysis (PSHA) program from USGS are classified into three types of seismic source zones, including [7]:

1. Subduction zone, which refers to the earthquake zone occurring near the boundary between an oceanic plate subducting beneath a continental plate. In this study, there is a subduction within 500 km from the research area
2. Fault zone, which refers to the zone of shallow crustal fault earthquakes with strike-slip, reverse, or normal mechanisms occurring along well-defined faults. Additionally, information on slip rates, dips, fault lengths, and locations is also provided. In this study, there are 2 faults on the Kalimantan Island and 18 faults of Sulawesi Island.

1.4 Characterization of Seismic Sources

Probabilistic seismic hazard analysis requires input parameter data, including a-b parameters, maximum magnitude, slip rate, fault mechanism type, and fault dimensions. The seismic source characterization used in this study is based on National Earthquake Study Center Team (PUSGEN 2017), as shown in Table 2 and Table 3.

Table 2. Subduction seismic source parameters within a 500 km radius of West Sulawesi [8].

Indeks	Structure Name		L (km)	W (km)	a	b	Mmax Geodesy
	Main	Segmen					
M15	Northern Sulawesi Thrust	Northern Sulawesi	480	120	4.28	0.91	8.5

Table 3. Fault seismic source parameters within a 500 km radius of West Sulawesi [8].

ID	Structure Name		Slip Rate	Sense Mecanism	Dip	Top	Bottom	L (Km)	M Max
	Main	Segmen							
				Kalimantan Island					
1	Mangkalihat		0.5	Strike-slip	90	3	18	111	7.0
2	Meratus		0.5	Reverse-slip	45 W	3	18	105	7.0
				Sulawesi Island					
3	Palu koro *fault*	Makassar Strait	10	Left-lateral SS	90	3	18	130	7.1
4	Palu koro *fault*	Palu	10	Left-lateral SS	90	3	18	31	6.8
5	Palu koro *fault*	Saluki	10	Left-lateral SS	90	3	18	44	6.9

6	Palu koro *fault*	Moa	10	Left-lateral SS	90	3	18	66	7.1
7	Matano	Kuleana	7	Left-lateral SS	90	3	18	22	6.6
8	Matano	Pewusai	7	Reverse-slip	45N	3	18	46	6.9
9	Matano	Matano	7	Left-lateral SS	90	3	18	35	6.8
10	Matano	Pamsoa	7	Left-lateral SS	90	3	18	44	6.9
11	Matano	Ballawai	77	Left-lateral SS	90	3	18	26	6.7
12	Matano	Garessa	7	Left-lateral SS	91	3	18	80	7.2
13	Batui Thrust		1	Reverse-slip	45SE	3	18	84	7.3
14	Balantak		0.5	Right-lateral SS	90	3	18	149	7.6
15	Makassar *Strait Thrust*	North	2	Reverse-slip	45E	3	18	100	7.1
16	Makassar *Strait Thrust*	Central	2	Reverse-slip	45E	3	18	170	7.3
17	Makassar *Strait Thrust*	Mamuju	2	Reverse-slip	45E	3	18	40	7.0
18	Makassar *Strait Thrust*	Somba	2	Reverse-slip	45E	3	18	80	7.3
19	Buton	B	0.1	Strike-slip	90	3	18	60	7.1
20	Peleng		1	Strike-slip	90	3	18	44	6.9
21	Ambelang		1	Reverse-slip	45	3	18	50	7.0
22	Sausu		1	Strike-slip	90	3	18	31	6.8
23	Tokoraru		1	Reverse-slip	45	3	18	80	7.2
24	Poso		0.1	Reverse-slip	45E	3	18	80	7.2
25	Loa		0.1	Strike-slip	90	3	18	86	7.3
26	Tomini		0.1	Reverse-slip	45W	3	18	59	7.1
27	Lawanopo		0.1	Strike-slip	90	3	18	130	7.5
28	Tolo *Thrust*		1	Reverse-slip	45W	3	18	120	7.4
29	Gorontalo	North	1	Strike-slip	90	3	18	74	6.9
30	Gorontalo	South	1	Strike-slip	90	3	18	70	6.8
31	Walanae		0.5	Strike-slip	90	3	18	130	7.1
				Nusa Tenggara Region					
32	Selayar *Fault*	West	0.5	Normal	60E	3	18	152	7.2
33	Selayar *Fault*	East	0.5	Normal	60W	3	18	167	7.2

IOP Conf. Series: Earth and Environmental Science **1525** (2025) 012004 doi:10.1088/1755-1315/1525/1/012004

1.5 Attenuation Function

In this study, the attenuation functions used are the same as those used in the 2017 Indonesia Seismic Source and Hazard Map [8], which are:

Table 4. Attenuation Function [8].

Seismic Source	Attenuation Function	Weight
Fault	Boore-Atkinson NGA [9].	1/3
	Campbell-Bozorgnia NGA [10].	1/3
	Chiou Young NGA [11].	1/3
Subduction	Geomatrix Subduction [12].	1/4
	Atkinson-Boore BC rock and global subduction [13].	1/4
	Zhao, with variable Vs30 [14].	1/2

1.6 Logic Tree

The logic tree model used is adjusted to the seismic source model employed. The use of a logic tree in the Probabilistic Seismic Hazard Analysis (PSHA) method is highly necessary due to the uncertainty factors in data management for seismic hazard analysis, which can introduce uncertainty. Therefore, the application of a logic tree is required as an approach or alternative solution.

1.7 Probabilistic Seismic Hazard Analysis (PSHA)

Probabilistic Seismic Hazard Analysis (PSHA) includes the probability of distance and magnitude, as well as considering the uncertainty of magnitude, distance, and exceeded intensity parameters. In this study, the seismic hazard analysis was conducted using the USGS PSHA 2007 software.

For earthquake calculations, the National Seismic Hazard Maps were utilized. In this calculation, it is assumed that the magnitude (M) and distance (R) are continuous independent random variables [9].

$$P[I \geq i] = \int_r \int_m P[I \geq i | M\ dan\ R] . f_M(m) . f_R(r) dm\ dr \tag{1}$$

where:

fM : distribution function of magnitude

fR : distribution function of hypocentral distance

P[I ≥ i | m dan r] : conditional probability of intensity I exceeding value i at a given location for an earthquake with magnitude M and hypocentral distance R.

2. Result and Discussion

The research results consist of the Peak Ground Acceleration (PGA) values and the acceleration response spectra for short-period (T = 0.2 seconds) and long-period (T = 1 second) at bedrock. The hazard analysis using a probabilistic approach is limited to a 2% exceedance probability within 50 years, equivalent to an earthquake return period of 2,475 years. The analysis in this study is focused on six locations such as North Mamuju Regency, Central Mamuju Regency, Mamuju Regency, Majene Regency, Mamasa Regency, and Polewali Mandar Regency.

2.1 Peak Ground Acceleration (PGA) in West Sulawesi

The PGA analysis results for at the bedrock in West Sulawesi Province show a range of values between 0.05 g - 0.9 g (for PGA), 0.2 – 2.5 g (for T = 0.2 seconds) and 0.05 – 0.7 g (for T = 1 seconds) as displayed on the peak ground acceleration (PGA) map at bedrock in **Figure 2**, **Figure 3** and **Figure 4**.

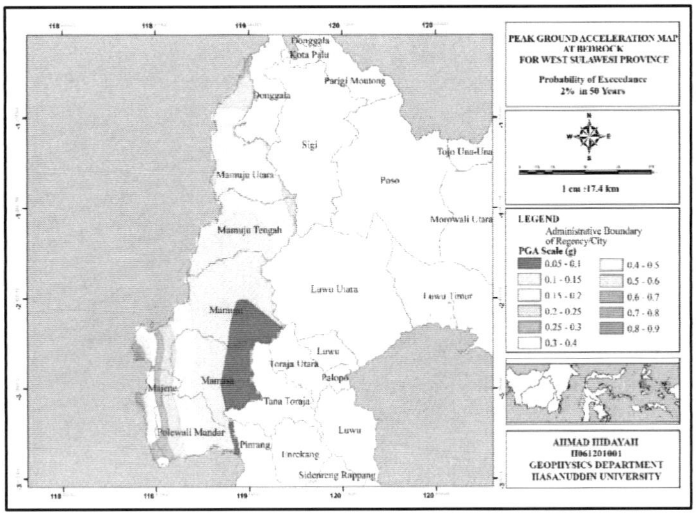

Figure 2. Peak Ground Acceleration (PGA) Map at Bedrock of West Sulawesi Province for a 2% Exceedance Probability in 50 Years

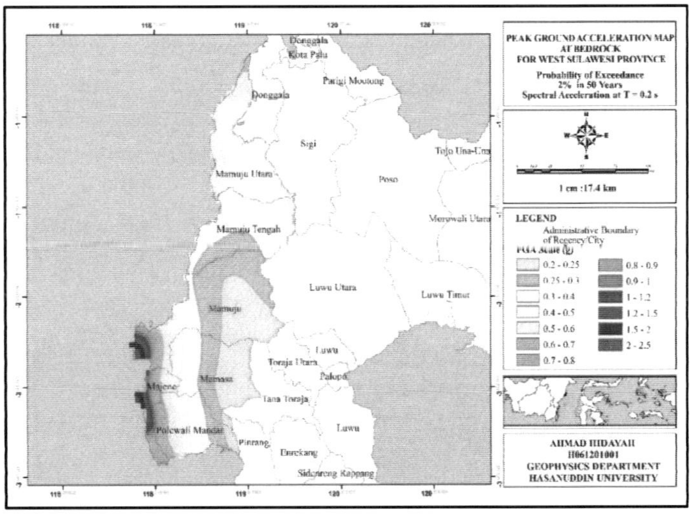

Figure 3. Peak Ground Acceleration (PGA) Map at Bedrock of West Sulawesi Province at Spectral Conditions of T = 0.2 Seconds for a 2% Exceedance Probability in 50 Years

IOP Conf. Series: Earth and Environmental Science **1525** (2025) 012004 doi:10.1088/1755-1315/1525/1/012004

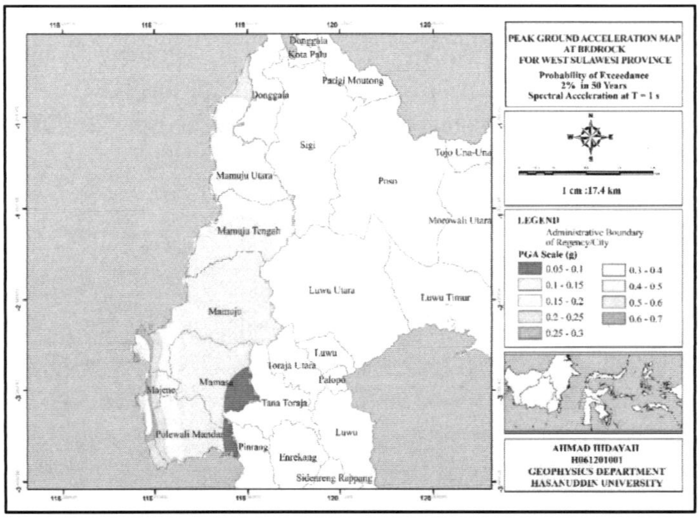

Figure 4. Peak Ground Acceleration (PGA) Map at Bedrock of West Sulawesi Province at Spectral Conditions of T = 1 Seconds for a 2% Exceedance Probability in 50 Years

Figure 2, 3 and 4 shows that parts of North Mamuju Regency, Mamuju Regency, Majene Regency and Polewali Mandar Regency have higher PGA values compared to other areas. This is due to these regions being closest to the Palu Koro Fault and the Makassar Strait Fault seismic sources.

2.2 Comparison of Peak Ground Acceleration (PGA) values

The Peak Ground Acceleration (PGA) values in West Sulawesi Province in this study are lower compared to the PUSGEN 2017 map. However, both maps exhibit a similar pattern, where North Mamuju, Central Mamuju, Polewali Mandar, and Majene regencies have higher PGA values compared to other areas in West Sulawesi. The difference in PGA values is due to the seismic sources and the earthquake data used, which comes from different time periods; this study uses data from 1973 to 2023, while PUSGEN covers the period from 1907 to 2016.

Table 5. Comparison of Peak Ground Acceleration (PGA) value

Period (second)	PGA Scale (g)		Difference (%)
	This Research	PUSGEN 2017	
PGA	0.05-0.9	0.25-1.2	45
0.2	0.2-2.5	0.6-2.5	14.81
1	0.05-0.7	0.1-1.2	73.33

2.3 Seismic Hazard Curve

The seismic hazard curve represents the relationship between the average annual exceedance rate and the level of hazard (ground motion acceleration) that occurs. Through the seismic hazard curve, it is possible to identify the seismic sources that have the most influence on a seismically vulnerable area. This study will generate the seismic hazard curve for all regencies in West Sulawesi province.

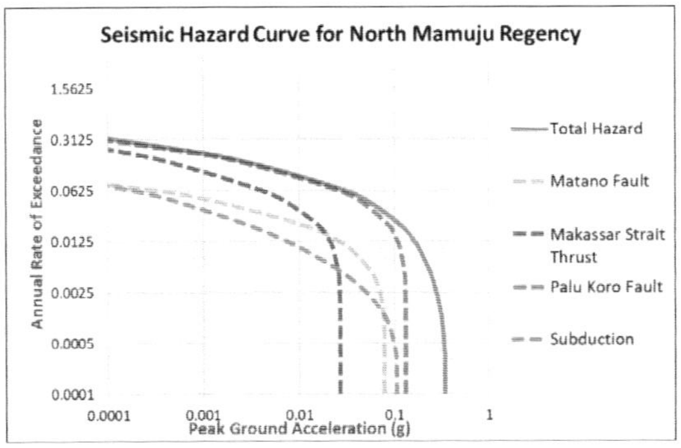

Figure 5. Seismic Hazard Curve for North Mamuju

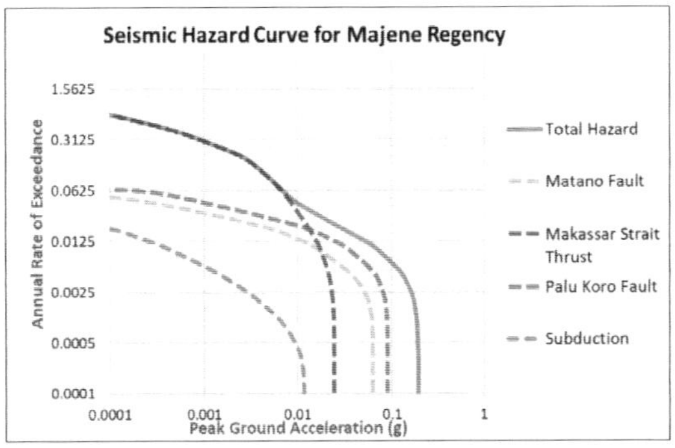

Figure 8. Seismic Hazard Curve for Majene

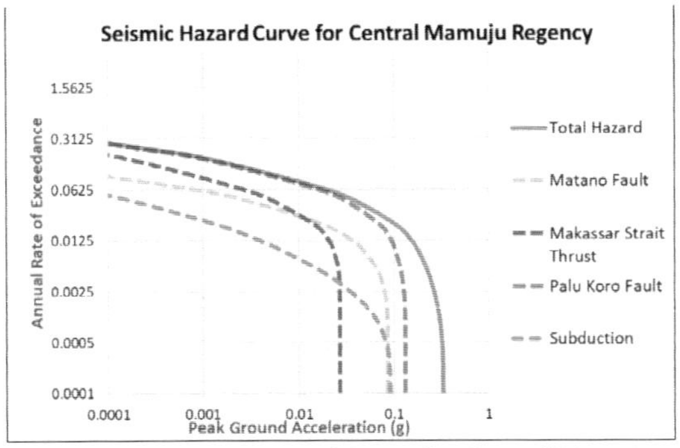

Figure 6. Seismic Hazard Curve for Central Mamuju

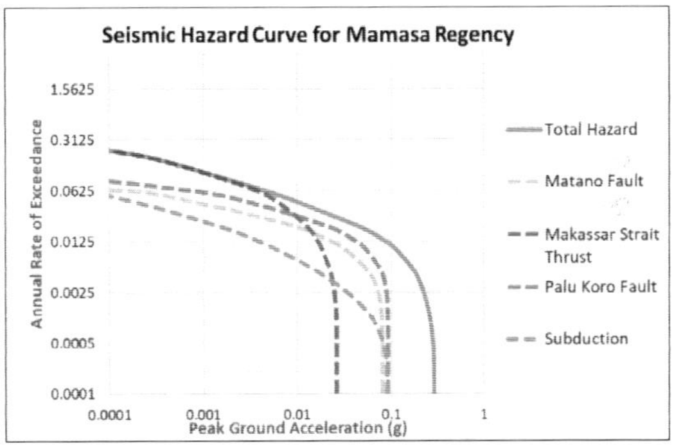

Figure 9. Seismic Hazard Curve for Mamasa

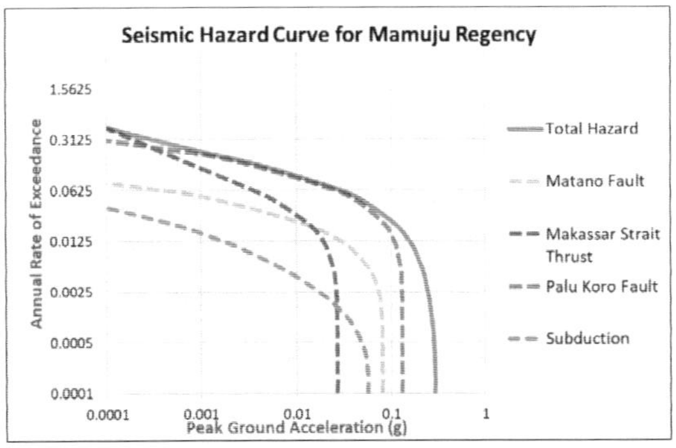

Figure 7. Seismic Hazard Curve for Mamuju

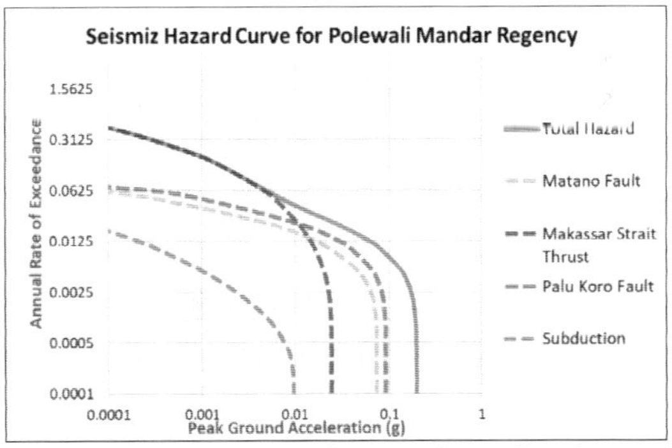

Figure 10. Seismic Hazard Curve for Polewali Mandar

The seismic hazard curve generated in the bedrock of North Mamuju and Central Mamuju Regency, as shown in **Figure 5** and **Figure 6**, indicates that the seismic source with the highest annual exceedance rate is the Palu Koro fault. This exceedance rate reaches a value of 0.48 per year, indicating a relatively high frequency of occurrence. Additionally, seismic source with the highest hazard value is the Palu Koro fault, which reaches a value of 0.12 g. The Palu Koro fault is the closest seismic sources to North Mamuju and Central Mamuju Regency. Meanwhile, the seismic hazard curve generated in the bedrock of Mamuju, Majene, Mamasa and Polewali Mandar Regency, as shown in **Figure 7**, **Figure 8**, **Figure 9** and **Figure 10** indicates that the seismic source with the highest annual exceedance rate is the Makassar Strait fault. This exceedance rate reaches a value of 0.48 per year, indicating a relatively high frequency of occurrence. Additionally, seismic source with the highest hazard value is the Palu Koro fault, which reaches a value of 0.12 g. The Palu Koro fault and Makassar Strait Thrust is the closest seismic sources to these four regencies in West Sulawesi.

3. Conclusion

1. The seismic hazard analysis in bedrock for a 2% exceedance probability over 50 years (return period of 2475 years) in West Sulawesi yields a range of acceleration values of 0.1 – 0.9 g for PGA, 0.25 – 2.5 g for short period (T = 0.2 seconds), and 0.05 – 0.7 g for long period (T = 1 second). The highest PGA or hazard values are found in North Mamuju, Mamuju, Majene and Polewali Mandar. This is due to the proximity of these four regencies to the Palu Koro fault and the Makassar Strait fault.
2. Seismic hazard curve in bedrock for North Mamuju, Central Mamuju, Mamuju, Majene, Mamasa and Polewali Mandar indicates that generally fault sources dominate or significantly contribute to seismic hazards.

References

[1] Indri, R., Taunaumang, H., & Reynol Tumimomor, F. (2022). *Analisis Bahaya Gempa Bumi Menggunakan Metode Probabilistic Seismic Hazard Analysis Di Wilayah Likupang, Minahasa Utara* **3**(1).

[2] Baillie, P., & Decker, J. (2022). Enigmatic Sulawesi: The tectonic collage. *Berita Sedimentologi*, **48**(1) 1–30. https://doi.org/10.51835/bsed.2022.48.1.388

[3] Hartono, D., Khoirudin Apriyadi, R., Winugroho, T., Aprilyanto, A., Hadi Sumantri, S., Wilopo, W., & Surya Islami, H. (2021). Analisis Sejarah, Dampak, Dan Penanggulangan Bencana Gempa Bumi Pada Saat Pandemi Covid-19 Di Sulawesi Barat. *PENDIPA Journal of Science Education*, *5*(2), 218–224. https://doi.org/10.33369/pendipa.5.2.218-224

[4] Joshi, R., Bhadauria, S. S., & Kushwaha, S. S. (2020). Probabilistic seismic hazard analysis of Madhya Pradesh (Central India) using alternate source models: a logic tree approach. *Asian Journal of Civil Engineering*, **21**(8) 1399–1414. https://doi.org/10.1007/s42107-020-00286-4

[5] Irsyam, M., Sengara, W., Aldiamar, F., Widiyantoro, S., Triyoso, W., Hilman, D., Kertapati, E., Meilano, I., Asrurifak, M., & Ridwan, M. (2010). *Ringkasan Hasil Studi Tim Revisi Peta Gempa Indonesia 2010.*

[6] Sunardi, B. (2013). *Peta Deagregasi Hazard Gempa Wilayah Jawa dan Rekomendasi Ground Motion di Empat Daerah*. Universitas Islam Indonesia.

[7] Wells, D. L., & Coppersmith, K. J. (1994). New empirical relationships among magnitude, rupture length, rupture width, rupture area, and surface displacement. *Bulletin of the Seismological Society of America*, **84**(4) 974–1002. https://doi.org/10.1785/BSSA0840040974

[8] Tim Pusat Studi Gempa Nasional. (2017). *Peta Sumber Dan Bahaya Gempa Indonesia Tahun 2017.* Pusat Penelitian dan Pengembangan Perumahan dan Permukiman Badan Penelitian dan Pengembangan Kementerian Pekerjaan Umum dan Perumahan Rakyat.

[9] Asrurifak, M., Budiono, B., Triyoso, W., & Hendriyawan. (2010). Development of Spectral Hazard Map for Indonesia with a Return Period of 2500 Years using Probabilistic Method. *Civil Engineering Dimension*, **12**(1) 52–62.

IOP Conf. Series: Earth and Environmental Science **1525** (2025) 012005 doi:10.1088/1755-1315/1525/1/012005

Aquifer Mapping Using Geoelectric Method in Bulukumba Regency, South Sulawesi

Abdul Mun'im Thariq[1]*, Hamsu Abdul Gani[1], Darmawang[2]

[1]Departement of Vocational Engineering Education, Universitas Negeri Makassar, 90222, Indonesia.

[2]Departement of Automotive Engineering Education, Universitas Negeri Makassar, 90224, Indonesia.

* Email: abdu.thq@gmail.com

Abstract. This research uses the geoelectric method to analyze the aquifer depth in Bulukumba Regency, South Sulawesi. The geoelectric method was applied to identify subsurface characteristics and determine the depth of aquifers. Based on measurements taken at several locations, the study results showed variations in aquifer depth ranging from 49 to 63 meters below the surface. The distribution of aquifer depth indicated significant variation, with the Ujung Loe area and its surroundings having deeper aquifers than other areas, such as Kajang and Bulukumba. This study provides essential information on the potential groundwater resources in Bulukumba Regency, which can serve as a reference for further exploration and groundwater resource management in the region.

1. Introduction

The Bulukumba District in South Sulawesi is defined by its geological formations and groundwater potential, which are critical for sustainable water resource management. Identifying aquifers in this region can be effectively accomplished using geo-electrical methods, particularly the resistivity technique. This approach allows for delineating subsurface geological structures and assessing groundwater potential, which is essential for agricultural irrigation and other uses. Agussalim (2022)[1] conducted a study in Bontorannu Village, Bulukumba, using the Schlumberger array of the resistivity method to identify groundwater potential. Their findings highlighted the area's suitability for irrigation, emphasizing the importance of groundwater in a predominantly rainfed agricultural landscape. Many align with the broader application of geo-electrical methods in groundwater exploration, as demonstrated by Dey et al. (2010)[2], who noted the effectiveness of vertical electrical sounding and other techniques in mapping groundwater potential in challenging terrains. The resistivity method is particularly advantageous due to its non-invasive nature and ability to provide clear contrasts in resistivity, indicating the presence of groundwater [2].

In Bulukumba, the geological characteristics, including various lithological layers, play a significant role in aquifer identification. Resistivity data can reveal the thickness and extent of aquifers, as evidenced by studies in other regions, such as Kaltungo, Nigeria, where the resistivity method successfully delineated multiple geoelectric layers indicative of groundwater zones [2]. Additionally, the application of 2D electrical resistivity imaging has become increasingly popular for addressing hydrological challenges, allowing for a comprehensive understanding of subsurface conditions [3].

Integrating geo-electrical methods with geological surveys enhances the reliability of aquifer identification. For example, in Banda Aceh, Indonesia, the resistivity method was used to map groundwater layers, contributing to a better understanding of aquifer distribution [4]. This method can be similarly applied in Bulukumba to ascertain aquifer characteristics and their potential for a sustainable water supply.

Moreover, the geological framework of Bulukumba District, which may include sedimentary and volcanic formations (Figure 1), influences the resistivity obtained during surveys. Variations in resistivity can indicate different geological materials, which are crucial for understanding groundwater flow and storage [5]. Findings from such studies can inform local water management strategies, ensuring that groundwater resources are utilized while maintaining ecological balance.

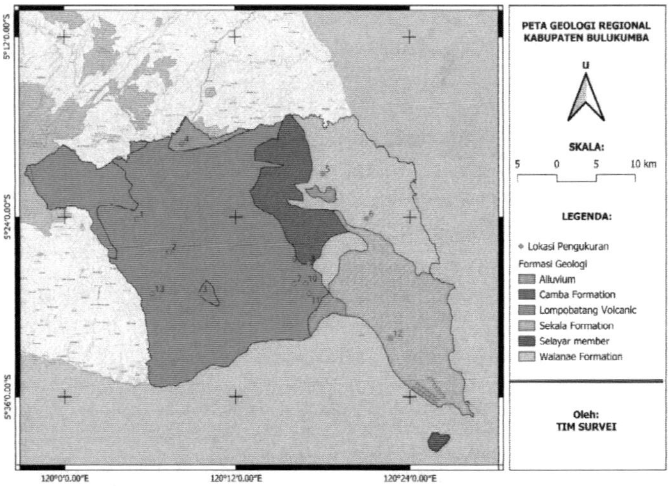

Figure 1. Geological formation of Bulukumba.

In conclusion, geo-electrical methods, particularly resistivity techniques, are vital for aquifer identification in Bulukumba District, South Sulawesi. By leveraging these methods, researchers can gain insights into the subsurface geology and groundwater potential, thereby supporting sustainable water resource management in the region.

2. Material and Method

The research method utilizing the 1D Schlumberger configuration geoelectric technique follows several systematic steps to obtain accurate data regarding subsurface characteristics and groundwater potential [6]. The first step involves selecting measurement locations based on preliminary studies and regional geological analysis to identify points with potential groundwater sources (e.g. 13 observation point in Fig.1). Once the locations are determined, electrodes are set up in the field using the Schlumberger configuration [7], where the current electrodes AB with distance 200 m are positioned symmetrically at increasing distances from the central potential electrodes (M and N). Increasing the distance of the current electrodes allows for deeper penetration of the electrical current into the subsurface, enabling the identification of deeper geological layers.

Subsequently, soil resistivity measurements by a resistivity meter are used on two electrodes. These measurements record the voltage generated when the current flows through the current electrodes, and the recorded values are processed to obtain apparent resistivity. This measurement process is repeated at each predefined electrode interval to acquire resistivity variations at different depths. The resistivity data obtained is then analyzed using geoelectric interpretation software to produce a subsurface profile in the form of a 1D model [7], [8], [9]. This model illustrates the resistivity variations at various depths, allowing for the identification of the

IOP Conf. Series: Earth and Environmental Science **1525** (2025) 012005 doi:10.1088/1755-1315/1525/1/012005

presence and thickness of aquifers or other significant geological layers. Following these steps, the acquired data are interpreted to accurately determine the potential and depth of the aquifers. The configuration geoelectrical method used in this research is Schlumberger configuration. For mapping aquifers in the Bulukumba Region, there are thirteen points of data acquisition, and IPI2Win software is used for data processing. The interpretation results are saved in the .csv format used for the aquifer mapping. Finally, interpolation conducted using the inverse distance weight (IDW) method in QGIS software for aquifer mapping is applied.

3. Results and Discussions

This section presents and interprets the findings of the geoelectric surveys conducted to map aquifers in Bulukumba Regency, South Sulawesi. The data obtained are evaluated regarding their implications for groundwater management and potential future applications. Table 1 and the geological formations of the Bulukumba region (Figure 1) serve as references for interpreting the results.

Table 1. Interpretation results of vertical electrical sounding (VES).

ρ(Ωm)	Height	Depth	Interpretation
vertical electrical sounding 1			
72.9	1.26	1.26	Topsoil
22	17.3	18.6	Tuff
330	18.9	37.5	Volcanic rock
1.31			Aquifer
vertical electrical sounding 2			
65.1	1.19	1.19	Topsoil
196	3.31	4.51	Sandstone/fluvial
18.0	46.6	50.9	Aquifer/sandstone
47.6			Sandstone/conglomerate
vertical electrical sounding 3			
98.8	1.95	1.95	Topsoil
35.2	8.55	10.5	Sandstone/fluvial
7.84	46.3	56.8	Aquifer/sandstone
28.3			Sandstone/conglomerate
vertical electrical sounding 4			
19.2	0.18	0.18	Topsoil
82.9	8.31	8.49	silt
13.2	41.2	49.7	Aquifer/sandstone
59.7			conglomerate
vertical electrical sounding 5			
8.31	0.314	0.314	Topsoil
273	1.75	2.06	
13.2	47.1	49.1	Aquifer/sandstone
63			conglomerate
vertical electrical sounding 6			
44.3	0.1	0.1	Topsoil
15	8.33	8.33	Tuff
2.38	44.8	53.3	Aquifer/sandstone
13.1			conglomerate
vertical electrical sounding 7			
12.8	0,73	0.73	Topsoil
185	7.22	7.95	tuff
13.9	52.2	60.2	Aquifer/batupasir
93.6			conglomerate

IOP Conf. Series: Earth and Environmental Science **1525** (2025) 012005 doi:10.1088/1755-1315/1525/1/012005

vertical electrical sounding 8			
ρ(Ωm)	**Height**	**Depth**	**Interpretation**
25.9	0.22	0.22	Topsoil
35.4	3.31	3.52	Fluvial/tuff
2.21	58.0	61.6	aquifer/sandstone
8.68			conglomerate
vertical electrical sounding 9			
40,6	2,72	2,72	Topsoil
72,9	32,2	35,1	Fluvial/tuff
4,02	35,9	71	Aquifer/sandstone
148			Conglomerate
vertical electrical sounding 10			
4,98	0,56	0,56	Topsoil
358	7,02	7.57	Fluvial
33,9	52,9	57,9	Aquifer/sandstone
108			conglomerate
vertical electrical sounding 11			
41,1	2,82	2,82	Topsoil
317	2,21	5,03	Fluvial
1,71	52,9	57,9	Aquifer/sandstone
0,012			conglomerate
vertical electrical sounding 12			
18.1	1.98	1.98	Topsoil
53.5	7.55	9.54	Fluvial
10.7	44.1	53.7	Aquifer/sandstone
13.2			conglomerate
vertical electrical sounding 13			
57.90	2,05	2,05	Topsoil
4.16	16,5	18,6	Fluvial
15.9	34,1	52,7	Aquifer/sandstone
81.3			conglomerate

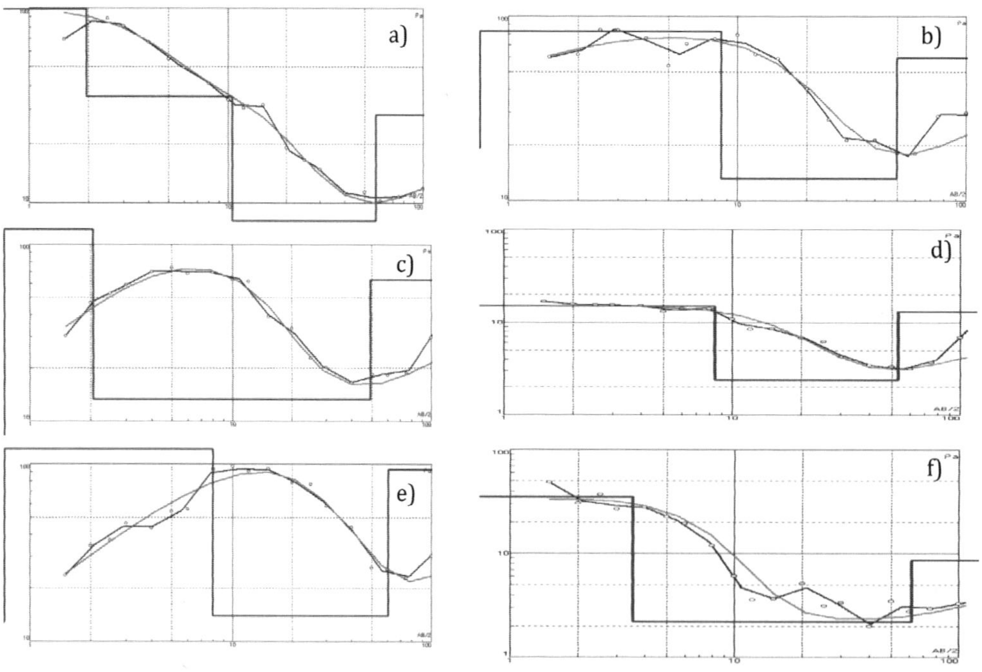

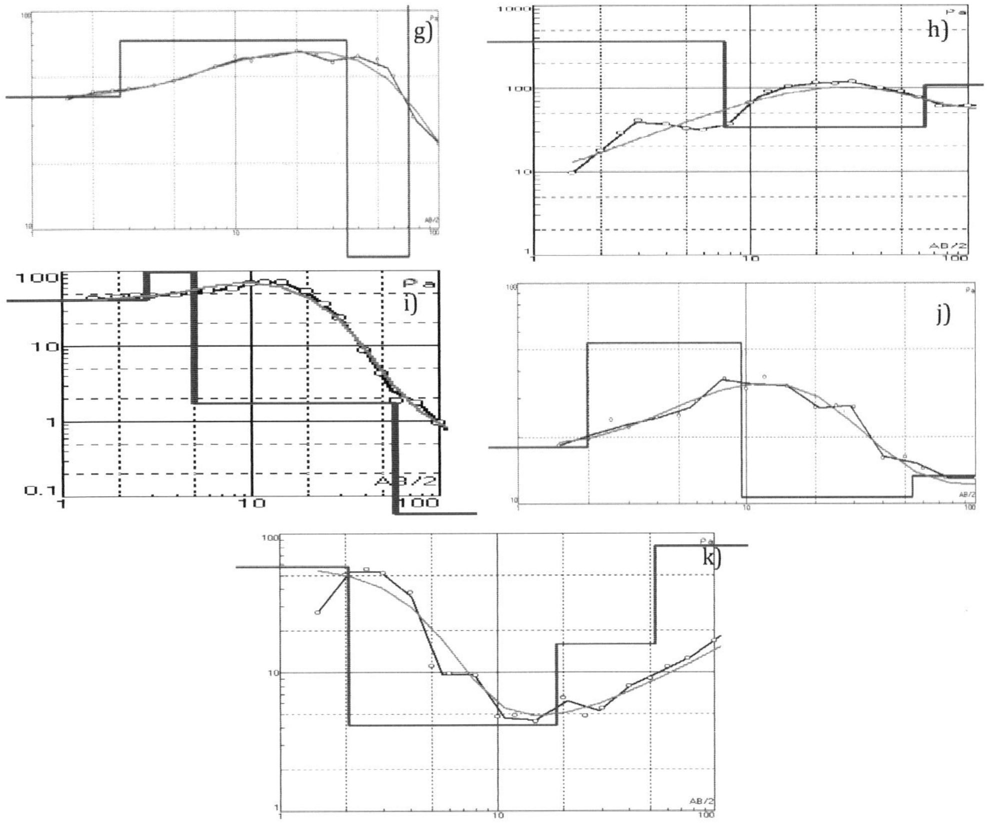

Figure 2. a.) Inversion result of 1.) VES 1, b.) VES 2, c.) VES 3, d.) VES 4, e.) VES 5, f.) VES 6, g.) VES 7, h.) VES 8, i.) VES 9, j.) VES 10, k.) VES 11. l.) VES 12, and k.) VES 13.

The discussion on aquifer distribution in Bulukumba highlights the significant role of geological factors in determining the potential and depth of aquifers. Bulukumba Regency consists of various geological formations, including sedimentary, volcanic, and metamorphic rocks, which influence the variation in soil resistivity detected during geoelectric measurements. Sedimentary rock layers, such as sandstone and conglomerate, found at intermediate to deep depths, are identified as productive aquifer layers due to their high porosity and permeability, as depicted in Figure 2. In the Ujung Loe area, deeper aquifers were identified, likely due to more complex geological structures, such as folds and faults, which can impact groundwater flow and storage. Conversely, in areas such as Kajang and the central part of Bulukumba Regency, aquifers were located at shallower depths, indicating differences in lithological composition, where more impermeable rock layers or those with lower water storage capacity are present. This aquifer distribution mapping emphasizes the importance of understanding local geological variations to optimize groundwater resource utilization, as each area exhibits distinct geological characteristics that affect groundwater availability and sustainability.

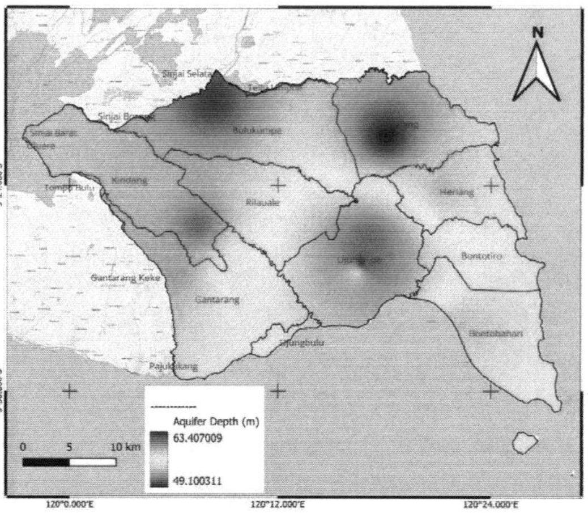

Figure 3. Aquifer mapping in Bulukumba District.

The geoelectric measurements conducted at several locations in Bulukumba Regency revealed significant variations in aquifer depths, ranging from 49 to 63 meters below the surface (Figure 3). In the Ujung Loe area, aquifers were detected at greater depths than in other regions, such as Kajang and the central part of Bulukumba Regency, indicating differences in groundwater distribution influenced by variations in local lithology and geological structures. The resistivity data, processed using IPI2Win software, produced subsurface profiles showing several geoelectric layers that could be interpreted as topsoil, sandstone, and conglomerate. These layers indicate a significant aquifer potential, particularly in the sandstone and conglomerate layers at intermediate to deep depths.

These depth variations reflect the geological diversity across the region, where sedimentary formations, such as sandstone and conglomerates, may provide more favorable conditions for aquifer development in specific locations, as indicated by previous studies on aquifer characteristics in similar lithological environments [3]. Moreover, the depth variations of aquifers align with findings from groundwater exploration studies, where deeper aquifers often correspond to more fractured or permeable bedrock formations, which allow for more significant water storage potential [2]. The water table may be more susceptible to seasonal fluctuations and recharge processes in areas with shallower aquifers, a pattern observed in other hydrological studies of rainfed agricultural zones [4].

This mapping provides a critical visual tool for guiding water resource management in Bulukumba. It underscores the need for location-specific groundwater exploration strategies that consider the depth and distribution of aquifers in different areas, in line with recommendations from previous research on groundwater management in heterogeneous terrains [5]. Consequently, this information is essential for supporting sustainable water resource development, particularly in the context of agricultural irrigation and domestic water needs in the region.

The discussion emphasizes the importance of understanding local geological characteristics for sustainable groundwater management. By applying the inverse distance weight (IDW) method in QGIS software for aquifer mapping, this study successfully visualized the distribution and thickness of aquifers across the study area, as depicted in Figure 3. The interpretation shows that regions with deeper aquifers, such as Ujung Loe, hold more significant potential for developing sustainable bore wells. On the other hand, the variations in depth and thickness of the layers highlight the need for location-specific management approaches to optimize groundwater exploration and utilization in Bulukumba. These findings provide crucial information that can serve as a reference for groundwater resource planning and management to support agricultural activities and local community needs.

4. Conclusion

The findings of this study demonstrate that the Schlumberger configuration geoelectric method effectively identifies the depth and distribution of aquifers in Bulukumba Regency. The variation in aquifer depths, ranging from 49 to 63 meters, indicates differences in geological characteristics across locations, particularly in the Ujung Loe area, which has deeper aquifers. The resistivity analysis successfully depicted detailed subsurface layers, including sandstone and conglomerate layers, which have the potential to serve as productive aquifers. Aquifer mapping using the IDW approach in QGIS software effectively visualized the distribution and thickness of aquifers, providing a basis for more efficient groundwater management. This study offers critical information supporting the sustainable exploration and management of groundwater resources in Bulukumba, serving as a reference for planning agricultural irrigation needs and local community water requirements.

References

[1] A. Agussalim and N. I. AS, "Identification of Groundwater Potential using Schlumberger Array of Resistivity Method in Bulukumba, South Sulawesi," *INTEK: Jurnal Penelitian*, vol. 8, no. 2, 2022, doi: 10.31963/intek.v8i2.3016.

[2] S. Dey, C. Debbarma, P. Sarkar, and S. Paul, "Geo-electrical mapping and groundwater potential zoning in some selected pockets of Baromura hill of Tripura (India)," *Journal of Applied and Natural Science*, vol. 2, no. 2, 2010, doi: 10.31018/jans.v2i2.121.

[3] A. G. Mohammed and A. Ibrahim, "2D Geo-electrical Resistivity Imaging of Clay Deposit in Agbonmwoba Village, Edo State, Nigeria," *Journal of Environment and Earth Science*, 2020, doi: 10.7176/jees/10-10-06.

[4] D. Sugiyanto, I. Rusydy, M. Marwan, D. M. Hidayati, and A. Asrillah, "A Preliminary Study On Aquifer Identification Based on Geo-Electrical Data In Banda Aceh, INDONESIA," *Jurnal Natural*, vol. 18, no. 3, 2018, doi: 10.24815/jn.v18i3.11204.

[5] A. Puttiwongrak, S. Keo, S. Vann, T. Arpornthip, T. Suteerasak, and P. Sukontasukkul, "Geological and Climatic Factors Affecting the Correlation between Electrical Resistivity and SPT N-Value in Sandy Soils of Phuket, Thailand," *Geosciences (Switzerland)*, vol. 13, no. 6, 2023, doi: 10.3390/geosciences13060185.

[6] D. R. Rahmawati, Supriyadi, N. P. Aryani, and M. A. Naufal, "Groundwater potential prediction by using geoelectricity method a case study in Simpang Lima and around it," in *Journal of Physics: Conference Series*, 2018. doi: 10.1088/1742-6596/983/1/012003.

[7] A. ila Sahmila, "Identifikasi Akuifer Air Tanah Berdasarkan Data Geolistrik Vertical Electrical Sounding (VES) Konfigurasi Schlumberger (Studi Kasus: Kelurahan Liang Kecamatan Kota Bangun)," *GEOSAINS KUTAI BASIN*, vol. 6, no. 1, 2023, doi: 10.30872/geofisunmul.v6i1.987.

[8] K. A. Wikranta Setiawan, "Identifikasi Jenis Akuifer Menggunakan Metode Vertical Electrical Sounding Pada Daerah CAT Ampibabo Kabupaten Parigi - Moutong, Sulawesi Tengah," *Jurnal Multidisiplin West Science*, vol. 2, no. 09, 2023, doi: 10.58812/jmws.v2i09.522.

[9] A. Wijaya and A. Kusmiran, "Identifikasi Jenis Akuifer Air Tanah Menggunakan Vertical Electrical Sounding Konfigurasi Schlumberger," *JFT: Jurnal Fisika dan Terapannya*, vol. 8, no. 1, 2021, doi: 10.24252/jft.v8i1.21210.

IOP Conf. Series: Earth and Environmental Science **1525** (2025) 012006 doi:10.1088/1755-1315/1525/1/012006

Open Pit Coal Mine Failure Model Based on Statistical Data

Mulyadi Sannang[1,2*], Hendra Pachri[1] and Ilham Alimuddin[1]

[1] Department of Geological Engineering, Hasanuddin University, 92171, Indonesia.
[2] Geotechnical Specialist Services, PT. GroundProbe Indonesia, 76114, Indonesia.

*E-mail: ma.sannang@geologist.com

Abstract. Slope stability is paramount in open-pit safety and mine operations, where unexpected failures can have significant consequences. This necessitates a deeper understanding of specific slope behaviour to guide geotechnical professionals in taking preventive measures. The primary objective of the study is to create an empirical model of the recorded failures to enhance failure prediction accuracy using the inverse velocity method and to define optimal alarm parameters, ensuring timely warnings before slope failures occur. This study investigates 14 slope failure events categorized into original in-situ slope failure, subsoil failure, and disposal failure. The failures are monitored using three GroundProbe Slope Stability Radar (SSR) monitoring systems that facilitate real-time data delivery with submillimetre accuracy, operated at an open-pit coal mine in South Kalimantan. A total of 15 parameters of different slope behaviours obtained from failure back analysis have been tested using the Anderson-Darling fitting test where the mean values are used as an empirical model of each failure category. The model demonstrates that the parameters of each failure category have different fitting test values that best fit with either normal or lognormal distribution types. The model was also compared to the actual failure data by analysing the distribution and variance of the data. The result indicates that most of the failure parameters are effectively represented by the empirical model proposed in this study, with certain parameters are also observed to be not closely aligned with the model.

1. Introduction

Slope failure in open-pit mining often occurs unexpectedly despite detailed slope design analyses. The complexities of rock mass properties, variations in geological conditions, and different slope behaviour under certain conditions can contribute to the unpredictability of slope failures. A study of about 58 mine slope failures in several coal mining in Kalimantan [1] indicated that slope monitoring using Slope Stability Radar (SSR) technology has been reliable in detecting mine slope failures and reducing the risk of fatality, injury, and equipment damage. However, while comprehensive monitoring systems can significantly predict failure occurrences, they are accompanied by uncertainties that must be addressed to improve prediction accuracy and operational safety [2]. This study aims to fill the existing research gaps by generating an empirical model of slope deformation behaviour prior to the failure that is defined statistically using Anderson-Darling fitting method.

2. Methodology

2.1. Slope Stability Radar (SSR)

The SSR slope monitoring concept is based on radar interferometry technique [3]. The SSR scans the slope by emitting an electromagnetic signal toward an area on the slope and records signal phase change of the reflected signal between each consecutive scan (Figure 1). A collection of radar scan data over a specific monitoring period will illustrate deformation trends that can be further analysed to determine slope behaviour and generate proper alarm threshold [4].

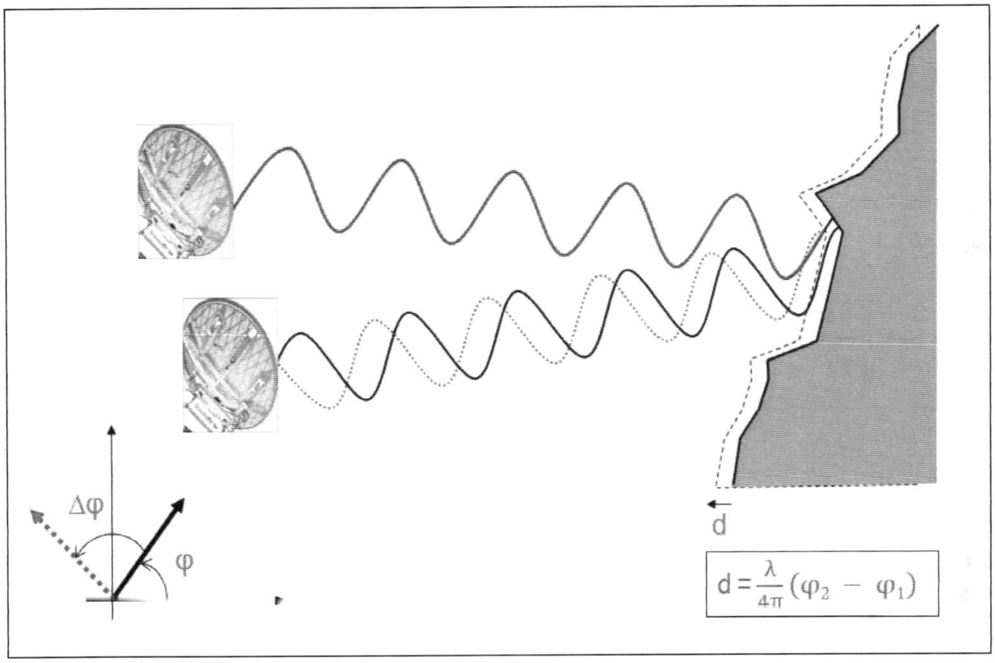

Figure 1. Interferometry techniques of GroundProbe slope stability radar [5].

Slope behaviour over time can be categorized into three typical types [6]: transitional deformation, regressive deformation, and progressive deformation (Figure 2). Transitional deformation is characterized by the wall moving outward at a constant velocity. In contrast, regressive deformation involves the wall moving outward but with a decreasing velocity over a specified time interval. Progressive deformation, on the other hand, is marked by the wall moving outward with an increasing velocity during a given period.

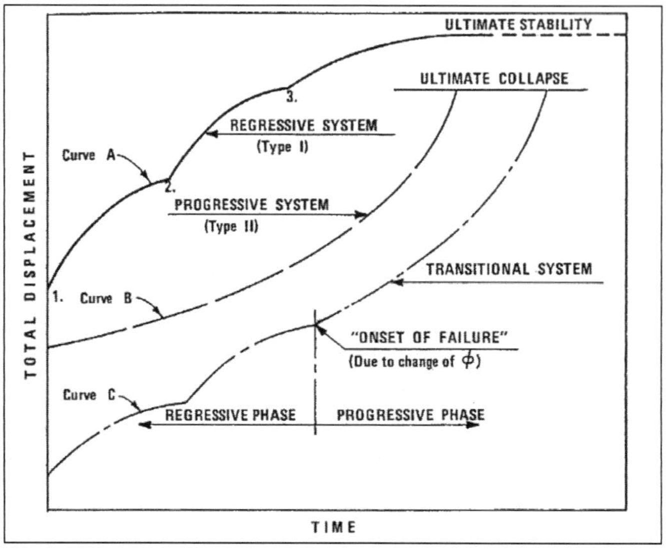

Figure 2. Three types of slope deformation behaviour [6].

A total of fourteen mine slope failures monitored by GroundProbe SSR are divided into three categories based on its material; disposal, subsoil, and original *in-situ* material and are back-analysed to obtain up to fifteen parameters from different slope behaviours such as velocity, duration time of each slope behaviour, accumulated deformation, minimum inverse velocity, and the coherence value at the time of failure. Coherence values indicate the consistency of amplitude and range with values ranging from 0 to 1. Shown in figure 3, a coherence value of 1 signifies that the monitored slope remains intact and undisturbed. On the contrary, any disturbance in the monitored slope area such as rockfall or collapse will result in a coherence drop lower than 1 [7].

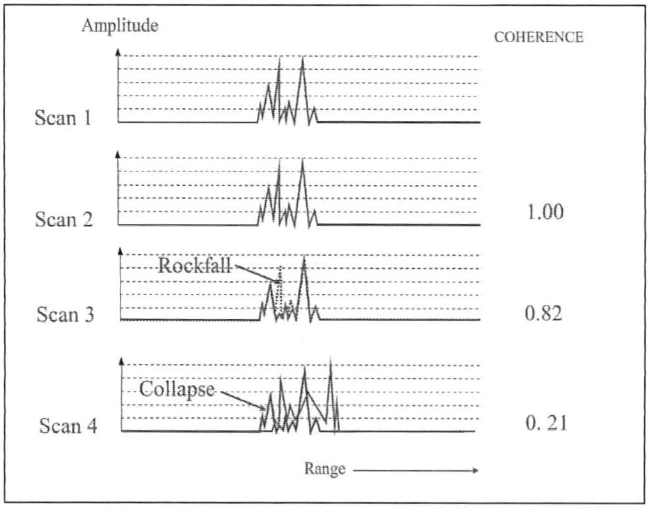

Figure 3. Simplified concept of coherence measurements [5, 7].

IOP Conf. Series: Earth and Environmental Science **1525** (2025) 012006 doi:10.1088/1755-1315/1525/1/012006

2.2. Anderson-Darling Fitting Test

The Anderson-Darling (A-D) fitting test method is one of the most powerful tests [8]. In the test, two values are used to assess the goodness of fit of a data distribution: the AD score and the P-value. The types of data distributions used in this fitting test are normal distribution, log-normal distribution, and gamma distribution. P-value less than the significance level (α=0.05) is indicating that the data do not follow a specific distribution type, hence the hypothesis is rejected. On the contrary, if the P-value is greater than the significance level (α=0.05) across all three distribution tests, the distribution with the smallest AD score is considered the best fit for the data [9]. Although the A-D test can be effectively applied to smaller samples, similar to other statistical tests, its robustness increases with larger datasets [10]. The general form of the A-D test can be found in Equations (1) below [11].

$$W_n^2 = -n - \frac{1}{n}\sum_{j=1}^{n}(2j-1)[\log u_j + Log\,(1-u_{n-j+1})] \quad (1)$$

The critical value can also be utilized to assess whether the data meets the hypothesis. If the calculated AD value is greater than or equal to the critical value, the hypothesis is rejected, indicating a significant difference between the sample data and the theoretical distribution. For normal and lognormal data distributions, the critical value is derived from Equation (2), where n represents the total number of data, while a, b, and d are obtained from Table 1 [12]. The significance level (α=0.05) used in Equation (2) is yellow highlighted in Table 1.

$$Critical\ value = \alpha\left(1 - \frac{b}{n} - \frac{d}{n^2}\right) \quad (2)$$

Table 1. Parameter for calculating critical value in one-sample Anderson-Darling test at each significance level (normal and log-normal distribution).

α	0.005	0.01	0.025	0.05	0.1	0.2
a	1.1578	1.0348	0.8728	0.7514	0.6305	0.5091
b	1.063	1.013	0.881	0.795	0.75	0.756
d	1.34	0.93	0.94	0.89	0.8	0.39

This study employs fourteen mine slope failure datasets that categorized based on its failed material. Specifically, the dataset includes five failures categorized into disposal material, three failures categorized into subsoil material, and six failures categorized into original in-situ material. Considering the significance level of α=0.05 and different sample size of each failure category, resulting in different critical values for normal and log-normal data distributions as specified in Table 2. Any AD value equal to or greater than the critical value of each failure category for normal and log-normal distributions will lead to the rejection of the hypothesis.

IOP Conf. Series: Earth and Environmental Science **1525** (2025) 012006 doi:10.1088/1755-1315/1525/1/012006

Table 2. Critical value of each failure category for normal and log-normal distributions

Failure Category	Critical Value
Disposal	$Critical\ value = 0.7514\ (1 - \frac{0.795}{5} - \frac{0.89}{5^2}) = 0.659$
Subsoil	$Critical\ value = 0.7514\ (1 - \frac{0.795}{3} - \frac{0.89}{3^2}) = 0.627$
Original in-situ	$Critical\ value = 0.7514\ (1 - \frac{0.795}{6} - \frac{0.89}{6^2}) = 0.670$

Table 3 is utilized to ascertain the critical value for the gamma distribution. This table employs the shape parameter k to establish the critical value at each significance level [12].

Table 3. Parameter for calculating critical value in one-sample Anderson-Darling test at each significance level for gamma distribution type.

$k \setminus \alpha$	0.005	0.01	0.025	0.05	0.10	0.25
1	1.227	1.092	0.917	0.786	0.657	0.486
2	1.190	1.062	0.894	0.768	0.643	0.477
3	1.178	1.052	0.886	0.762	0.639	0.475
4	1.173	1.048	0.883	0.759	0.637	0.473
5	1.170	1.045	0.881	0.758	0.635	0.472
6	1.168	1.043	0.880	0.757	0.635	0.472
8	1.165	1.041	0.878	0.755	0.634	0.471
10	1.164	1.040	0.877	0.754	0.633	0.471
12	1.162	1.038	0.876	0.754	0.633	0.471
15	1.162	1.038	0.876	0.754	0.632	0.470
20	1.161	1.037	0.875	0.753	0.632	0.470
>	1.159	1.035	0.873	0.752	0.631	0.470

After completing the fitting test and selecting the distribution type, the mean and standard deviation of each parameter were calculated using the equations provided in Table 4 [13].

Table 4. Mean and standard deviation equations for each distribution type.

Distribution type	Mean	Standard Deviation
Normal	$M = \frac{\sum x}{n}$	$SD = \sqrt{\frac{\sum_i^n (x_i - \bar{x})^2}{n-1}}$
Log-Normal	$M = e^{\mu + \frac{1}{2}\sigma^2}$	$SD = \sqrt{e^{2\mu+\sigma^2}(e^{\sigma^2}-1)}$
Gamma	$M = \alpha\beta$	$SD = \alpha\beta^2$

n: amount of data, $\bar{x}$: mean, x: value of data μ: location, σ: scale, α: shape, and β: scale

3. Result and Discussion

3.1. Statistical Analysis: Anderson-Darling Fitting Test

A total of fifteen parameters derived from the back-analysis of each failure category were tested using the Anderson-Darling method. Each parameter was found to conform to at least one distribution type. The lowest AD values for each parameter indicates the best distribution fit. All identified best distribution fits are statistically reliable, as evidenced by P-values exceeding the significance level (α = 0.05) and AD values falling below the critical value for each failure category, as specified in Tables 5 – 7 and highlighted in yellow.

Table 5. Anderson-Darling (AD) values, p-values, and critical values for each back-analysis parameter of disposal failure category.

Parameter			Normal		Lognormal		Gamma	
			AD	P-Value	AD	P-Value	AD	P-Value
Transitional deformation	Duration		0.654	0.036	0.292	0.443	0.429	>0.25
	Avg. Velocity 1h		0.738	0.020	0.286	0.461	0.457	>0.25
Progressive deformation	Time to failure		0.958	<0.005	0.465	0.135	0.715	0.077
	Accumulated Deformation		0.530	0.086	0.373	0.258	0.387	>0.25
	Maximum Velocity - pre ambiguous	60	0.381	0.244	0.437	0.164	0.505	0.223
		180	0.504	0.104	0.573	0.064	0.665	0.092
		360	0.582	0.060	0.675	0.031	0.778	0.050
		720	0.482	0.119	0.628	0.043	0.705	0.080
		1440	0.315	0.381	0.399	0.215	0.417	>0.25
	Minimum Inverse Velocity - pre ambiguous	60	0.511	0.099	0.459	0.141	0.570	0.165
		180	0.603	0.051	0.578	0.061	0.701	0.080
		360	0.855	0.009	0.540	0.081	0.667	0.096
		720	0.200	0.584	0.209	0.547	0.266	>0.25
		1440	0.529	0.087	0.398	0.217	0.526	0.215
Coherence at the Time of Failure			0.277	0.490	0.279	0.484	0.327	>0.25

IOP Conf. Series: Earth and Environmental Science **1525** (2025) 012006 doi:10.1088/1755-1315/1525/1/012006

Table 6. Anderson-Darling (AD) values, p-values, and critical values for each back-analysis parameter of subsoil failure category.

Parameter			Normal		Lognormal		Gamma	
			AD	P-Value	AD	P-Value	AD	P-Value
Transitional deformation	Duration		0.480	0.060	0.485	0.058	0.624	0.127
	Avg. Velocity 1h		0.207	0.555	0.232	0.479	0.287	>0.25
Progressive deformation	Time to failure		0.198	0.590	0.193	0.614	0.252	>0.25
	Accumulated Deformation		0.256	0.394	0.303	0.268	0.369	>0.25
	Maximum Velocity - pre ambiguous	60	0.191	0.623	0.220	0.515	0.267	>0.25
		180	0.460	0.072	0.468	0.067	0.592	0.138
		360	0.299	0.279	0.348	0.184	0.425	>0.25
		720	0.192	0.617	0.200	0.584	0.249	>0.25
		1440	0.193	0.615	0.201	0.579	0.249	>0.25
	Minimum Inverse Velocity - pre ambiguous	60	0.277	0.334	0.222	0.509	0.307	>0.25
		180	0.476	0.063	0.470	0.066	0.600	0.132
		360	0.396	0.121	0.352	0.178	0.469	>0.25
		720	0.248	0.420	0.200	0.583	0.274	>0.25
		1440	0.254	0.403	0.200	0.582	0.276	>0.25
Coherence at the Time of Failure			0.208	0.551	0.192	0.621	0.253	>0.25

Table 7. Anderson-Darling (AD) values, p-values, and critical values for each back-analysis parameter of original in-situ failure category.

Parameter			Normal		Lognormal		Gamma	
			AD	P-Value	AD	P-Value	AD	P-Value
Transitional deformation	Duration		1.173	<0.005	0.343	0.348	0.601	0.149
	Avg. Velocity 1h		0.232	0.658	0.429	0.198	0.389	>0.25
Progressive deformation	Time to failure		0.503	0.119	0.432	0.194	0.529	0.206
	Accumulated Deformation		0.311	0.427	0.330	0.378	0.353	>0.25
	Maximum Velocity - pre ambiguous	60	0.532	0.100	0.655	0.044	0.711	0.069
		180	0.511	0.115	0.634	0.050	0.682	0.082
		360	0.284	0.504	0.452	0.169	0.451	>0.25
		720	0.265	0.544	0.395	0.248	0.384	>0.25
		1440	0.287	0.496	0.510	0.116	0.470	>0.25
	Minimum Inverse Velocity - pre ambiguous	60	0.416	0.155	0.398	0.177	0.479	0.241
		180	0.271	0.450	0.270	0.451	0.345	>0.25
		360	0.690	0.034	0.475	0.145	0.609	0.129
		720	0.632	0.051	0.391	0.255	0.516	0.215
		1440	0.702	0.026	0.456	0.143	0.594	0.139
Coherence at the Time of Failure			0.319	0.407	0.221	0.703	0.281	>0.25

The result shows that the parameters follow either normal or lognormal distribution types, none is fit with the gamma distribution type. After a distribution type is chosen for each parameter, the mean and standard deviation can be calculated as specified in Table 4.

Tables 8 – 10 presents a summary of the calculated mean and standard deviation for each parameter across the different failure categories. The relative minimum value is calculated by subtracting the standard deviation from the mean and the relative maximum value is calculated by adding the standard deviation to the mean.

Table 8. Anderson-Darling (AD) values, p-values, and critical values for back-analysis parameters of disposal failure category.

	Parameter		Distribution	Mean	Std. Dev.	Rel. Min	Rel. Max
Transitional deformation	Duration (Hh:Mm:Ss)		Lognormal	47:53:48	38:37:34	9:16:14	86:31:21
	Avg. Velocity 1h (mm/h)		Lognormal	0.82	0.75	0.06	1.57
Progressive deformatio n	Time to failure (Hh:Mm:Ss)		Lognormal	23:49:38	25:30:30	1:40:52	49:20:08
	Acc. Deformation (mm)		Lognormal	429.69	864.97	-435.28	1294.67
	Maximum Velocity (mm/h) - pre ambiguous	60	Normal	26.77	16.87	9.90	43.64
		180	Normal	23.47	16.08	7.39	39.55
		360	Normal	19.76	14.40	5.36	34.16
		720	Normal	14.42	11.09	3.33	25.51
		1440	Normal	9.07	7.99	1.09	17.06
	Minimum Inverse Velocity (h/mm) - pre ambiguous	60	Lognormal	0.06	0.05	0.01	0.12
		180	Lognormal	0.09	0.09	0.00	0.18
		360	Lognormal	0.40	0.93	-0.53	1.33
		720	Normal	0.05	0.01	0.04	0.05
		1440	Lognormal	0.38	0.49	-0.11	0.87
Coherence at the Time of Failure			Normal	0.47	0.33	0.14	0.80

Table 9. Anderson-Darling (AD) values, p-values, and critical values for back-analysis parameters of subsoil failure category.

	Parameter		Distribution	Mean	Std. Dev.	Rel. Min	Rel. Max
Transitional deformation	Duration (Hh:Mm:Ss)		Normal	32:39:50	24:57:36	7:42:14	57:37:26
	Avg. Velocity 1h (mm/h)		Normal	0.88	0.23	0.25	1.11
Progressive deformation	Time to failure (Hh:Mm:Ss)		Lognormal	4:44:57	2:54:01	0:31:57	7:38:58
	Acc. Deformation (mm)		Normal	42.30	15.67	17.40	57.97
	Maximum Velocity (mm/h) - pre ambiguous	60	Normal	22.76	9.87	10.09	32.63
		180	Normal	11.95	4.47	5.16	16.42
		360	Normal	7.30	3.01	3.40	10.30
		720	Normal	5.03	2.42	2.35	7.45
		1440	Normal	2.95	1.52	1.47	4.46
	Minimum Inverse Velocity (h/mm) - pre ambiguous	60	Lognormal	0.05	0.04	0.02	0.09
		180	Lognormal	0.10	0.07	0.03	0.16
		360	Lognormal	0.17	0.11	0.06	0.28
		720	Lognormal	0.25	0.17	0.11	0.42
		1440	Lognormal	0.44	0.31	0.22	0.75
Coherence at the Time of Failure			Lognormal	0.66	0.42	0.19	1.08

Table 10. Anderson-Darling (AD) values, p-values, and critical values for back-analysis parameters of original in-situ failure category.

Parameter			Distribution	Mean	Std. Dev.	Rel. Min	Rel. Max
Transitional deformation	Duration (Hh:Mm:Ss)		Lognormal	194:04:33	708:50:19	514:45:45	902:54:52
	Avg. Velocity 1h (mm/h)		Normal	1.54	0.66	0.88	2.20
Progressive deformatio n	Time to failure (Hh:Mm:Ss)		Lognormal	12:58:42	12:20:06	0:38:36	25:18:48
	Acc. Deformation (mm)		Normal	141.60	102.40	39.20	244.00
	Maximum Velocity (mm/h) - pre ambiguous	60	Normal	34.24	13.20	21.04	47.44
		180	Normal	23.18	11.09	12.09	34.27
		360	Normal	16.16	8.43	7.74	24.59
		720	Normal	10.89	6.40	4.49	17.29
		1440	Normal	6.90	4.04	2.86	10.93
	Minimum Inverse Velocity (h/mm) - pre ambiguous	60	Lognormal	0.02	0.01	0.01	0.04
		180	Normal	0.03	0.00	0.03	0.04
		360	Lognormal	0.09	0.07	0.02	0.16
		720	Lognormal	0.15	0.12	0.03	0.27
		1440	Lognormal	0.15	0.11	0.05	0.26
Coherence at the Time of Failure			Lognormal	0.50	0.33	0.16	0.83

3.2. Empirical Model

Based on the summary provided in Tables 8 – 10, the empirical model of each failure category is illustrated in Figure 4 - 6. These models utilize one hour (1h) velocity calculation period, which is commonly used for an accurate failure prediction. The empirical models of different failure categories serve as a critical tool for establishing appropriate alarm thresholds, providing inverse velocity reference value during failure prediction, and guiding decision-making processes in response to the potential risk of slope failures.

The failure model for disposal material initiates with a transitional deformation trend approximately 47 hours and 53 minutes with average velocity around 0.82 mm/hour and further turns into a progressive deformation trend, with maximum velocity reaching 26.77 mm/hour within approximately 23 hours and 49 minutes prior to the failure. Disposal failure exhibits a minimum inverse velocity of 0.06 hour/mm and a coherence value of 0.47 with accumulated deformation up to 429.69 mm during the progressive deformation stage (Figure 4).

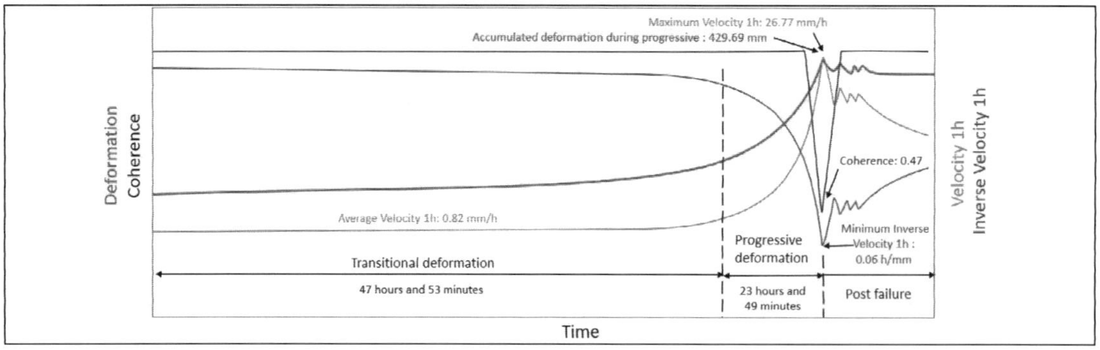

Figure 4. Empirical model of the disposal failure (not to scale).

IOP Conf. Series: Earth and Environmental Science **1525** (2025) 012006 doi:10.1088/1755-1315/1525/1/012006

The failure model for subsoil material also starts with a transitional deformation trend, lasting approximately 32 hours and 39 minutes with average velocity around 0.88 mm/hour before evolving into a progressive deformation trend, with velocity increases to a maximum of 22.76 mm/hour within approximately 4 hours and 44 minutes before the failure. Subsoil failure exhibits a minimum inverse velocity of 0.05 hour/mm and a coherence value of 0.66 with accumulated deformation up to 42.30 mm during the progressive deformation stage (Figure 5).

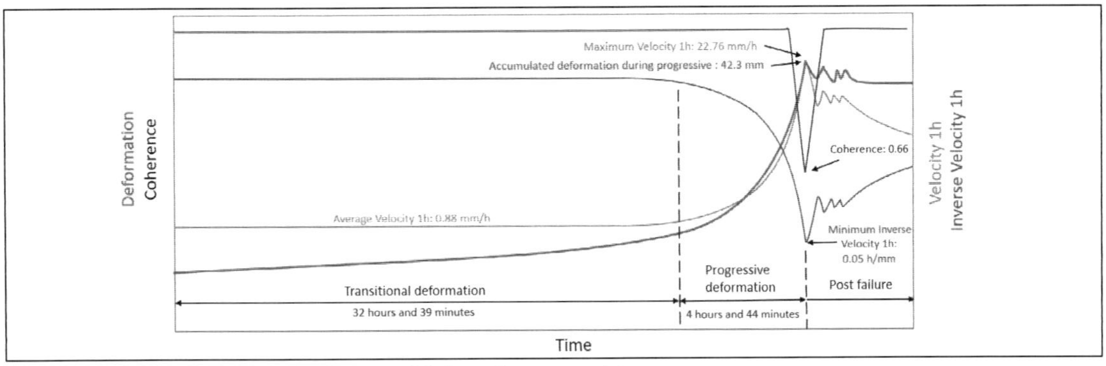

Figure 5. Empirical model of the subsoil failure (not to scale).

The failure model for original in-situ material initiates with the longest duration of the transitional deformation trend approximately 194 hours with average velocity around 1.54 mm/hour and turned into a progressive deformation trend with a maximum velocity up to 34.24 mm/hour within approximately 12 hours and 58 minutes prior to the failure. The failure of the original in-situ material exhibits a minimum inverse velocity of 0.02 hour/mm and a coherence value of 0.5 with accumulated deformation up to 141.6 mm during the progressive deformation stage (Figure 6).

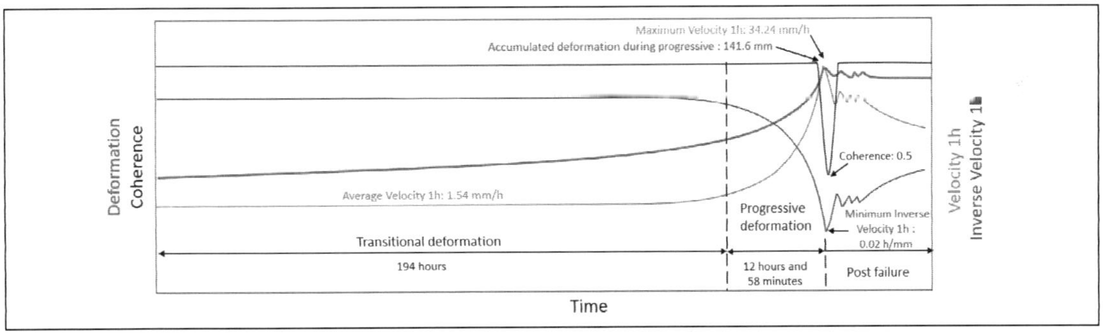

Figure 6. Empirical model of the original in-situ failure (not to scale).

3.3. *Case Study*

A failure event of original in-situ material that occurred on December 7, 2022, is used for this case study and considered to resemble the empirical model of the original in-situ failure as illustrated in Figure 7. The transitional deformation was initially observed at 2:43 PM on December 1, 2022, and lasted for 121 hours and 48 minutes with an average velocity (1h) around 1.99 mm/hour. The deformation further developed into a progressive deformation trend with accumulated deformation of 247 mm and velocity (1h) increased to approximately 45.54 mm/hour over a time span of about 15 hours and 53 minutes prior to the failure. The failure also exhibits a minimum inverse velocity (1h) of 0.02 hour/mm and a coherence value of 0.4.

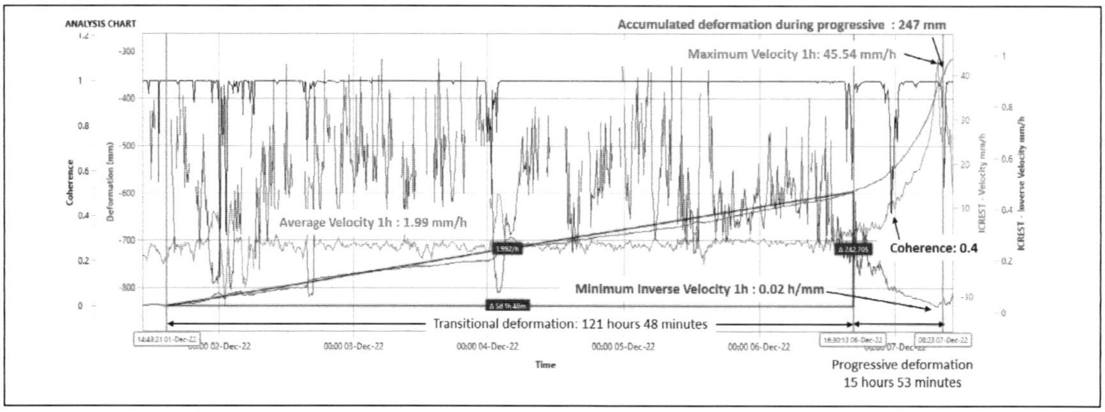

Figure 7. Back analysis of an original in-situ failure event on December 7, 2022.

3.4. Data Comparison

The empirical model was compared to the actual failure data by analysing the distribution and variance. Variance values were calculated as the square of the standard deviation for each failure parameter. A high variance value in Table 11 indicates a random data distribution that does not represent the empirical model while a low variance suggests a good representation. The result indicates that most of the failure parameters exhibit low variance values below 100. However, high variance data is observed on several parameters across all failure categories (highlighted in yellow). In the disposal failure category, high variance is observed over five parameters: accumulated deformation, and maximum velocity over 60, 180, 360, and 720 minutes. For the original in-situ failure category, four parameters indicated high variance: duration of transitional deformation, accumulated deformation, and maximum velocity over 60, and 180. In the subsoil failure category, high variance is observed only in accumulated deformation.

Table 11. Data variance of all failure parameters.

Parameter			Data Variance		
			Disposal	Subsoil	Original
Transitional deformation	Duration		2.59	1.08	872.31
	Avg. Velocity 1h		0.57	0.05	0.43
Progressive deformation	Time to failure		1.13	0.01	0.26
	Acc. Deformation		748180.46	245.55	10485.76
	Maximum Velocity - pre ambiguous	60	284.60	97.46	174.24
		180	258.57	19.95	122.99
		360	207.36	9.04	70.98
		720	122.99	5.87	40.97
		1440	63.76	2.30	16.29
	Minimum Inverse Velocity - pre ambiguous	60	0.00	0.00	0.00
		180	0.01	0.00	0.00
		360	0.87	0.01	0.01
		720	0.00	0.03	0.01
		1440	0.24	0.10	0.01
Coherence at the Time of Failure			0.11	0.18	0.11

Data distribution analysis was also performed to complement the variance data. A failure parameter is considered to satisfy the empirical model if its value is distributed between the relative minimum and maximum values (Tables 8 – 10) and is accompanied by a low variance value. Figure 8 shows an example of a failure parameter that meets this criterion. The parameter for average velocity (1h) of transitional deformation for disposal failure indicates that only one outlier observed in its distribution and has a low variance value of 0.57, therefore this parameter is considered to conform to the empirical model.

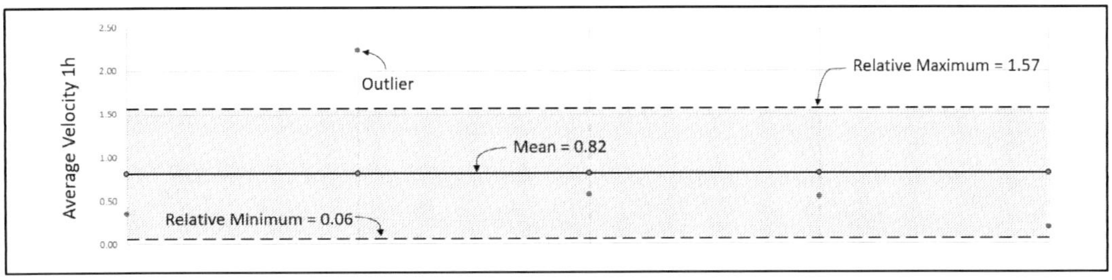

Figure 8. Data distribution of the average velocity (1h) during the transitional deformation for disposal failure.

4. Conclusion

1. The Anderson-Darling fitting test indicates that empirical models of each failure category start with a transitional deformation trend and further develop into a progressive deformation trend prior to the failure. Parameters of each failure category show different fitting test values that best fit with either normal or lognormal distribution types.
2. The empirical model for disposal failure suggests that the transitional deformation trend with an average velocity (1h) of 0.82 mm/hour lasts approximately 47 hours and 43 minutes before developing into a progressive deformation. During the progressive deformation, the velocity (1h) increases to a maximum of 26.77 mm/hour with accumulated deformation reaching up to 429.69 mm within approximately 23 hours and 49 minutes prior to the failure. The failure exhibits a minimum inverse velocity of 0.06 hour/mm with the coherence value dropped up to 0.47 at the failure time.
3. The empirical model for subsoil failure suggests that the transitional deformation trend with an average velocity (1h) of 0.88 mm/hour, lasts approximately 32 hours and 39 minutes before developing into a progressive deformation. During the progressive deformation, the velocity (1h) increases to a maximum of 22.76 mm/hour with accumulated deformation reaching up to 42.30 mm within approximately 4 hours and 44 minutes prior to the failure. The failure exhibits a minimum inverse velocity of 0.05 hour/mm with the coherence value dropped up to 0.66 at the failure time.
4. The empirical model for original in-situ failure suggests that the transitional deformation trend with an average velocity (1h) of 1.54 mm/hour lasts approximately 194 hours before developing into a progressive deformation. During the progressive deformation, the velocity (1h) increases to a maximum of 34.24 mm/hour with accumulated deformation reaching up to 141.6 mm within approximately 12 hours and 58 minutes prior to the failure. The failure exhibits a minimum inverse velocity of 0.02 hour/mm with the coherence value dropped up to 0.5 at the failure time.

5. Most failure parameters align with the empirical model since their values fall between the relative minimum and maximum values and exhibit low variance. However, high variance data was observed in five parameters for disposal failure, four parameters for original in-situ failure, and one parameter for subsoil failure.
6. The empirical model generated in this study requires a continuous update and development in the future as new additional data is collected from each different failure category of subsoil, disposal, and original in-situ. Involving more specific geotechnical conditions and material properties will provide a more comprehensive perspective in understanding the open pit coal mine failure model.

Acknowledgment

The authors express their sincere gratitude to the Department of Geophysics and the Department of Geological Engineering, Hasanuddin University for organizing ICOGEST 2024. The authors also extend their deepest appreciation to GroundProbe and the associated coal mining company for facilitating the collection of failure data. Finally, the authors thank everyone who contributed to the completion of this study.

References

[1] Musa R H, Syafriya I, Risyana T, Rahmad. 2013. *Geotechnical Risk Management Using Slope Stability Radar for Open Pit Coal Mines Indonesia.* Indonesian Rock Mechanics Society (IRMS) annual conference.
[2] Cabrejo A G. 2013. *Analysis offailures in open pit mines and consideration of the uncertainty when predicting collapses.* The International Symposium on Rock Slope Stability in Open Pit Mining and Civil Engineering. Australian Centre for Geomechanics pp 483-497.
[3] Noon D, and Harries N. 2003. *Slope Stability Radar for Managing Rock Fall Risks in Open Cut Mines.* Sixth Large Open Pit Mining Conference. Australian Institute of Mining and Metallurgy. pp 93-98.
[4] Sannang M A, Musa R H, Manaf A. 2017. *Analisa Balik dan Penentuan Ambang Batas Alarm Menggunakan Slope Stability Radar (SSR) Studi Kasus Longsor Pada Batuan Keras.* TPT XXVI PERHAPI Proceeding.
[5] GroundProbe Geotechnical Support Services. 2024. *SSR Geotechnical Training Modules.* Internal Document GroundProbe Geotechnical Support Services.
[6] Broadbent C D, Zavodni Z M. 1982. *Influence of Rock Structure on Stability.* Stability in Surface Mining. Society of Mining Engineers. Volume 3. Chapter 2.
[7] Cahyo F, Dwitya R, Musa R H. 2020. *New Approach to Detect Imminent Slope Failure by Utilising Coherence Attribute Measurement on Ground-based Slope Radar.* Slope Stability Symposium, Australian Centre for Geomechanics. pp 1401-1412.
[8] Razali N, Wah Y B. 2011. *Power Comparisons of Shapiro-Wilk, Kolmogorov-Smirnov, Lilliefors and Anderson-Darling Tests.* Journal of Statistical Modelling and Analytics. Volume 2, pp 21-33.
[9] Minitab. 2017. *How the Anderson-Darling statistic is used to assess the distribution fit.* Minitab Support. Available from: https://support.minitab.com/en-us/minitab/help-and-how-to/quality-and-process-improvement/capability-analysis/supporting-topics/distributions-and-transformations-for-nonnormal-data/anderson-darling-and-distribution-fit/. Accessed on October 21, 2024.
[10] Smith J. 2012. *Testing for normality: A Tale of Two Samples by Anderson-Darling.* The Minitab Blog. Available from: https://blog.minitab.com/en/fun-with-statistics/testing-for-normality-a-tale-of-two-samples-by-anderson-darling. accessed on October 21, 2024.
[11] Anderson T W, Darling D A. 1954. *A test ofgoodness offit.* J. Am. Stat. Assoc., 49, pp 765–769.
[12] Zaiontz C. 2022. *One-Sample Anderson-Darling Test,* Available from: https://real-statistics.com/non-parametrictests/goodness-of-fit-tests/anderson-darling-test/, accessed October 21, 2024.
[13] Putra G Z M, Sadewa L P, Putuhena Y A. 2023. *Mine Waste Dump Failure Behaviour Model Based on Statistical Data,* TPT XXXII PERHAPI Proceeding.

IOP Conf. Series: Earth and Environmental Science **1525** (2025) 012007 doi:10.1088/1755-1315/1525/1/012007

Identification of Metal Mineral Potential in the Form of Iron Ore through Principal Component Analysis (PCA) and X-ray fluorescence (XRF) Testing in the Laeya Region, Southeast Sulawesi

Abd. Hadid Rahman Ulfa[1*], Erwin Anshari[1], and Muhammad Chaerul[2]

[1]Department of Mining Engineering, Halu Oleo University, Kendari Southeast Sulawesi 93232, Indonesia

[2] Department of Infrastructure and Environmental Engineering, Fajar University, Makassar South Sulawesi 90231, Indonesia

*E-mail: abd.hadid@uho.ac.id

Abstract. Buton Island is a fragment from the rifting of the Australian-New Guinea plate and has a complex tectonic history involving rifting, drifting, and collision during the Neogene period. This evolution has enriched the island with abundant natural and mineral resources. This study aims to identify the potential for metal mineralization, particularly iron ore, in the Laeya region, North Buton Regency, Southeast Sulawesi. The analysis was conducted using Landsat 8 OLI TIRS satellite imagery and Principal Component Analysis (PCA) to delineate prospective areas containing iron ore. PCA results indicate potential metal mineralization, specifically iron ore, in certain areas, primarily in open regions not covered by vegetation. The morphology of the study area consists of hills and karst plains, with river flow patterns controlled by geological structures. The lithology of the region is dominated by limestone and marble. Results are supported by X-ray fluorescence (XRF) testing of rock samples, which reveals an average iron (Fe) content of 1.4%. Although the iron content is relatively low, these findings provide initial indications of metal mineral potential in the area and may serve as a basis for further exploration research.

1. Introduction

In the context of iron ore exploration, remote sensing technology has become one of the most effective methods. This technology enables researchers to identify and map mineral deposits based on the spectral properties of surface materials [1,2]. Each material has unique characteristics in reflecting or emitting electromagnetic waves, which can be used to differentiate various types of minerals. This technology is highly beneficial in supporting more efficient and targeted exploration efforts [3,4,5,6].

In this study, we will employ remote sensing technology by applying two processing techniques: the Image Ratio technique, represented by the Iron Ratio, and Principal Component Analysis (PCA). Landsat imagery is chosen for its effectiveness and cost-efficiency in the early exploration of mineral resources such as iron ore. It provides multispectral data with sufficient spatial resolution to distinguish rock types and mineral alterations. Additionally, we will use supervised classification to obtain data that accurately reflects field conditions. The X-ray fluorescence (XRF) method will be used to verify the success of the image data processing.

2. Study Site

The Laeya area is located within the administrative region of Wakorumba District, North Buton Regency, Southeast Sulawesi Province, Indonesia (Fig. 1). Buton Island, like many other islands in the Banda Sea region, is believed to be a fragment resulting from the rifting of the Australian-New Guinea continent. The geology of Buton Island reflects a complex tectonic history, including pre-rift, rifting, drifting, and collision events during the Neogene period [7]. This geological background contributes to the richness of natural resources such as asphalt and mineral deposits on the island. The Laeya area has a morphology characterized by hills and karst plains, with elevations ranging from 0 to 1,055 mdpl (Fig. 2).

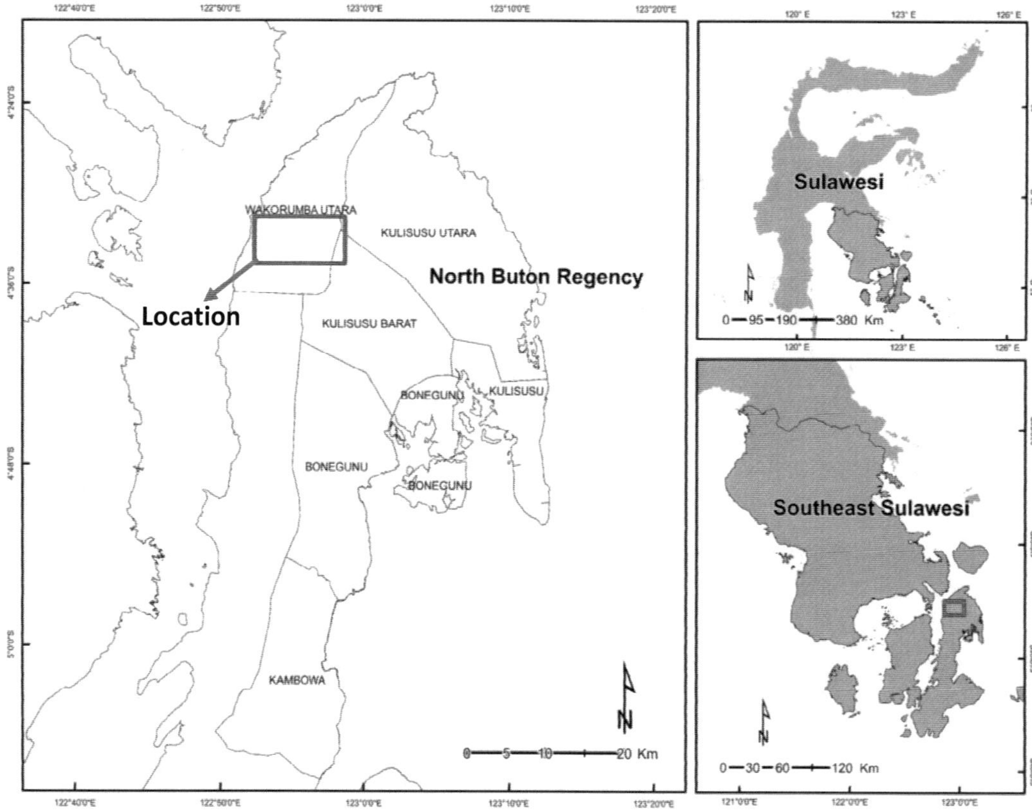

Figure 1. The study location is in the Laeya area of North Buton Regency, Southeast Sulawesi Province.

3. Data and Methods

3.1 Data

This study utilizes Landsat 8 OLI_TIRS to produce data from the spectrum of the rock reflection and Sample test results obtained from the field using the X-ray fluorescence (XRF) method. The Landsat 8 OLI_TIRS produces a record of the spectral reflectance of objects on the Earth's surface, where the reflectance spectrum of rocks is a combination of their constituent minerals [8].

IOP Conf. Series: Earth and Environmental Science **1525** (2025) 012007 doi:10.1088/1755-1315/1525/1/012007

Minerals possess characteristic spectra that enable remote identification, with metals such as iron often appearing as anomalies known as "gossan," a mixture of hydrated iron oxides [9,10]. Sample data, which serve as evidence of the remote sensing results, were collected from several outcrops detected through remote sensing processing.

Figure 2. Morphological Overview of the Laeya Area

3.2 Methods

X-ray fluorescence (XRF) technology was employed in geochemical analysis to determine the elemental composition of rock samples by measuring the fluorescence emitted from elements like iron (Fe) [14].

The processing of Landsat 8 OLI_TIRS imagery aims to enhance quality by removing noise, such as cloud interference, sharpening the image, and correcting georeferencing issues. This process includes atmospheric correction to reduce noise and geometric correction to align the imagery accurately with its actual geographic location [11, 12]. Following this, the Image Ratio technique is applied between band 4 (red) and band 2 (blue), known as the Iron ratio, to detect materials like iron oxides on the Earth's surface, which helps highlight high-contrast areas for further analysis [4,13]. The Principal Component Analysis (PCA) method is also used to reduce the number of image bands while retaining essential information, generating principal components from Landsat 8 OLI_TIRS multispectral bands for more efficient analysis [3,5]. Essentially, PCA is a transformation of the spectral channels or bands in an image. For example, if there are consecutive bands named B1, B2, B3, and B4, the linear combination would be as follows:

$$A = C1B1 + C2B2 + C3B3 + C4B4 \tag{1}$$

Where A is the new 'PCA channel' produced, and C1, C2, C3, and C4 are coefficients applied to each channel

IOP Conf. Series: Earth and Environmental Science **1525** (2025) 012007 doi:10.1088/1755-1315/1525/1/012007

4. Result and Discussion

Using Landsat 8 OLI_TIRS imagery data, iron ore was identified through the Image Ratio technique with the Iron Ratio formula (where the red band is divided by the blue band). The resulting image, shown in Figure 3A, highlights lighter tones, interpreted as gossan [15,10]. In the Principal Component Analysis (PCA) method, an RGB composite was generated using PCA 1, PCA 2, and PCA 3, as shown in Figure 3B. Iron ore appears in a pale yellow color [3, 5], though the distribution pattern requires further analysis. Both processing techniques showed consistent similarities in texture, tone, and color. The Supervised Classification technique was then applied to create an iron ore distribution map, which shows a concentration of deposits along riverbanks (Fig. 4B). However, due to the area's heavy vegetation cover, the Image Ratio and PCA results were mainly influenced by vegetation reflectance, limiting iron ore identification to river and riverbank areas.

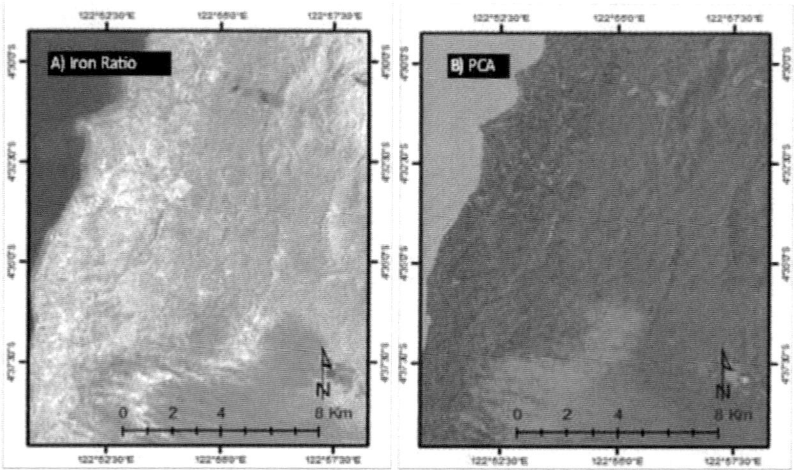

Figure 3. Results of Landsat 8 OLI_TIRS Image Processing: (A) Iron Ratio, (B) PCA

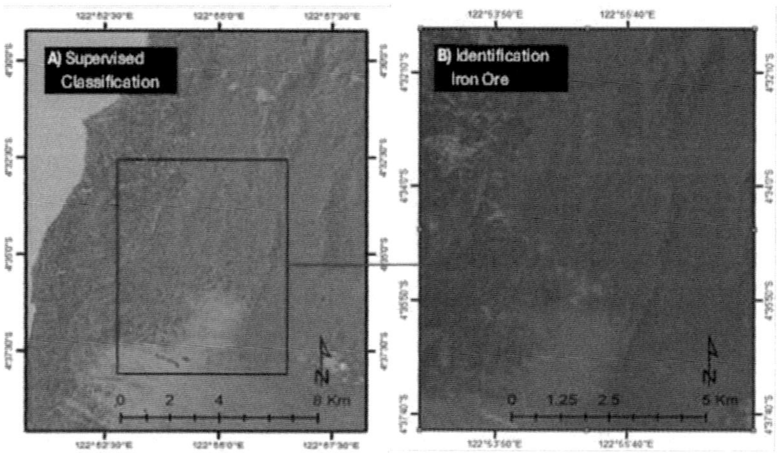

Figure 4. (A) Supervised Classification and (B) Iron ore identified along river flow patterns

X-ray fluorescence (XRF) testing on outcrop samples from the Laeya region indicated that CaO was the dominant compound, followed by SiO_2 and MgO, with metals such as Fe, Al, and Mn appearing in low percentages (Table 1). Some oxide mineral nodules, potentially hematite, were identified as weathering products of fine-grained clastic sedimentary rocks (Fig. 5). The low iron content (6.42%) suggests that significant iron oxide formation has yet to occur.

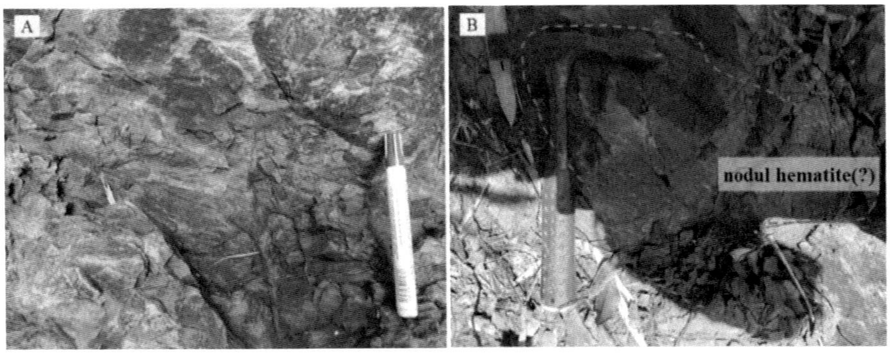

Figure 5. Indications of iron oxide presence as a weathering product in (A) marl and (B) mudstone in the Laeya area, North Buton Regency

Table 1. XRF analysis results of elemental content in rock samples

Sample ID	Lithology	CaO (%)	SiO_2 (%)	MgO (%)	Fe_2O_3 (%)	Al_2O_3 (%)	Cr_2O_3 (%)	MnO (%)
ST03 / LYA	Mudstone	5,61	53,41	13,59	5,73	9,45	0,26	<0,01
ST10 / LYA	Carbonate sandstone	5,93	46,72	28,18	6,42	2,49	0,34	0,05
ST12 / LYA	Marl	14,30	43,06	24,40	6,38	1,34	0,40	0,33
ST16 / LYA	Mudstone	2,88	50,01	16,45	5,87	12,61	0,14	0,07

5. Conclusion

The potential presence of metallic minerals is indicated in certain areas based on the results of Principal Component Analysis (PCA) on Landsat 8 OLI_TIRS imagery, with the primary distribution located in open areas not covered by vegetation. The distribution of metallic mineral indications is influenced by river flow patterns controlled by geological structures. This condition leads to the detection of metallic minerals, particularly iron ore, along riverbanks. Observations and analyses of mudstone, carbonate sandstone, marl, and mudstone rocks indicate the presence of iron ore, although in insignificant quantities.

References

[1] Amigun, J., Faruwa, R. and Komolafe, A. (2015) Integrated Landsat Imagery and Geophysical Exploration for Groundwater Potential Evaluation of Okene and Its Environs, Southwestern Nigeria. *International Journal of Geosciences*, **6**, 209-229. doi: 10.4236/ijg.2015.63015.

[2] El Atillah, A., El Morjani, Zea, & Souhassou, M. (2018). Use of Multispectral Image for Exploration and Research of Mineral Resources: State of Knowledge and Proposal of a Processing Model. European Scientific Journal, ESJ, 14 (24), 350. https://doi.org/10.19044/esj.2018.v14n24p350.

[3] Alasta A. F., (2011). Using Remote Sensing data to identify iron deposits in central western Libya. International Conference on Emerging Trends in Computer and Image Processing (ICETCIP2011), Bangkok, December 2011.

[4] Allo D. K., Hantchi, K. D., and Idi K. L. (2024). "Contribution of LANDSAT 8 Imagery in the Mapping of Iron Ore of the Continental Terminal 3 Formation of the Arewa Region, South-West of the Iullemmeden Basin in Niger". International Journal of Science and Research Methodology (IJSRM) Vol 27, Issue 8, August 2024, pp 23-28.

[5] Ghoneim, S.M., Salem, S.M., El-Wahid, K.H.A. (2022). Application of remote sensing techniques to identify iron ore deposits in the Central Eastern Desert, Egypt: a case study at Wadi Karim and Gabal El-Hadid areas. *Arab J Geosci* **15**, 1596. https://doi.org/10.1007/s12517-022-10871-3.

[6] Saepuloh A. (2020). Prinsip dan Aplikasi Pengindraan Jauh Geologi Gunung Api Pendekatan Fisis dan Observasi Geologi Lapangan. Bandung; Institut Teknologi Bandung.

[7] Zakir M. N. A, Kadir W. G. A and Wahyudi E. J. (2023). Interpretasi Bawah Permukaan dan Delineasi Struktur Cekungan Muna-Buton Berdasarkan Data Gravitasi Satelit. Jurnal Rekayasa Geofisika Indonesia. Vol. 05 No. 03, Desember 2023 (118-31).

[8] Zhao., Huijie., Ziwei Wang., Guorui Jia., Jia Tian., Shuliang Jin., Shuneng Liang., and Yumeng Liu. (2023). "The Impact and Correction of Sensitive Environmental Factors on Spectral Reflectance Measured In Situ" *Remote Sensing* 15, no. 22: 5332. https://doi.org/10.3390/rs15225332.

[9] Husain A. Alanazi, Habes A. Ghrefat. (2013). *Spectral Analysis of Multispectral Landsat 7 ETM + and ASTER Data for Mapping Land Cover at Qurayah Sabkha, Northern Saudi Arabia., 41(4), 833–844.* https://doi.org/10.1007/s12524-013-0291-2.

[10] Ozdemir A., Sahinoglu A. (2018) Important of Gossans in Mineral Exploration: A Case Study in Northern Turkey. Int J Earth Sci Geophys 4:019.

[11] Abdullah H., Roshanak Darvishzadeh, Andrew K. Skidmore, and Marco Heurich. (2019). "Sensitivity of Landsat-8 OLI and TIRS Data to Foliar Properties of EarlyStage Bark Beetle (*Ips typographus*, L.) Infestation" *Remote Sensing* 11, no. 4: 398. https://doi.org/10.3390/rs11040398.

[12] Wulder, M. A., Loveland, T. R., Roy, D. P., Crawford, C. J., Masek, J. G., Woodcock, C. E., & Zhu, Z. (2019). Current status of Landsat program, science, and applications. *Remote sensing of environment, 225*, 127-147.

[13] Gopinathan, P., Parthiban, S., Magendran, T., Al-Quraishi, A. M. F., Singh, A. K., & Singh, P. K. (2020). Mapping of ferric (Fe^{3+}) and ferrous (Fe^{2+}) iron oxide distribution using band ratio techniques with ASTER data and geochemistry of Kanjamalai and Godumalai, Tamil Nadu, south India. *Remote Sensing Applications: Society and Environment, 18*, 100306.

[14] Abd Halim, M. H., Mokhtar, N. A. M., Masnan, S. S. K., Talib, N. K., Jusoh, A. F., & Saidin, M. (2023). X-ray fluorescence (XRF) analysis of iron ore at ancient Kedah iron smelting site, Sungai Batu archaeological complex, Bujang Valley, Kedah, Malaysia. *Heliyon, 9*(4).

[15] Abubakar, Jabir, Zhaochong Zhang, Zhiguo Cheng, Fojun Yao, and Abdoul-Aziz Bio Sidi D. Bouko. (2024). "Advancing Skarn Iron Ore Detection through Multispectral Image Fusion and 3D Convolutional Neural Networks (3D-CNNs)" *Remote Sensing* 16, no. 17: 3250. https://doi.org/10.3390/rs16173250.

Characteristic of Iron Ore (Fe) Metallic Mineral Potential in Laeya Village, North Buton Regency, Southeast Sulawesi

Deniyatno[1*], Erwin Anshari[1], Abd. Hadid Rahman Ulfa[1], Marwan Zam Mili[1], Masri[2], La Ode Muhamad Iradat Salihin[3], Al Rubaiyn[4]

[1*]Department of Mining Engineering, Halu Oleo University, Kendari Southeast Sulawesi 93232, Indonesia.
[2]Department of Geology Engineering, Halu Oleo University, Kendari Southeast Sulawesi 93232, Indonesia.
[3]Department of Geography, Halu Oleo University, Kendari Southeast Sulawesi 93232, Indonesia.
[4]Department of Geophysics Engineering, Halu Oleo University, Kendari Southeast Sulawesi 93232, Indonesia.

*E-mail: Deniyatno@uho.ac.id

Abstract. This study aims to assess the potential of metallic minerals, particularly iron ore (Fe), in the Laeya region, North Buton Regency, Southeast Sulawesi Province. The methods used include petrography, XRF, and magnetic susceptibility to analyze 16 samples collected from various research sites. Petrographic analysis reveals that the rocks in the Laeya area consist of wackestone, carbonate sandstone, and siltstone. XRF analysis indicates variations in iron (Fe) content in the rock samples, with the highest concentration reaching 7.15% in the siltstone. Magnetic susceptibility analysis supports this research, with the highest recorded susceptibility value of 414.5×10^{-8} m^3/kg. The geological structures identified include quartz veinlets and fault brecciation in the carbonate sandstone outcrops, which, along with indications of iron ore (Fe) mineralization, provide preliminary evidence of potential metallic mineral accumulation in the area.

1. Introduction

Iron (Fe) is an element found in almost all types of rocks, and its presence in large quantities with economic value involves various geological processes related to mineralization zoning [1, 2]. Iron ore (Fe) is very important because it is one of the common elements frequently found as a raw material in the production of iron metal, which is essential in various fields [2].

This research focuses on preliminary to identify the characteristic of iron ore in the Laeya region, North Buton Regency, Southeast Sulawesi, Indonesia. The Laeya area is part of Buton Island, which has complex formation processes. Buton Island is located in the Southeast Sulawesi region and is believed to be part of the Australian continental fragment that drifted northeastward, separating from the Australian continent [3]. The dynamics of this continental fragment later merged with the Sunda microcontinent during the Miocene epoch, triggering the formation of geological structures in Sulawesi [4]. Generally, Buton Island consists of island clusters formed due to intense tectonic processes. Several tectonic events that influenced the formation of geological structures on Buton Island include pre-rift, rift, drift phases, and Neogene collision, with a predominance of geological structures formed during the most recent collision [4,5,6].

2. Research Site

This research focuses on the Laeya Village area in North Buton Regency, Southeast Sulawesi, Indonesia (Fig. 1A). The geomorphology of this area is part of the North Buton zone, dominated by lowlands and karst mountains (Figure 1C). The river flow patterns in this region show geological structure control, contributing to weathering processes and the formation of secondary minerals. The general trend of the mountains in this area is Northwest–Southeast [3, 4].

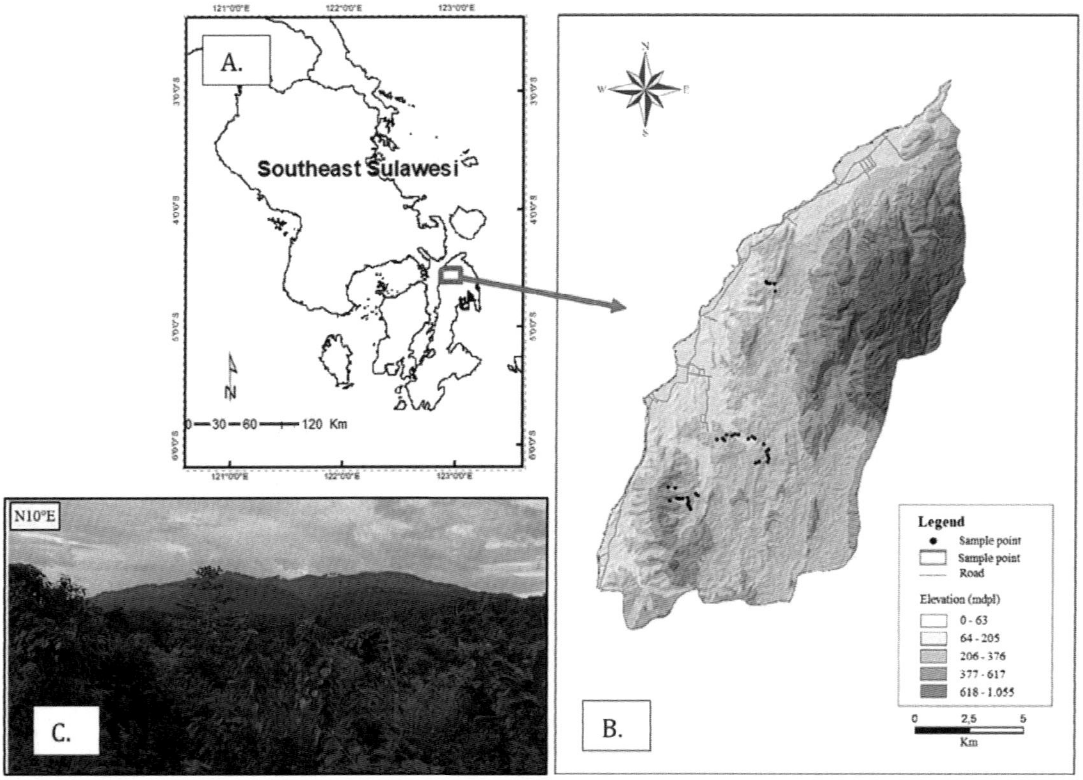

Figure 1. (A) The Research location is in the Laeya of North Buton Regency, Southeast Sulawesi Province (B), Elevation of the area, dan (C) Morphology of the area.

3. Materials and Methods

Samples were collected directly in the field using the grab sampling method. Rock outcrops were also observed to identify surface geological characteristics, such as structure, texture, and rock composition. Each rock sample was analyzed using methods suited to its characteristics to obtain accurate and relevant data. Sample testing was conducted in the laboratory, including petrographic analysis, XRF analysis, and magnetic susceptibility testing. Petrographic analysis aims to identify the mineral composition of the rock samples and the presence of alteration

IOP Conf. Series: Earth and Environmental Science **1525** (2025) 012008 doi:10.1088/1755-1315/1525/1/012008

minerals that may have formed during geological processes. X-Ray Fluorescence (XRF) analysis is used to detect and identify economically valuable metal compounds within the samples, while magnetic susceptibility testing measures the degree of magnetization in the sample material [5,7,8]. Magnetic susceptibility values can provide insights into the presence of oxide minerals, as each type of oxide mineral has different magnetic properties [2].

4. Result and Discussion

The geological structure in the study area is characterized by the presence of folds, faults, and quartz veins (Fig. 2B). The folds have an axis that extends from north to south, forming a gently sloping syncline of the open fold type with a plunging axis. The faults are characterized by a fault plane oriented at N140°E/45°SW, with the strike of the lineation having a dip angle of 56°, forming a pitch of 40° relative to the fault plane (Fig. 2A). The lineation indicates that this is a thrust fault type. Quartz veins are only sporadically found cutting through the layers of carbonate sandstone, with an opening width of 3–10 mm and an orientation of N65°E/60°SE.

Figure 2. (A) Fault plane and lineation on carbonate sandstone, and (B) quartz vein cutting through the layers.

In the research area, the Tondo Formation is characterized by clastic sediments consisting of sandstone, siltstone, and shale. The Tondo Formation is one of the formations found on Buton Island [4,6,9]. The sandstone is yellowish-gray in color, with good sorting, closed packing, and a carbonate matrix predominantly composed of rounded quartz and lithic fossil fragments. Meanwhile, the shale is dark gray, with open packing, carbonate composition, and a brittle texture (Fig. 3).

The results of the X-Ray Fluorescence (XRF) analysis indicate that the iron (Fe) content in the tested samples is not significant (Fig. 4). Among the analyzed samples, the average iron content was found to be 6.31%. The siltstone sample exhibited the highest iron (Fe) content at 7.15% (Table 1).

Figure 3. (A) Carbonate sandstone ST10 (4861) with an abundance of quartz and a carbonate matrix. (B) Dark gray shale ST11 (4663), carbonate.

Table 1. The results of the magnetic susceptibility analysis and XRF analysis based on variations in different lithologies.

Sample Id	Litology	Fe (%)	Suceptibility (10^{-8} m^3/kg)
ST01 (4853)	wackestone	2,34	10,0
ST03 (4857)	Siltstone	5,73	29,7
ST04 (4859)	wackestone	2,68	15,2
ST08 (4854)	wackestone	2,67	6,7
ST09 (4862)	wackestone	2,89	18,6
ST10 (4861)	Carbonate sandstone	6,42	414,5
ST11 (4856)	Siltstone	7,15	76,3
ST11 (4863)	Marl	4,09	11,2
ST12 (4860)	Marl	6,38	108,7
ST14 (4855)	Reef limestone	2,25	7,8
ST16 (4858)	Siltstone	5,87	24,6

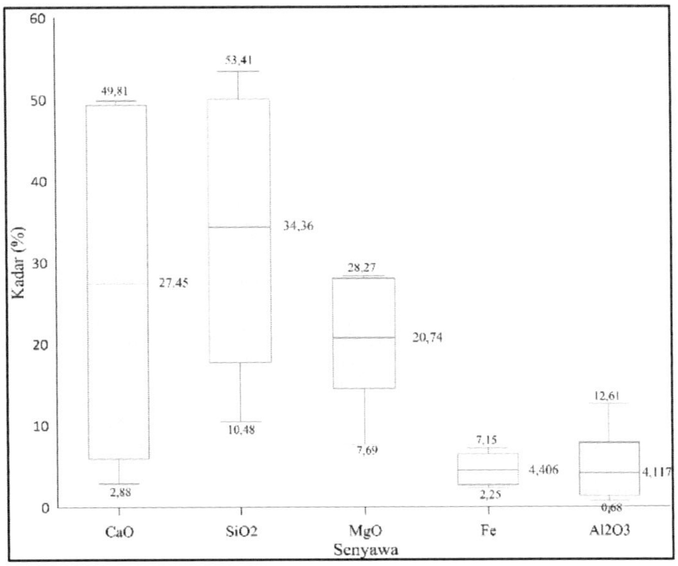

Figure 4. The distribution of major compound contents in the rock samples, consisting of minimum values, maximum values, and average values.

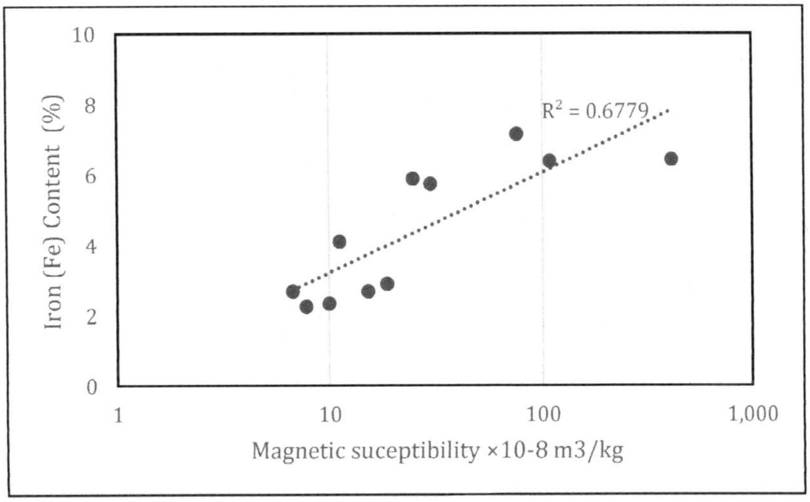

Figure 5. Correlation of magnetic susceptibility values with iron (Fe) content in the rock samples.

Magnetic susceptibility values can differentiate rocks containing oxide minerals from surrounding rocks, especially in rocks with high mica content [2,7,10]. This measurement is often used in mineral to identify potential areas containing metallic minerals, particularly iron (Fe) [2]. High magnetic susceptibility values often indicate concentrations of iron oxide minerals. Measurement results show a variation in magnetic susceptibility values ranging from 6.7 to 414.5 × 10^{-8} m^3/kg (Table 1), with the lowest values found in limestone and the highest in carbonate sandstone, which is closely related to iron oxide content based on XRF analysis results (Fig. 5).

5. Conclusion

This study evaluates the iron (Fe) mineral potential in Laeya Village using petrography, X-Ray Fluorescence (XRF), and magnetic susceptibility methods. Petrographic analysis shows that the rocks consist of wackestone, carbonate sandstone, and siltstone. Through XRF analysis, the highest iron content was found in siltstone, reaching 7.15%. Magnetic susceptibility testing indicates the highest magnetic values in carbonate sandstone, suggesting the presence of iron oxide minerals. The results of these three methods conclude that Laeya has significant iron mineralization potential, especially in zones with high magnetic susceptibility values.

References

[1] Abraham, M., Gift, R., Hennery, S., Asuman, G., Rita, A., Dexter, D. and Joseph, M. (2020). The Prospects of Uganda's Iron Ore Deposits in Developing the Iron and Steel Industry. *Journal of Minerals and Materials Characterization and Engineering*, **8**, 316-329. doi: 10.4236/jmmce.2020.84019.

[2] Roy, Subhnit K.; Nayak, Deepak; Rath, Swagat S.. (2020). A review on the enrichment of iron values of low-grade Iron ore resources using reduction roasting-magnetic separation. Powder Technology, 367, 796–808. https://doi:10.1016/j.powtec.2020.04.047.

[3] Burhanuddin, M. , Chiyonobu, S. , Ando, T. , Yusriani, A. , Husain, R. and Asfar, S. (2024). Hydrocarbon Potential of the Triassic-Jurassic Sediments in Southeast Sulawesi, Indonesia, Based on Lithofacies and Geochemical Analysis. *Open Journal of Geology*, **14**, 723-745. doi: 10.4236/ojg.2024.148031.

[4] Davidson, J.W. (1991). The Geology and Prospectivity of Buton Island, SE Sulawesi, Indonesia, Proc. 20th Ann. Con. Indon. Petroleum Assoc.

[5] Nugraha, A.M.S. and Hall, R. (2022). Neogene Sediment Provenance and Paleogeography of SE Sulawesi, Indonesia. Basin Research, 34, 1714-1730.

[6] Surono and Hartono, U. (2013) Geology of Sulawesi. Centre of Geological Survey, Ministry of Mineral and Energy Resources of Indonesia. LIPI Press.

[7] Zhang, B., Shi, J., Jiang, Y.R., Zhou, J.G. and Li, S.Q. (2021) XRF Analysis of Impurity Metal Elements in Iron Base Alloy. *Open Access Library Journal*, **8**, 1-5. doi: 10.4236/oalib.1107846.

[8] El-Bahi, S., Sroor, A., Arhoma, N. and Darwish, S. (2013). XRF Analysis of Heavy Metals for Surface Soil of Qarun Lake and Wadi El Rayan in Faiyum, Egypt. *Open Journal of Metal*, **3**, 21-25. doi: 10.4236/ojmetal.2013.32A1003.

[9] Jabłońska, Mariola, Marzena Rachwał, Małgorzata Wawer, Mariola Kądziołka-Gaweł, Ewa Teper, Tomasz Krzykawski, and Danuta Smołka-Danielowska. (2021). "Mineralogical and Chemical Specificity of Dusts Originating from Iron and Non-Ferrous Metallurgy in the Light of Their Magnetic Susceptibility" *Minerals* 11, no. 2: 216. https://doi.org/10.3390/min11020216.

[10] Müller-Huber, E. and Börner, F. (2017). Multi-parameter petrophysical characterization of variscan greisen rocks from the southern bohemian batholith (austria) and the eastern erzgebirge volcano-plutonic complex (germany). Austrian Journal of Earth Sciences, 110 (1). hal. 78-100. https://doi.org/10.17738/ajes.2017.0006.

IOP Conf. Series: Earth and Environmental Science **1525** (2025) 012009 doi:10.1088/1755-1315/1525/1/012009

Geochemical Properties of Landslide Materials, West Sinjai District, South Sulawesi Province

A. M. Imran*[1], Busthan[1], Safruddim[1], Karina Ipanin[1], and Asmita Ahmad[2]

[1] Department of Geology, Engineering Faculty, Hasanuddin University, Makassar, Indonesia
[2] Department of Soil Science, Faculty of Agriculture,Hasanuddin University, Makassar, Indonesia

*E-mail: imran-am@eng.unhas.ac.id

Abstract. Malino - Sinjai road section is the main access connecting Sinjai and Malino, the mountain tourism destination. As a tourism destination, the access road to the Malino area frequently experiences landslides, especially in the rainy season. The area consists of volcanic rocks that are highly weathered. Some factors controlling the landslides are the expansive mineral contents, steep slopes, and human activities. A further detailed study of soil characteristics is needed to mitigate the area. Grain size distribution and soil geochemical analysis are conducted on landslide and non-landslide materials. The grain size distribution denotes poorly sorted distribution at the scarp and head zone. Meanwhile, it is well sorted in the body and foot zone. The particle size distribution is dominated by medium – fine-sized sand ranging from 30% to 67%. Montmorillonite develops at the foot zone (41%) but is limited at the scarp zone (1,4%) as an expansive clay mineral well. In contrast to that one, vermiculite well develops at the scarp (24,8%), but it restrictedly presents at the foot zone (4,8%).

1. Introduction

A landslide is a geological hazard frequently occurring along the Manipi – Malino road. The area is a mountain tourism destination in South Sulawesi Province, Indonesia, which is well known as a part of Mount Bawakaraeng. Geologically, the area consists predominantly of semi-consolidated volcanic rocks from the Lompobattang Formation of the Pleistocene age. The huge landslide material due to mass movement came from the collapse of Mount Bawakaraeng caldera on March 26th, 2004, resulting in around 200–300 million m^3 sediments. [1] steep slopes, soil properties, and high rainfall triggered landslides in the Malino region. The study area is located along the road of the Malino – Manipi segment, where landslides occur in the rainy season (Figure 1).

Several types of research have been conducted around Mount Bawakaraeng. Stated that the landslide occurred along the Malino-Manipi road section due to discontinuities between tuff, volcanic breccia, and lava [2]. Furthermore, [3][4] explained that the weathering of volcanic rocks was the main factor in landslides on that road section. Ahmad [5] found that clay minerals within soil mineral particles triggered soil movement (landslide) in the Tombolo area, Gowa South Sulawesi. In general, soil's chemical and physical properties, especially expansive and grain size distribution, play an important role in landslides. Hence, soil assessments such as mineralogy, texture, level of weathering, and mineral fluid interface processes should be analyzed to predict landslide potential.

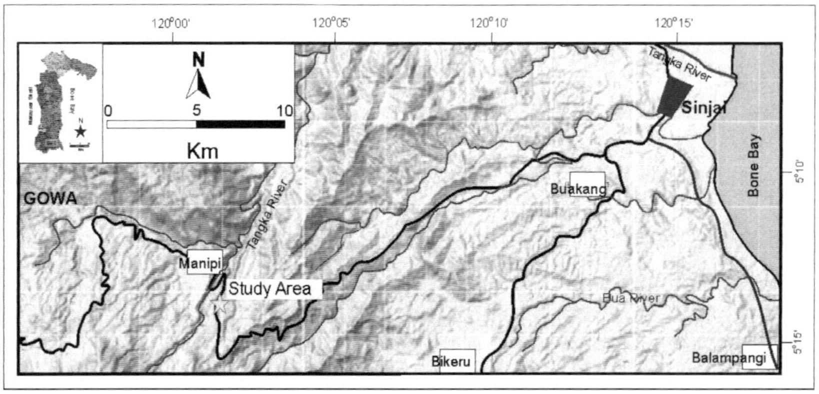

Figure 1. Study area map showing the research location in Maniipi, Sinjai, South Sulawesi. The map includes major rivers such as the Tangka River and Bua River, as well as nearby settlements including Bukakang, Bikeru, and Balampangi. The inset map provides a regional context with a scale bar and a north arrow for orientation.

Clay minerals have extensive properties that are directly observable in the field. Its physical properties shrink in dry conditions but swell in wet conditions. The expansive nature of clay depends on its grain size and mineralogy [6]. The swelling factor of clay is due to a void that may be filled by water. This study aims to perform control of grain size and mineral alteration of volcanic rock in tropical areas. The survey of several landslide materials in Taiwan by Marc et al. [7] found that grain size material depends upon the depth of the rock parent that has been moved. The coarser size of the material means the deeper rocks experience landslides. The swelling and shrinking properties of clay are controlled by voids within the clay mineral [8].

2. Geological Background

The study area lies on the eastern flank of Lompobattang Mountain, composed of volcanic rocks. The area is a mountainous region with slope ranges between 14 – 550 and heights ranging from 75 – 500m. The highest point is around 1000 m above sea level (Fig. 2). The rocks experience weathering and form soil 50 – 150 cm thick. Several normal faults are identified by the topographic liniment (Fig. 3).

Figure 2. Mountain landscape with a steep slope (a). Meanwhile, people use it as a rice field despite its steep slope (b).

Regionally, the study area is covered by Camba Volcanics (Tmcv) of Miocene age, Baturappe-Cindako Volcanics (Tpbv) of Pliocene age, and Lompobattang Volcanics (Qlv) of Plio-Pleistocene age (Fig. 3). Camba Volcanics dominates the western part of the study area. It consists of tuffs and a small amount of volcanic breccia. The tuff is generally dipping downstream about 40O. Baturape-Cindako Volcanics mainly consists of volcanic breccia. The rock is partly distributed in the northeast part of the study area. Fragment sizes of breccia range from 15 – 65 cm, consisting of basalt and andesite. A geological explanation will be provided for Lompobattang Volcanics, where the landslide occurred (Fig. 3).

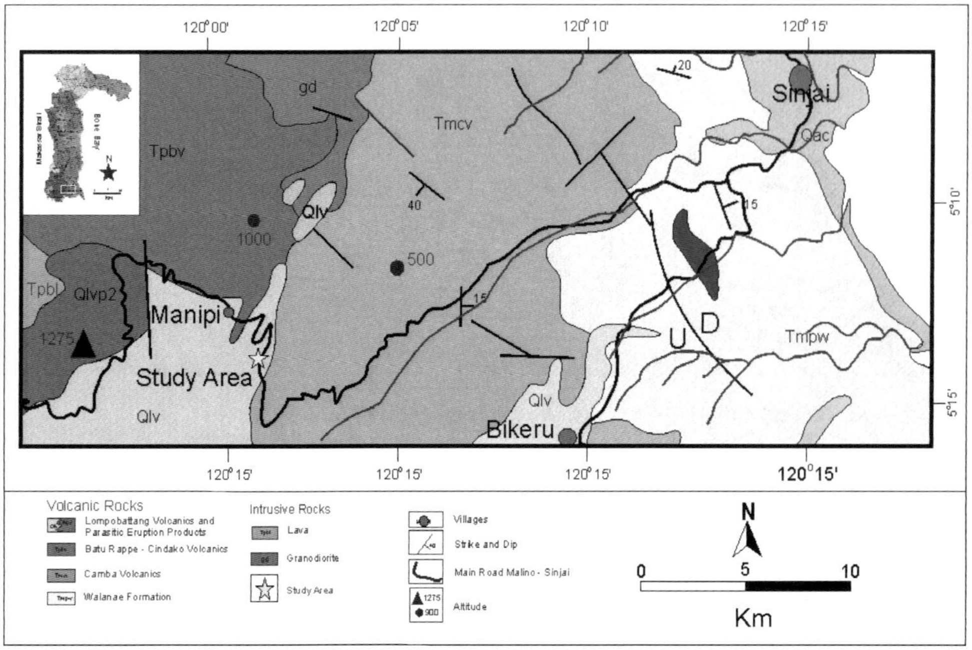

Figure 3. Regional geology of the study area, adapted from [9]. The map illustrates the research location's geological framework, highlighting major lithological units, structural features, and significant geological formations. Key tectonic and stratigraphic elements provide insight into the broader geological setting of the study area.

Lompobattang volcanic rocks are extrusive rocks with an age of Plio-Pleistocene [10]. It is composed of alternating lava flow and pyroclastic breccia. It is occasionally associated with dikes and lava domes (Fig. 4). The volcanic rocks have a silica-saturated shoshonitic composition as the first type, and another type is a more basic member, which has a silica-undersaturated alkali potassic composition.

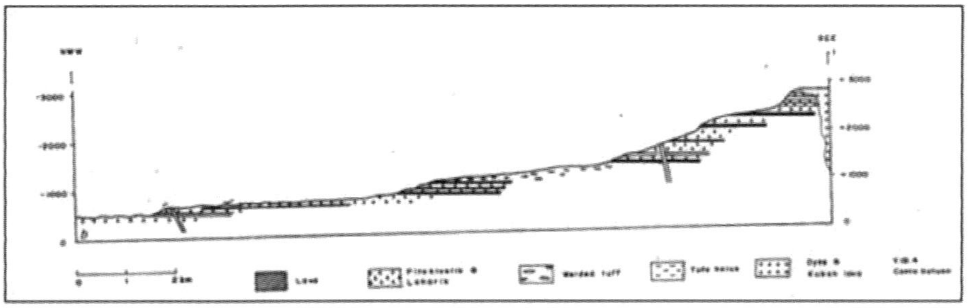

Figure 4. Stratigraphy of the Lompobattang Volcanic Rocks at elevations above 1000 m asl, adapted from [11]. The figure illustrates the lithological composition and stratigraphic sequence of volcanic deposits in the high-altitude region, providing insights into the geological evolution of the Lompobattang area.

The fragments of volcanic breccia generally consist of alkaline igneous rock with a size ranging between 45 – 150 mm (Fig. 5a), and the matrix consists of coarse tuff ranging from 1/16 – 2mm. These rocks are generally exposed on river cliffs and cutting roads. The breccia alternately presents with fine grain size tuff, which is highly weathered (Fig. 5b). Basaltic lava (Fig. 5c) presents in two types, silica-undersaturated alkali potassic rocks and silica-saturated shoshonitic rocks [10]. The first type tends to be a basic composition; however, the second type is more acidic.

Figure 5. Outcrops of Lompobattang Volcanic Rocks showcasing volcanic breccia, tuff, and basaltic lava. The figure highlights the primary lithological units exposed in the study area, representing different phases of volcanic activity and deposition

3. Material and Methods

Surveys were carried out on cliffs that experienced landslides and their material deposits. The distance of observation points from the head was 50 m to the scarp, while to the body of the landslide was 10 m and to the foot was around 14 m.

We performed 4 test pits at different parts of the landslide deposit, specifically at the scarp, head, body, and foot of the landslide (Fig. 6). The geometry of the test pit is 100 cm3 and divided into three parts: upper (0 – 33 cm), middle (33 – 66 cm) and lower parts (66 – 100 cm). Each part in the test pit was compositely sampled for laboratory analyses. Twelve samples were used for grain size analyses, and four samples were placed in the lower layer, each test pit for XR-D analysis.

Grain size distribution was obtained by sampling each layer of the test pit and developed to measure the vertical distribution of landslide materials based on basic classification tests of sieving [12].

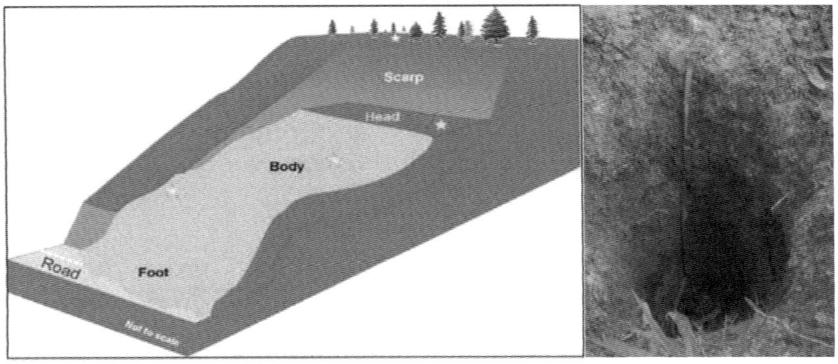

Figure 6. Generalized profile of landslide material along the Malino–Manipi road (a) and test pit conducted at the scarp (b). Yellow stars indicate the locations of test pits. The figure provides an overview of the landslide deposits and subsurface conditions, aiding in the assessment of slope stability in the area.

IOP Conf. Series: Earth and Environmental Science **1525** (2025) 012009 doi:10.1088/1755-1315/1525/1/012009

4. Result and Discussion

The occurrence of landslides along the Malino – Manipi road is controlled by discontinuity due to bedding and fracture [2]. Another factor controlling landslides in the study area is weathering, steep slopes, and faults [4].

The geometry of landslide material shows a sliding plane forming a scarp with slump-type movement (Fig. 7). The landslide material consists of head, body, and foot zones. The width of the material is 15 m and spans up to 75 m. The original slope is 39O, and the cliff's height is around 40 m from the edge of the foot zone that crosses the road.

Figure 7. Profile of landslide material along the Malino–Manipi road. The figure illustrates the stratigraphic composition and distribution of landslide deposits, providing insights into the geological and geotechnical characteristics influencing slope stability in the area.

4.1. Grain Size Distribution

The grain size distribution of soil depends on generated minerals and depth of weathering, where a high degree of weathering produces thick residual soil [13] [14]. Moreover, the degree of weathering generates variation in pore-size distribution [15].

The grain size distribution at the lower part of the 4 test pits shows two graphs that have poorly graded type (at the scarp and head of landslide material). The graphs are gap-graded distribution. Meanwhile, the other two graphs (at the main body and foot of the landslide material) show a well-graded distribution. Generally, the grain size of landslide materials looks homogeneous, but it does not work in different lithologies [7]. The study area comprises volcanic breccia, so we used an area-weighted sum for grain size analysis.

The sampled soil is in situ and classified into clayey sand based on the Unified Soil Classification System (USCS) classification. It comprises 43,71% coarse sand, 52,92% fine sand, and 3,36% silt clay. Grain size analysis of the test pit at scarp shows poorly graded distribution with coefficient gradation (Cc) of 0,81. The grain size distribution tends to increase to the upper layer (Fig.8). This graph indicates that plasticity index values rise to the upper layer with decreasing particle size.

Figure 8. Grain size distribution of soil at the scarp zone. The figure presents the particle size distribution curve, highlighting the proportions of gravel, sand, silt, and clay. This analysis provides insights into the soil's textural properties and its influence on slope stability.

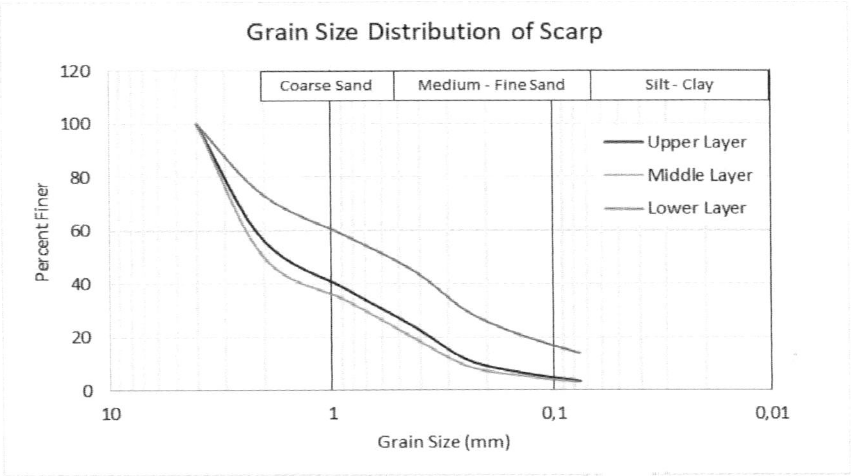

Figure 9. Grain size distribution of the Main Scarp from bottom to top layer. The figure illustrates the variation in particle size distribution, with a particular focus on the 0.4 mm – 1 mm range (coarse sand), which exhibits a gap-graded pattern. This characteristic indicates possible selective deposition or post-failure soil reworking processes affecting slope stability.

Figure 9 shows grain size distribution that gives gap or poorly distributed material between 0,4 to 1 mm. The strongest segregation was found in the lower layer, which is close to the parent rocks. The lower layer (ll) of the materials has fine grains (silt-clay) 4 times more than that of the upper layer (ul) without significantly changing the graph form of distribution.

A grain size analysis of the studied landslide materials is provided in Figure 10, which explains the grain size distribution of clay–silt and sand fractions. The landslide materials are classified as clayey-silty sand texture and divided into three parts (head, body, and foot). The grain size distribution of the materials was also conducted at each test pit's lower, middle, and upper layers. The discriminant of material was poorly distributed. Coarse material (sand) dominates the landslide head with a proportion between 30,15 – 67,39%. Coarse sand dominates at the middle and lower layers; conversely, medium-fine sand dominates the upper layer (Fig. 10). Silt-clay materials occur more abundantly at the upper layer, although they are <5%.

The physical properties of the landslide materials in Fig. 10 show a close similarity in their general characteristics of body and foot landslides. Grain size tends to increase the silt-clay size to the upper layer but decreases medium-fine sand. Meanwhile, the coarse sand increases from the body to the upper layer. The materials of landslide heads consist of coarse sand ranging from 43 - 67%, medium - fine sand ranging from 30 - 53%, and silt clay ranging from 2,5 - 4%. The materials of the landslide body contain coarse sand ranging from 51 - 56%, medium - fine sand ranging from 37 - 44%, and silt clay ranging from 4,5 - 6%. The materials of foot landslide consist of coarse sand ranging from 21 - 24%, medium - fine sand ranging from 66 - 67%, and silt-clay ranging from 9 - 13%.

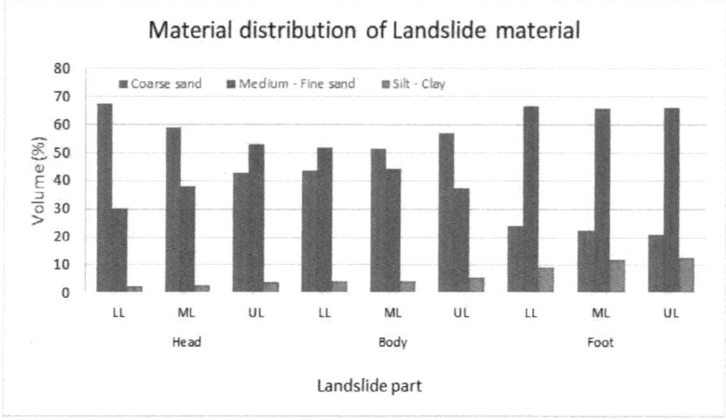

Figure 10. Grain size distribution of landslide material.

Material distribution changes from head to foot of the landslide, in which coarse sand dominantly develops at the head. In contrast, silt-clay size gradually increases to the landslide foot (Fig. 10). In line with fine-size material, medium-fine sand tends to increase in volume corresponding to distance from the scarp. This is an anomaly of coarse sand distribution, in which its volume increases from LL to UL at the landslide body compared to the head and foot of the landslide. Based on the gradation coefficient, the grain size distribution of landslide material is poorly graded with a value above 3, except at the landslide foot, the well-graded distribution from the lower to the upper layer with a gradation coefficient (Cc) is 1,29 (Fig. 11).

4.2. Clay Minerals

The weathering of volcanic rock produces residual soil commonly consisting of clay, silt, and fine sand [14]. Moreover, [16] those who studied the Late Cretaceous volcanic rocks found that the weathering products are kaolinite, smectite (mainly montmorillonite), and illite. A tropical climate zone like Indonesia has high rainfall and humidity, triggering rapid chemical weathering processes [17]. Hence, clay minerals are widely spread in that area and are present as the main element in soil. Common clay minerals are kaolinite, chlorite, illite, and smectite. [18] clay minerals' type, quantities, and distribution are important in landslide occurrence.

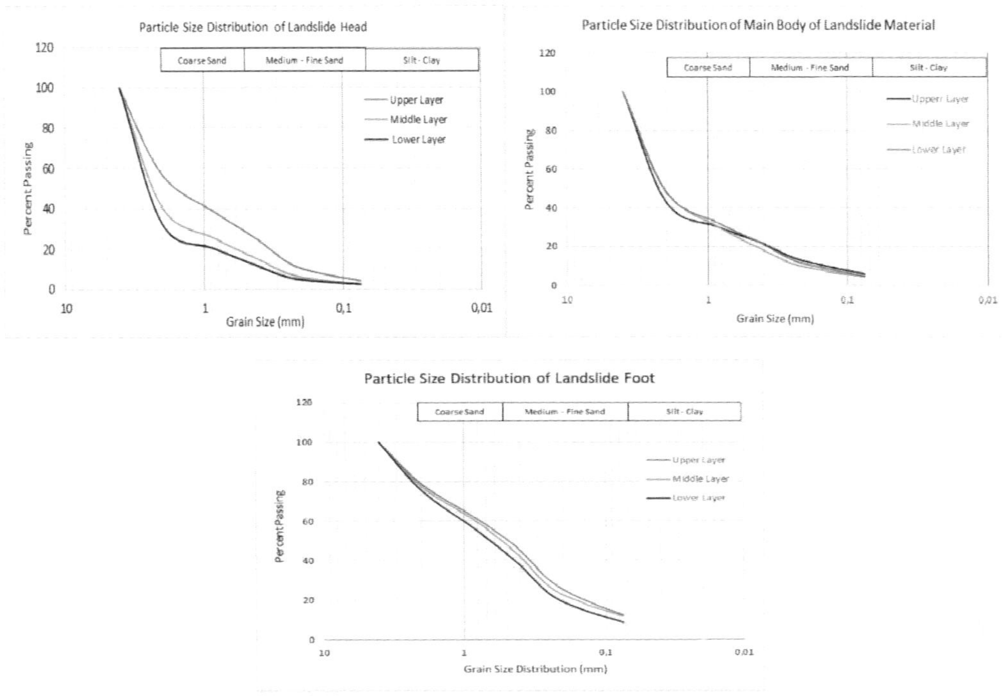

Figure 11. Distribution and grading of landslide material. The figure illustrates the spatial variation in grain size and grading characteristics within the landslide deposits, providing insights into sediment transport, deposition processes, and material stability in the affected area

We examined four soil samples at different parts of the landslide materials. All sampled materials were carried out from the lower layer of the test pits. There are 7 (seven) types of clay minerals were identified through XR-D analysis (Fig. 12). Montmorillonite, which is one of the expansive clay minerals, occurs more abundant at the foot zone (41,8%) than other parts of the landslide (Fig. 14). This mineral significantly increases toward the foot zone. It comprises 1,4% at the scarp zone, 24,3% at the head, and 30,3% at the body zones. Another expansive clay mineral (nontronite) is sparingly present in all landslide materials (<6%), but it slightly increases at the foot zone (5,8%).

Non-expansive clay minerals developing in the study area are halloysite and kaolinite. Other limited expansion minerals are vermiculite and chlorite. Halloysite is relatively evenly distributed at all landslide parts ranging 12 – 15% from scarp to foot zone. The existence of kaolinite decreases toward the foot zone (12% at scarp and 5,2% at foot zones). Vermiculite dominant comprises the scarp zone (24,8%); meanwhile, it presents ≤ 5% at other zones. Illite restrictedly appears at the foot zone (9,8%) but is significantly present at other zones (>25%).
Compared to the clay mineral distribution from the scarp zone, in situ soil, and landslide materials from the foot zone, expansive and non-expansive minerals have an inverse composition relation. Montmorillonite developed limitedly at the scarp zone (1,4%), but it developed well at the foot zone (41%). In contrast to that one, vermiculite is well developed at the scarp (24,8%) but restrictedly present at the foot zone (4,8%).

Expansive soil is a kind of clay rich in montmorillonite and other strong hydrophilic minerals, which has the characteristics of swelling in wet conditions and shrinking in dry conditions. In the natural state, expansive soil is generally in the hard plastic state with high strength. However, the expansive soil will soften with the increased water content, decreasing its strength.

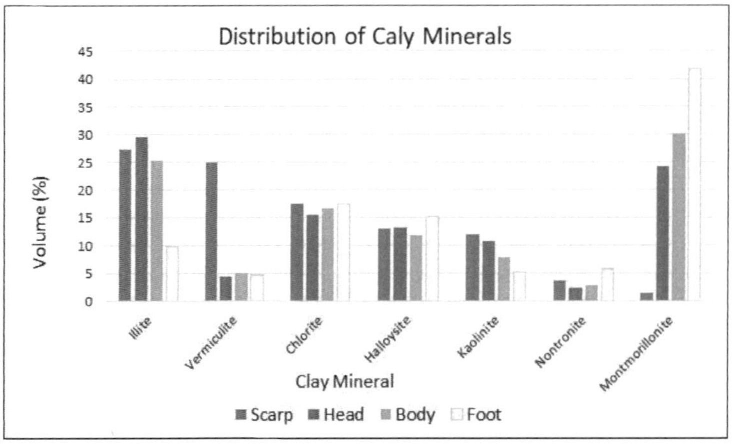

Figure 12. Distribution of clay minerals in the lower part of the landslide material.

Figure 13 shows a correlation between clay mineral content and grain size. The more fine grain size, the higher the content of montmorillonite. Similarly to non-expansive minerals, the less kaolinite content, the less coarse grain content. It means that the decreasing tendency of silt-clay size material follows the increasing tendency of kaolinite. This data indicates that the swelling properties of the landslide materials reflect the presence of expansive smectite clay minerals (montmorillonite and nontronite).

5. Discussion

The data argue that significant amounts of expansive clay minerals (montmorillonite and nontronite) and increasing fine particle size in landslide material triggered the landslide in the study area. The presence of montmorillonite and nontronite at the foot zone, both of which have high swell properties, describes that the volcanic breccia has undergone weathering, generating minerals with high levels of swelling. [14] confirm the relationship between depths of weathering levels and engineering properties of residual soils. [19] concludes that high-Na smectites have a low swelling capacity, but smectite clays with Ca and Mg have initially rapid water adsorption.

The soil of weathered volcanic breccia shows a high level of weathering, as seen by the presence of clay minerals such as illite and vermiculite, which is> 20%. Although expansive minerals of montmorillonite and nontronite (smectite clays) are quite low, < 5%, they are starting to appear and increase significantly at the head zone (24,3%). The grain size in Fig. 9 is poorly distributed with a 0,8 gradation coefficient. Fig. 8 shows that medium-fine sand is a dominant material of three layers with an average of 53%, while coarse sand has an average of 40%. The chemical weathering process is characterized by the enrichment of smectite clays [20]. Many expansive clay minerals (montmorillonite: a smectite group) suggest that the landslide materials have high swell potential [21].

The study of grain size control to landslides has been conducted on 17 recent landslides [7]. They found that the deeper the landslide, the coarser the particle size. This case comes from the study area where the grain size at the lower layer accumulates more coarse sand (43,71%) than at the upper layer (26,72%). Furthermore, [21] states that mechanical and chemical weakening within soil due to the chemical weathering of unstable sand- to silt-sized materials at the transitional zone, an enrichment zone of kaolinite and smectite. Another finding around the study area considered that the primary mineral weathering process started at the cleavage planes and mineral fragment planes of clay minerals at volcanic rocks in the Tombolopao area, Gowa District, Indonesia [5]. Small-sized clay minerals (montmorillonite, illite, vermiculite) absorb more water because they have more voids [23].

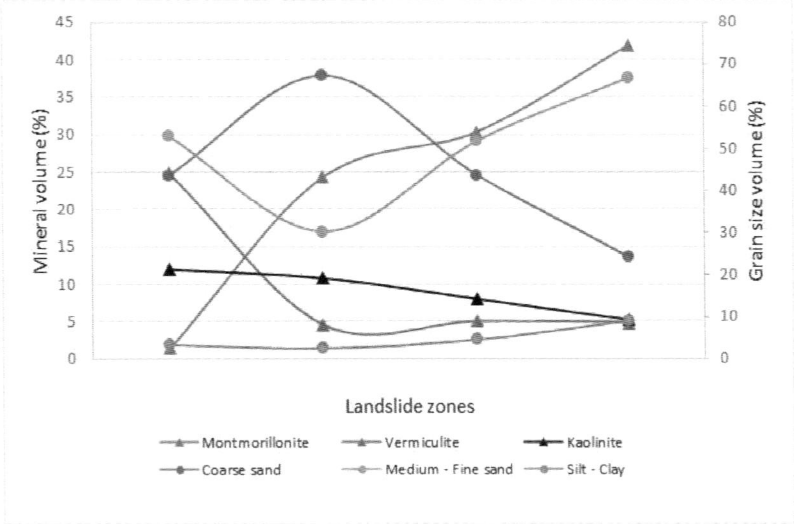

Figure 13. Relationship between clay minerals and grain size in the lower part of the landslide and scarp zone

The expansive minerals and nonexpansive minerals contrary occur in all zones (Fig. 14). Referred the statement, where the Lompobattang volcanic rock consists of 2 (two) types: shoshonitic type 1 is more acid condition and silica-saturated shoshonitic composition, and type 2 contains high kal-alkaline minerals [11]. Hence, the origin of montmorillonite in the study area is probably shoshonitic type 1 rock. [14] have tested the shear angle of volcanic breccia controlled by several factors, including the degree of weathering and the rock's mineral composition. The occurrence of kaolinite, illite, and smectite generated from andesitic–dacitic volcanic stones in Turkey was studied by [24]. Furthermore, [25] argued that andesitic volcanic rocks experiencing alteration produced smectite, kaolinite, illite, and montmorillonite.

The average soil water content in Gowa Regency, South Sulawesi, is 44%, with the average soil liquid limit in the landslides area range of 63.55%, while the no-landslide area is 55.65% [26]. Stated that the water content within the clay layer led the clay to swell [27]. In addition, clay mineral swelling type physicochemical swelling behavior is due to the interaction of water and minerals [28][8]. The effect of water on reducing the bond strength between soil particles has been proven through simulations by [29] [30]. They demonstrated that the weakening zone was due to the high content of water. Another study obtains that chlorite and illite precipitate at high water content [31]. The study of [30] suggests that the presence of clay minerals resulting from the weathering of dacite rocks in the Giresun Region of Türkiye triggers landslides. He found that the developed clay types were Illite and montmorillonite, both of which reduce the shear strength when saturation occurs.

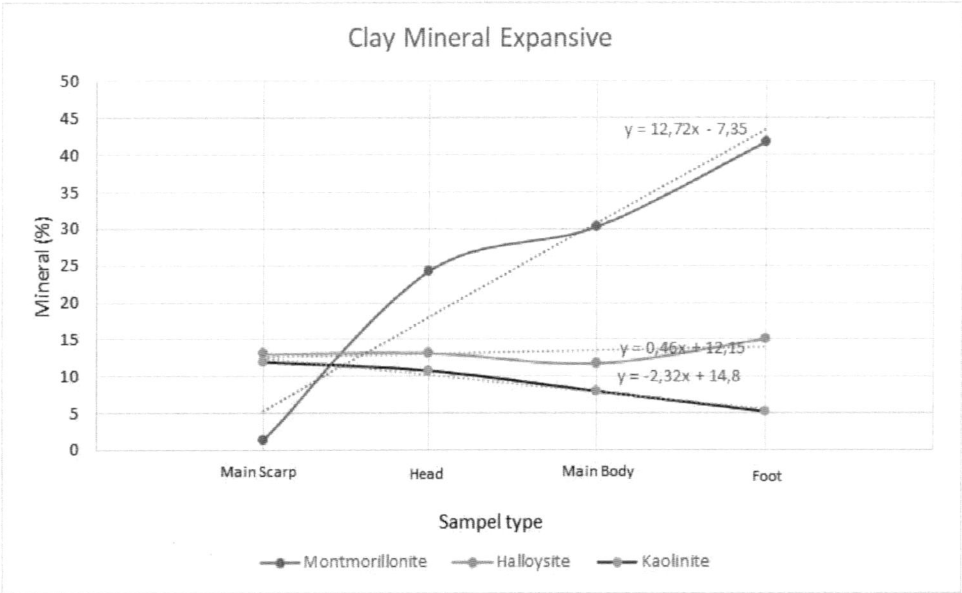

Figure 14. Distribution of clay minerals in the landslide material, including Montmorillonite (highly expansive), Halloysite (moderately changeable), and Kaolinite (less expansive). The figure highlights the varying swelling potential of these minerals, which significantly influence soil behavior, water retention, and slope stability.

6. Conclusion

Clay minerals in the landslide material are likely derived from the weathering of shoshonitic (Type 1) rocks of the Lompobattang Volcanics. The degree of weathering has resulted in a poorly sorted grain size distribution, predominantly composed of medium to fine sand, which enhances the soil's ability to absorb rainwater. Furthermore, the dominance of montmorillonite, a highly expansive clay mineral, suggests that the movement of landslide materials is strongly influenced by mineral expansion and contraction, making it a key factor in slope instability.

References

[1] Lias SA, Safaat M and Solle MS (2020). The soil characteristics of landslides in Manuju District, Gowa Regency. IOP Conf. Series: Earth and Environmental Science 486 012140 IOP Publishing. doi:10.1088/1755-1315/486/1/012140

[2] Imran AM, Azikin B and Sultan (2012). Peranan aspek geologi sebagai penyebab terjadinya longsoran pada ruas Jalan Poros Malino-Sinjai. Buletin Geologi Tata Lingkungan (Buletin of Enviromental Geology). 22 (3): 183-196. In Bahasa Indonesia

[3] Azikin B, Imran AM, Samang L, and Ramli M (2014)' Weathering stage control of tuff and influence to the malino-manipi landslide susceptibility, South Sulawesi-Indonesia, Journal South –East Applied Geology 6(1): 1-9.

[4] Azikin B (2015). Analysis of landslide susceptibility of rupture surface based on weathering stage of volcanic rock, Penelitian Fakultas Teknik Universitas Hasanuddin, Makassar.

[5] Ahmad A, Lopulisa C, Imran AM and Baja S (2018). Soil mineral as an indicator of soil stabilization in the sloping area: a case study in Tombolopao at Gowa District. Ecosolum 7 (1): 33-37. In Bahasa Indonesia.

[6] Yulianti H, Sarah D and Subowo E (2012). Pengaruh lempung ekspansif terhadap potensi amblesan tanah di Daerah Semarang, Riset Geologi Dan Pertambangan 22 (2): 93-103, Bandung. In Bahasa Indonesia.

[7] Marc O, Turowski JM and Meunier P (2021). Controls on the grain size distribution of landslides in Taiwan: the influence of drop height, scar depth, and bedrock strength; Earth Surf. Dynam 9: 995–1011. The European Geosciences Union.

[8] Gaspar TAV, Jacobsz SW, Smit G, Gens A, Toll DG, Osman AS (2023). Centrifuge modeling of an expansive clay profile using artificial fissuring to accelerate swell. Engineering Geology 312: 1-13. Elsevier; https://doi.org/10.1016/j.enggeo.2022.106928.
[9] Sukamto and Supriatna (1982). Geologi Lembar Ujungpandang, Benteng dan Sinjai Sulawesi, Pusat Penelitian dan Pengembangan Geologi Direktorat Jenderal Pertambangan Umum Departemen Pertambangan dan Energi, Bandung. In Bahasa Indonesia.
[10] Yuwono YS, Maury RC, Soeria-Atmadja R, Bellon H (1988). Tertiary and Quaternary geodynamic evolution of South Sulawesi: Constraints from studying volcanic units. Majalah Ikatan Ahli Geologi Indonesia 13 (1): 157 - 173.
[11] Yuwono YS (1989). Petrologi dan mineralogi Gunung Lompobattang, Sulawesi Selatan. Geologi Indonesia 12 (1): 483 - 509 Jakarta. In Bahasa Indonesia.
[12] ASTM (2017). ASTM D6913 / D6913M-17: Standard test methods for soil particle-size distribution (gradation) using sieve analysis. West Conshohocken, PA.
[13] Latib FW Mohd, Kasa A and Bachok Mohd F (2023). Geotechnical properties on residual soil of sedimentary rock. Journal of advanced research in applied sciences and engineering technology 30 (3): 182-191.
[14] Wibawa YS, Sugiarti K and Soebowo E (2018). Characteristics and engineering properties of residual soil of volcanic deposits, Global Colloquium on Geo Sciences and Engineering. IOP Conf. Series: Earth Environmental Science.118. 012041.
[15] Rahardjo H, Aung KK, Leong EC and Rezaur RB (2004). Characteristics of residual soils in Singapore as formed by weathering. Engineering Geology 73 (1-2): 157-169. Abstract. https://doi.org/10.1016/j.enggeo.2004.01.002.
[16] Çelik M, Karakaya N and Temel A (1999). Clay minerals in hydrothermally altered volcanic rocks, Eastern Pontides, Turkey. Clays and Clay Minerals 47 (6): 708-717.
[17] Yanti ED (2021). Mineralogy and characterization of clay mineral of West Lampung. IOP Conf. Series: Earth and Environmental Science 789 012072, IOP Publishing. doi:10.1088/1755-1315/789/1/012072
[18] Priyono KD (2012). Kajian mineral lempung pada kejadian bencana longsor lahan di Pegunungan Kulonprogo Daerah Istimewa Yogyakarta. Forum Geografi 26 (1): 53 – 64. In Bahasa Indonesia.
[19] Odom I E (1984). Smectite clay minerals: properties and uses. Phil. Trans. R. Soc. Land. 311: 391-409. Downloaded from https://royalsocietypublishing.org/ on 14 January 2024.
[20] Baldermann A, Dietzel M, Reinprecht V (2021). Chemical weathering and progressing alteration are possible controlling factors for creeping landslides. Science of the Total Environment 778: 1-13 Elsevier B.V. https://doi.org/10.1016/j.scitotenv.2021.146300.
[21] Gougazeha M and Al-Shabatat A (2013). Geological and geotechnical properties of soil materials at Tannur Dam, Wadi Al Hasa, South Jordan; Journal of Taibah University for Science 7: 216–224. Elsevier B.V. www.sciencedirect.com
[22] Wen B, Duzgoren- Aydin NS, Aydin A (2004). Geochemical characteristics of the slip zones of a landslide in granitic saprolite, Hong Kong: implications for their development and microenvironments, Environmental Geology 47:140–154.
[23] Isra N, Lias SA, Ahmad A (2019). The characteristics of grain sizes and soil clay minerals in landslide area (case study: Joncberang sub-watershed). Jurnal Ecosolum 8 (2): 62 -73. In Bahasa Indonesia
[24] Karakaya MC and Karakaya N (2001). Kaolin occurrences in the Erenler Dagi Volcanics, Southwest Konya Province, Turkey. International Geology Review 43: 711- 721. Winston & Son. Inc.
[25] Ece OI, Nakagawa Zenbe-E and Schroeder PA (2003). Alteration of volcanic rocks and genesis of kaolin deposits in the Şile Region, Northern Istanbul, Turkey. I: Clay mineralogy. Clays and Clay Minerals 51 (6): 675-688.
[26] Ahmad A, Lopulisa C, Baja S, and Imran AM (2019). The correlation of soil liquid limit and plasticity index for predicting soil susceptibility: A case study on landslides in South Sulawesi. IOP Conf. Series: Earth and Environmental Science 235 012007: 1 - 11. IOP Publishing doi:10.1088/1755-1315/235/1/012007.
[27] Husain R and Sultan (2023). Susceptibility of the surrounding soil and rock by the electrical resistivity method. Jurnal Geocelebes 7 (1): 29 -36.
[28] Al-Saeedi A Abd (2022). Clay basics and their physical and chemical properties: Review Paper: Iaetsd journal for advanced research in applied Sciences 10 (8): 13 – 18.
[29] Chen Xiao-ping, Zhu Hong-hu, Huang Jing-wu, Liu D (2015). Stability analysis of an ancient landslide considering shear strength reduction behavior of slip zone soil. Technical Note Landslides 9p. DOI 10.1007/s10346-015-0629-7. Springer-Verlag Berlin Heidelberg.
[30] Yalcin A (2007). The effects of clay on landslides: A case study. Applied Clay Science 38: 77–85. Elsevier B.V.
[31] Pusch R (1993). Evolution of models for conversion of smectite to non-expandable minerals. SKB Technical Report 93-33. Stockholm.

IOP Conf. Series: Earth and Environmental Science **1525** (2025) 012010 doi:10.1088/1755-1315/1525/1/012010

The provenance of quartz-rich Sandstone in Paleogen clastic rock of Mallawa Formation From The Southern Arm Of Sulawesi

Safruddim[1*], A. M. Imran[1], Ulva Ria Irfan[1], and Meutia Farida[1]

[1] Geological Engineering Department, Engineering Faculty, Hasanuddin University, Indonesia

*E-mail: safruddim@unhas.ac.id

Abstract. The sedimentary rocks of the Mallawa Formation found in the southern arm of Sulawesi have unique depositional environment characteristics that are highly interesting for research and remain controversial. Here, The physical, chemical, and biological characteristics of the sedimentary rocks of the Mallawa Formation are used to reconstruct the depositional environment. Initially, we provide a detailed overview of the physical attributes of several sedimentary rocks from different locations within the Mallawa Formation. The physical characteristics of the sedimentary rocks of the Mallawa Formation are characterized by constituent lithologies such as mudstone, feldspathic wacke, lithic arenite, and subarkose arenite, all of which exhibit clastic textures. Sedimentary structures include parallel lamination, convolute lamination, waviness, nodules, burrowing, flame structures, flasher bedding, and reverse-graded bedding. Interpretation of provenance is crucial in siliciclastic sedimentary rocks because siliciclastic minerals and rock fragments preserved within the sedimentary rocks provide important evidence of the lithology of the source rocks. Based on petrography analysis, the provenance of sandstone in the Mallawa Formation is derived from igneous rocks with alternative source rocks being Propylitized Volcanic Rocks and metamorphic rocks with alternative source rocks being the Bantimala and Barru basement complexes.

1. Introduction

As a rock formed from pre-existing rocks, sedimentary rock requires analysis of its origin, known as provenance. There are many ways to observe provenance, including mineral content with petrography analysis. The Mallawa Formation is found in the western part of South Sulawesi. It is believed to have been deposited in a terrestrial/marginal marine environment, transitioning upwards transgressively to shallow marine environments [1]. The sedimentary rocks of the Mallawa Formation hold abundant potential mineral resources, which can be utilized in today's modern industrial sector. This includes the presence of coal, potential for oil and gas, abundant quartz mineral potential for use in the renewable energy industry, and other potentials.

A formation that is younger than the Balangbaru Formation has been deposited, namely the Mallawa Formation, where this formation has an Early Paleocene - Middle Oligocene age with units in the form of interbedded mudstone with coal-insulated limestone [2]. Interpreted the deposition environment of the Mallawa Formation by considering the presence of coal [3]. The coal formation environment is in coastal marshes within the back barrier environment. Subsequently, a sea level transgression led to the influx of fine sediment materials into the basin, deposited and formed claystone as the base layer. Then, a slow regression occurred, reducing the direct impact of seawater on the marsh environment. This allowed for the development of various grasses and shrubs, forming in-situ organic deposits.

2. Material and Methods

This research utilizes rock samples from three sampling locations in South Sulawesi: Mallawa in Maros Regency, Massenrengpulu in Bone Regency, and Siloro in Pangkep District. These lithological samples are part of the sedimentary rocks of the Mallawa Formation scattered in the southern arm of Sulawesi. The regional stratigraphic arrangement in the research area, according to [4], is part of the Mallawa Formation deposited during the Eocene period and generally dominated by quartz-rich sandstone, conglomerate, claystone, and coal. The sedimentary rocks of the Mallawa Formation are characterized by abundant quartz content, with extensive variations in interbedding from one location to another. They contain channel deposits and coal seams, indicating that the Mallawa Formation is believed to have formed in a fluviatile environment. The widespread occurrence of fluviatile rocks, overlaying discordantly on the older Balangbaru Formation, locally containing lagoon deposits, suggests that the Mallawa Formation is believed to have formed in a very extensive coastal plain during the Early Eocene, underlying the carbonate rocks of the Eocene to Middle Miocene Tonasa Formation.

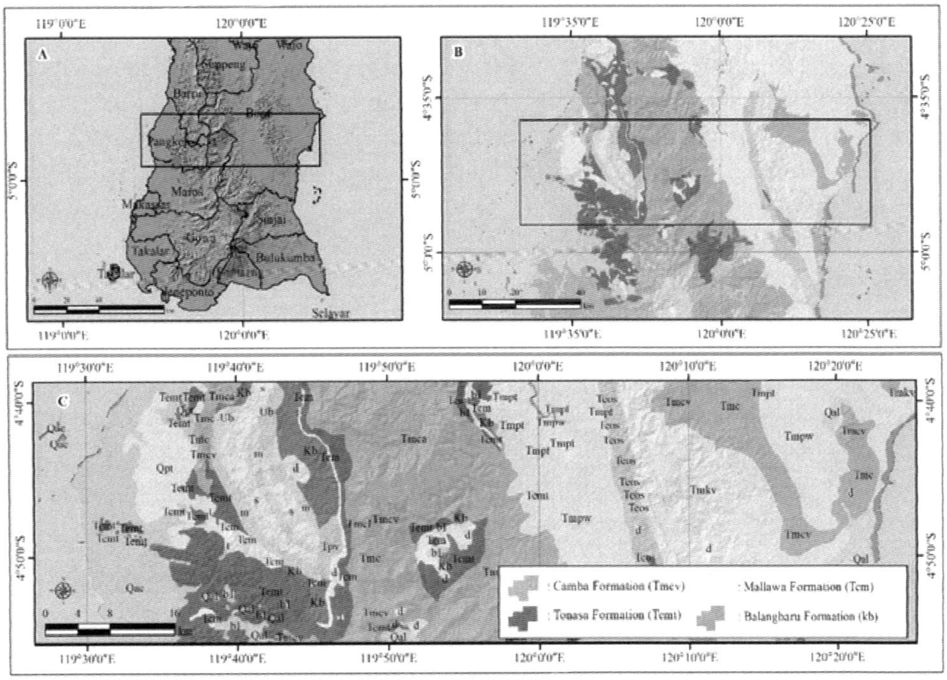

Figure 1. (A) Location map of the study area. Box delineates the region covered in Figure 1B. (B) Outcrop area of Regional Geology. Box delineates the region covered in Figure 1C. (C) Location of the three measuring sections used in this study.

This study's methodology consists of two main approaches: field observation and laboratory analysis. Field observation was conducted using stratigraphic section measurement methods to document the lithological characteristics, sedimentary structures, and stratigraphic relationships of the Mallawa Formation sandstone. This approach provided essential data to understand the geological context and depositional processes that formed the sandstone in the study area.

Additionally, laboratory analysis was carried out to identify the microscopic characteristics of the sandstone using petrographic methods. The thin-section analysis determined the mineral composition, primarily focusing on quartz, feldspar, and lithic fragments. This evaluation aimed to interpret the provenance and depositional environment of the sandstone, offering a clearer understanding of the origin and geological history of the studied sediments.

Figure 1 shows measurements were carried out in three different areas: Siloro region, Pangkep Regency, coordinates 119°39'10.8"S 4°47'44.2"E; Massenrengpula region, Bone Regency, coordinates 4°39'21.4"S 119°56'13.6"E; and Mallawa region, Maros Regency coordinates 4°50'47.7"S 119°53'04.4"E.

The samples taken are rock samples, systematically and progressively sampled in 3 different sections. The rock samples are then prepared based on the characteristics of the analysis. Petrographic samples are prepared in the sample preparation laboratory of the Department of Geological Engineering, Faculty of Engineering, Hasanuddin University.

3. Result and Discussion

The Mallawa Formation can be divided into two main lithofacies used to interpret the depositional environment: sandstone and mudstone facies. The sandstone facies dominate the lithology of the Mallawa Formation, primarily consisting of yellow-gray lithic quartz sandstone, litharenite, arkose, and pebbly sandstone. Some sandstone layers exhibit cross-bedding structures. Based on the overall characteristics of the deposits, this facies represents a delta front depositional environment, specifically within a sub-environment higher than the prodelta, precisely at the distributary mouth bar. Feldspathic Wacke characterizes the lithology in this facies and exhibits a thickening upward trend. The observation of the lowermost layer of the first unit indicates a transitional condition due to a significant increase in energy, shifting from a low-energy to a higher-energy environment.

The sediment structure in wavy laminations indicates deposition influenced by currents or waves due to tidal and pressure effects, forming ripples where material accumulates due to current activity during transportation. These ripple sand structures indicate shallow water depth [5]. Biogenic structures such as burrowing are also present in this layer, concluding that deposition occurred in a distributary mouth bar depositional environment. Distributary mouth bars are shoaling areas associated with a distributary channel's seaward terminus. They result from a decrease in velocity and the subsequent reduction in the river's carrying capacity as it leaves its confined channel and enters the open body of water. This leads to rapid sediment deposition [6].

Mudstone facies consist of mudstone with sandstone intercalations in some parts. Based on the overall deposit characteristics, it characterizes a delta plain depositional environment, specifically a sub-environment after the delta front, precisely located in the lower delta plain. The lithological change from mudstone to feldspathic wacke indicates a variation in grain size texture, with facies coarsening upwards, showing a transition from lower delta plain mud to sandier facies of the delta plain.

The sediment structure of parallel laminations, dominated by mudstone units, indicates deposition under low-pressure conditions and characterizes quiet depositional energy. Additionally, burrowing sediment structures indicate deposition under transitional conditions (from high-energy deposition to relatively quieter energy). The presence of massive (structureless) sediment structures also suggests stable depositional energy [7], and it is a primary characteristic of the absence of transportation traction, resulting in very rapid deposition from suspension solutions or deposition from high-concentration sediments [8]. It can be concluded that this deposition occurred in a lower delta plain depositional environment.

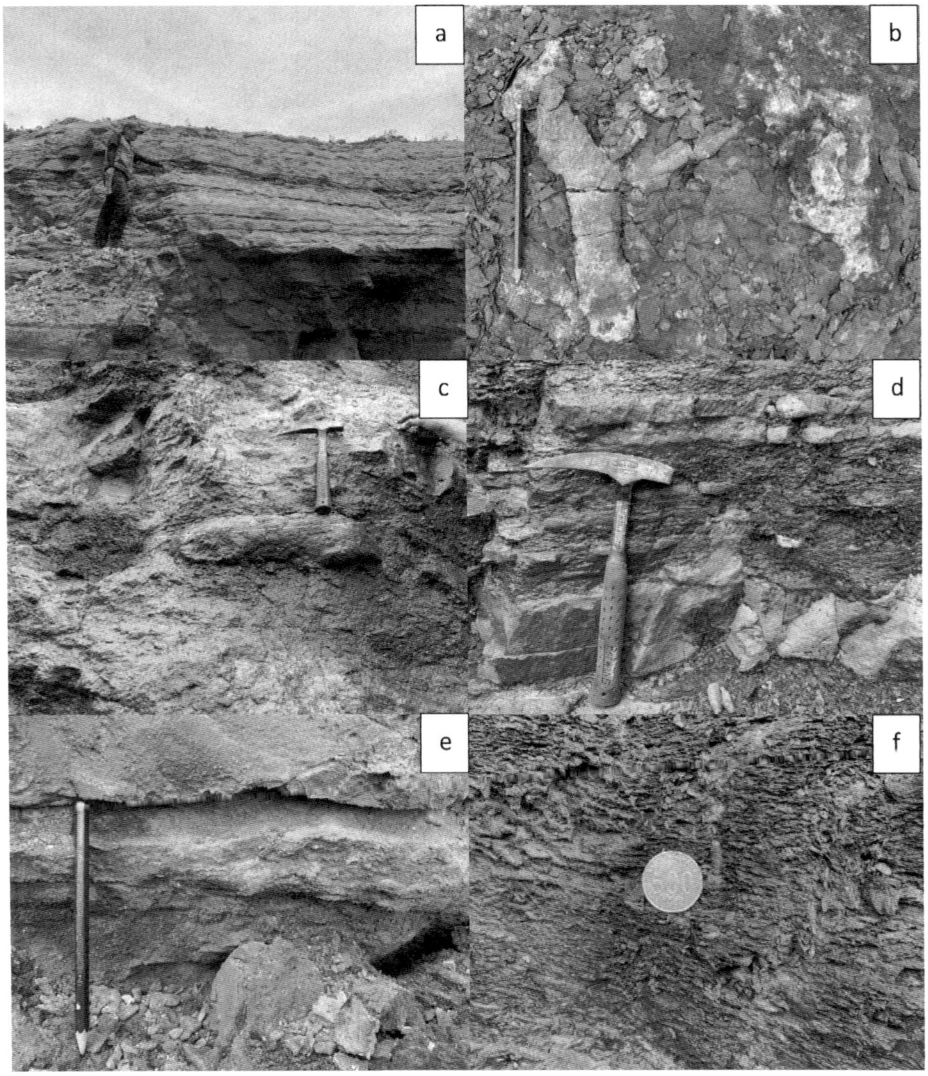

Figure 2. Photographs of sedimentary structures in Siloro Section, Mallawa Formation. (a). Sandstone with wavy structures interbedded with mudstone. A Researcher with a height of 165 cm. (b). Mudstone with burrow (bw) structure. The pencil is 15 cm long. (c). Mudstone nodule in sandstone. The hammer is 35 cm long. (d). Mudstone (Mst) is interbedded with sandstone (Sst). The hammer is 35 cm long. (e). Load cast structure in mudstone. The pencil is 15 cm long (f)—Laminasi in mudstone. The coin is 40 mm in diameter.

IOP Conf. Series: Earth and Environmental Science **1525** (2025) 012010 doi:10.1088/1755-1315/1525/1/012010

The rock naming utilizes the Pettijohn Classification, 1975, based on the abundance of quartz, feldspar, and rock fragments while considering the matrix content. The sandstone of the Mallawa Formation consists of feldspathic wacke, subarkose arenite, and lithic arenite.

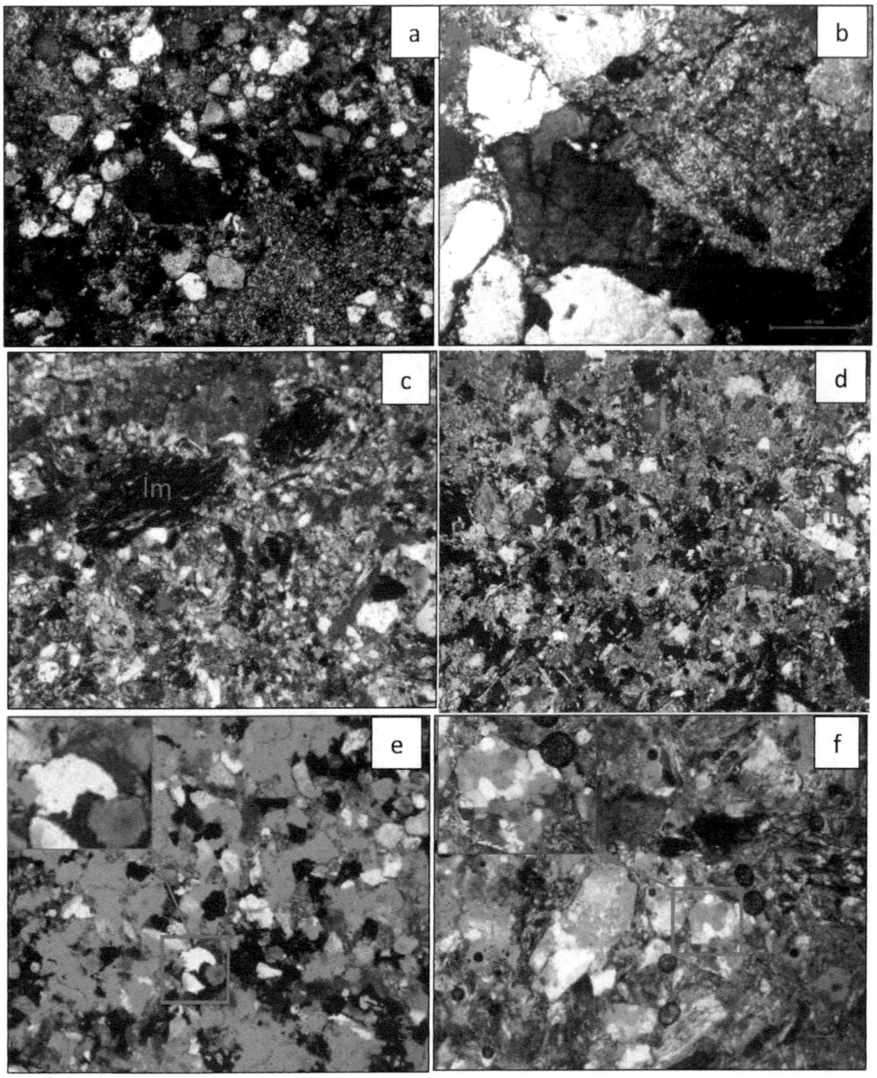

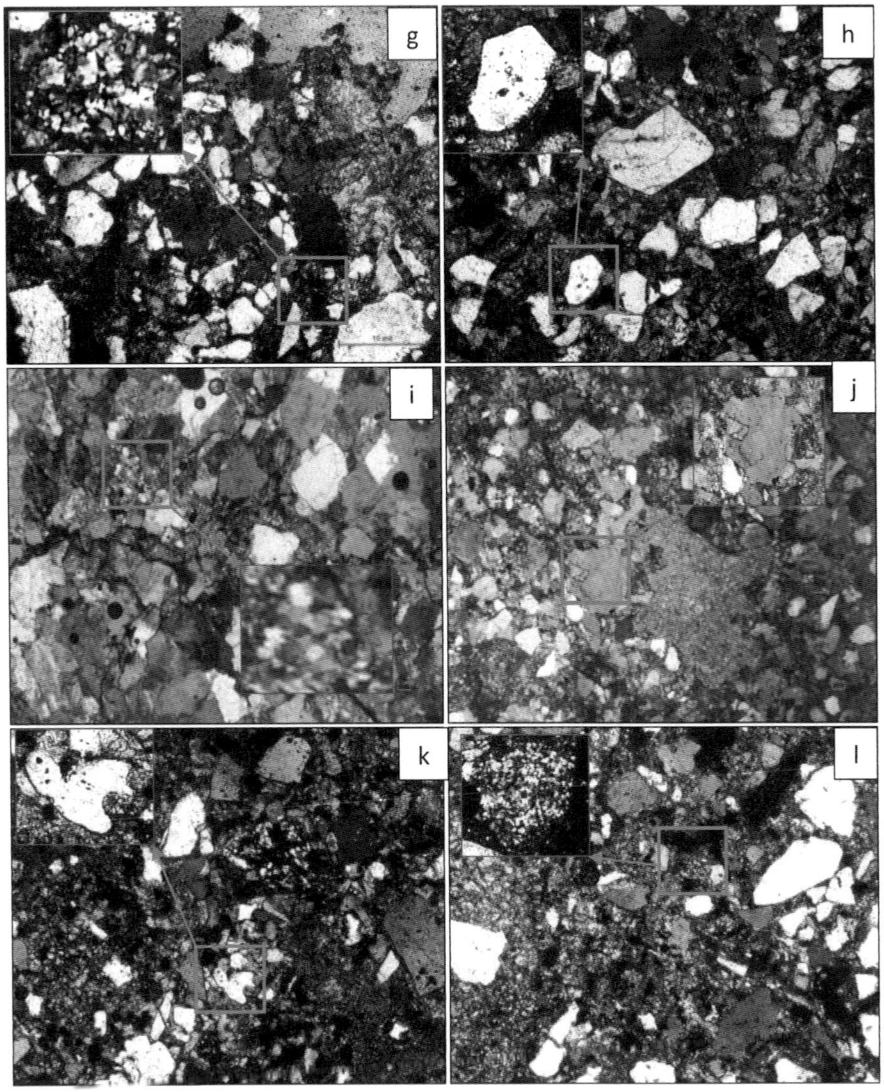

Figure 3. Representative cross-polarized light of sandstone from Mallawa formation. (a-b) Lithic rock fragmen from vulkanic rock (lv). (c-d) Lithic rock fragment from metamorphic rock (lm). (e). Monocrystalline quartz (volcanic quartz) typically originates from volcanic deposits. (f). Polycrystalline quartz (recrystallized metamorphic) typically originates from metamorphic deposits. (g) Polycrystalline quartz (Streched metamorphic) typically originating from metamorphic deposits. (h) Monocrystalline quartz (plutonic quartz) typically originates from plutonic rock deposits. (i) Polycrystalline quartz (recrystallized metamorphic) typically originates from metamorphic deposits; (j) Monocrystalline quartz (plutonic quartz) originates from plutonic rock deposits. (k) Monocrystalline quartz (volcanic quartz) originates from volcanic deposits. (l) Polycrystalline quartz (Streched metamorphic) typically originates from metamorphic deposits.

The Q-F-L, Qp-Ls-Lv, and Qm-K-P diagrams indicate that the sandstone of the Mallawa Formation in the study area is derived from a Recycled Orogen environment. This setting is typically associated with sedimentary sources that have undergone multiple cycles of erosion and deposition, reflecting the influence of tectonic uplift and reworking of older orogenic belts.

Furthermore, the Qp-Ls-Lv diagram suggests that the provenance of these sandstones is linked to a subduction zone and arc orogen. This implies that the sediments were sourced from tectonically active regions, such as forearc or accretionary complexes, where material from eroded volcanic arcs and uplifted orogenic belts was transported and redeposited. These findings align with the geological context of Sulawesi, which has undergone complex tectonic interactions involving subduction and orogenic recycling.

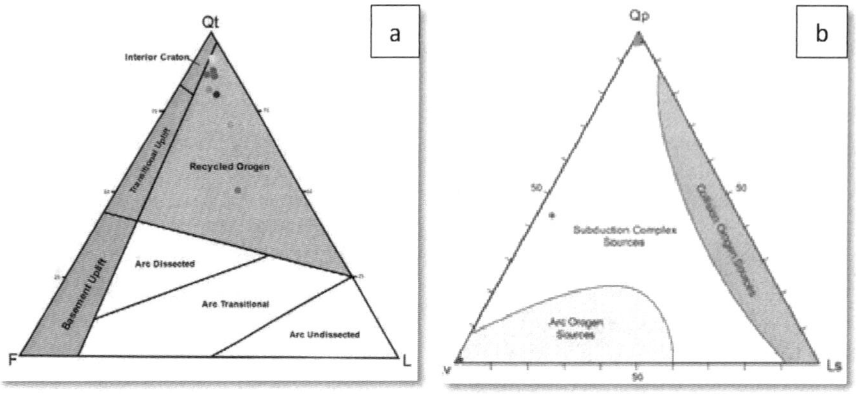

Figure 4. (a). Qt-F-L ternary diagram displays the relative abundance of framework grains in a sandstone, Qt: total quartz, F, feldspar, L: lithic rock fragments. The modal compositions of sandstone samples plot near the boundary of the "recycled orogenic" field in the QtFL diagram. (b) The plotting results on the Qp-Lv-Ls diagram in sandstone, Qp: Polycrystalline Quartz, Lv: Lithic Vulcanic, Lm: Lithic Metamorphic, indicate an association between Arc Orogen source and subduction complex sources. [9][10]

4. Conclusion

The analysis of sandstone and mudstone facies in the Mallawa Formation provides crucial insights into the origin and depositional environment of the formation. The dominance of sandstone facies, composed of lithic quartz sandstone, litharenite, arkose, and pebbly sandstone with cross-bedding structures, suggests a high-energy depositional setting characteristic of a delta front environment, specifically within the distributary mouth bar. The presence of Feldspathic Wacke and a thickening-upward trend further support this interpretation, indicating progradation influenced by sediment supply dynamics and hydrodynamic conditions.

In contrast, the mudstone facies, characterized by finer-grained deposits such as claystone and mudstone, represent lower-energy conditions typical of the prodelta or inter-distributary bay environments. These deposits suggest suspension settling in a relatively calm setting, marking a transition from the higher-energy sandstone deposits. The presence of sedimentary structures such as lenticular bedding, cross-lamination, convolute lamination, and flasher bedding further supports the interpretation of fluctuating energy conditions in a deltaic system.

Petrographic analysis reveals that the sandstone facies consist of lithic arenite, feldspathic wacke, and subarkose arenite, all exhibiting clastic textures. The composition and texture of these sandstones suggest that their provenance is primarily derived from igneous rocks, with additional contributions from altered volcanic rocks and metamorphic basement complexes, particularly from the Bantimala and Barru basement complexes.

The integration of sedimentological and petrographic data suggests that the Mallawa Formation was deposited in a Recycled Orogen setting associated with a subduction zone and arc orogen. This classification is consistent with the tectonic influences that shaped the sedimentary processes, highlighting the role of active geological processes in controlling sediment supply, transport, and deposition. The observed facies variations and stratigraphic relationships further indicate that the formation evolved within a dynamic deltaic system, influenced by fluctuations in sediment supply, hydrodynamic conditions, and tectonic activity over time.

References

[1] Wilson, M.E.J. dan Bosence, W.J. (1995). The Tertiary evolution of South Sulawesi: A record in redeposited carbonates of the Tonasa Limestone Formation. Geological Society 106, 365 – 389.

[2] Yusuf, H. A., Puteri, Y.A., Cifa, A. A., Sutadiwiria, Y., and Rendy., 2023, Geochemical Characteristics of the Mallawa Formation, and its Relationship with the History of Source Rock Formation in the South Makassar Basin, South Sulawesi. E3S Web of Conferences 500, 03038 (2024) https://doi.org/10.1051/e3sconf/202450003038 INTERCONNECTS 2023.

[3] Anshariah., Imran, A. M., Irfan, Ulfa Ria., Nawir, A., Budiman, A.A. 2019. Characterization And Reconstruction of Deposit Facies of Mallawa Coal Formation of South Sulawesi Based on Proximate and Petrography Analysis. IOP Conf. Series: Earth and Environmental Science 175 (2018) 012019.

[4] Sukamto, Rab., dan Supriyatna, Sam. 1982. Peta Geologi Lembar Pangkep, dan Watampone Bagian Barat. Pusat Penelitian dan Pengembangan Geologi, Bandung.

[5] Murakoshi, N & Masuda, F., (1992). Estuarine, barrier-island to strand-plain sequence and related ravinement surface developed during the last interglacial in Paleo-Tokyo Bay, Japan. Sedimentary Geology, 80, 167-184.

[6] Jones. R., Marcelissen. R., Fralick. P. 2022. Sedimentology and Stratigraphy of a Large, Pre-Vegatation Deltaic Complex. Frontiers in Earth Science. Doi: 10.3389/feart.2022.875838.

[7] Allen, G.P., and Chambers. J. L.C, 1998, Sedimentation in the Modern and Miocene Mahakam Delta, Indonesian Petroleum Association Proceeding, Jakarta, Indonesia.

[8] Boggs, S. (2006). Principles of Sedimentology and Stratigraphy 4th Edition. California: Pearson Prentice Hall.

[9] Augustsson, C., Aehnelt, M., Olivarius, M., Voigt, T., Gaupp, H., Hilse, U., 2023. Provenance from the geochemical composition of terrestrial clastic deposits — a review with case study from the intracontinental Permo-Triassic of European Pangea. Sedimentary Geology 456, 106496

[10] Ghasemlooytakantapeh, S., Zohdi. A., Lakirouhani, A., 2023. Petrography and geochemistry of the Miocene Upper Red Formation sandstones in NW Iran with an application to the origin and tectonic setting. Marine and Petroleum Geology 153 (2023) 106275.

Distribution of Tondo Clastics in the Wakoko and Wambona Rivers, Buton Island, Southeast Sulawesi

Harisma[1*], Muhammad Zamrun Firihu[2], Viska Inda Variani[2], Muliddin[3] and Erzam S. Hasan[3]

[1] Department of Geological Engineering, Halu Oleo University, Kendari, Indonesia
[2] Department of Physics, Halu Oleo University, Kendari, Indonesia
[3] Department of Geophysical Engineering, Halu Oleo University, Kendari, Indonesia

*E-mail: harisma@uho.ac.id

Abstract. The Buton Island is located in the the subduction zone between the Eurasian, Indo-Australian, and Pacific Plates. This island is exposed at the most southern part of southeast Sulawesi, adjacent to the Muna and the Tukang Besi Islands. Recent investigations have suggested that the Buton Island had experienced multiple tectonic events during the Neogene period, with the first collision occurred between Southeast Sulawesi/Muna and Buton and the second collision occurred between Buton and Tukang Besi. The objective of this study is to present the Neogene history of the Buton Island based on identification of the Tondo Formation. This study investigated the clastic of Tondo Formation in the Wakoko River (South Buton) and the Wambona River (North Buton) to identify the strata of stratigraphy, age and depositional environment, provenance, and evolution of the Tondo Formation in the Buton Island.

1. Background

The Buton Island is one of the best place in the eastern Indonesia to study about collision tectonics and stratigraphy. This island is attracted to be studied due to believed result of tectonic collision of microcontinents coming from Australian continent [1–3]. The structural evolution of the Buton Island is complex because experienced of multiple tectonic events occurred during Neogene [3]. The stratigraphy and structural style of this island is distinctly different than Southeast Sulawesi and Muna Island as adjacent islands. Deformed metamorphic, ophiolitic and Mesozoic sedimentary rocks from Triassic to Tertiary ages are found here [3]. Broad similarities are recognized between Buton and adjacent islands in Banda Arc, Timor, Seram, and Buru [3, 4].

Geologically, Buton Island is part of a Neogene collision zone. The Neogene tectonic evolution of this region has been dominated by a series of collision events. The Buton Island is believed to include two separate and distinct micro-continental fragments, one encompassing eastern Buton and Tukang Besi Islands, and the second covering western Buton and Muna Islands [3]. Collision of Muna and Buton Islands is recorded in the Early Miocene where resulted a sequence of thin skinned thrusts and folds developed and continued into the Late Miocene [3]. The next collision between Buton Island and Tukang Besi Island is recorded in the Late Pliocene where resulted in strike-slip and dip-slip movement with localized uplift and subsidence [3]. Smith et al [5] studied that Neogene strata cover most of Buton, all of the adjacent island of Muna, and much of the southern part of the Southeast Arm of Sulawesi. On Buton, late Miocene to recent folding, faulting, and uplift have exposed parts of the underlying collision complex in a number of areas. Therefore, the Neogene strata of Buton Island for the structurally complicated and lithologically diverse association [3].

The purpose of this study is to present a more comprehensive analyses of the Tondo Formation (figure 1A) as part of Neogene tectonic strata in the Buton Island in order to add our understanding on previous works. Many aspects of the geology of these microcontinent remain poorly understood and have been the subject of considerable debate which including: The tectonic event of Buton and Tukang Besi collision is remain unclear. [6] and [2] regarded eastern Buton and Tukang Besi as a single microcontintal fragment distinct from western Buton and Muna, while [2], [3] and recent work by [7] argued that Buton and Tukang Besi platform are two separate microcontinents which is Buton collided with Muna in early Miocene and Tukang Besi collided with Buton in late Pliocene.

This study area is located in the Wakoko and Wambona Rivers (figure 1B) of the Buton Island. The Wakoko River is lying in South Buton region, around 45 minutes from Pasar Wajo City, while the Wambona River is lying in North Buton region, it takes around 90 minutes from the Labuan Ferry Port of North Buton.

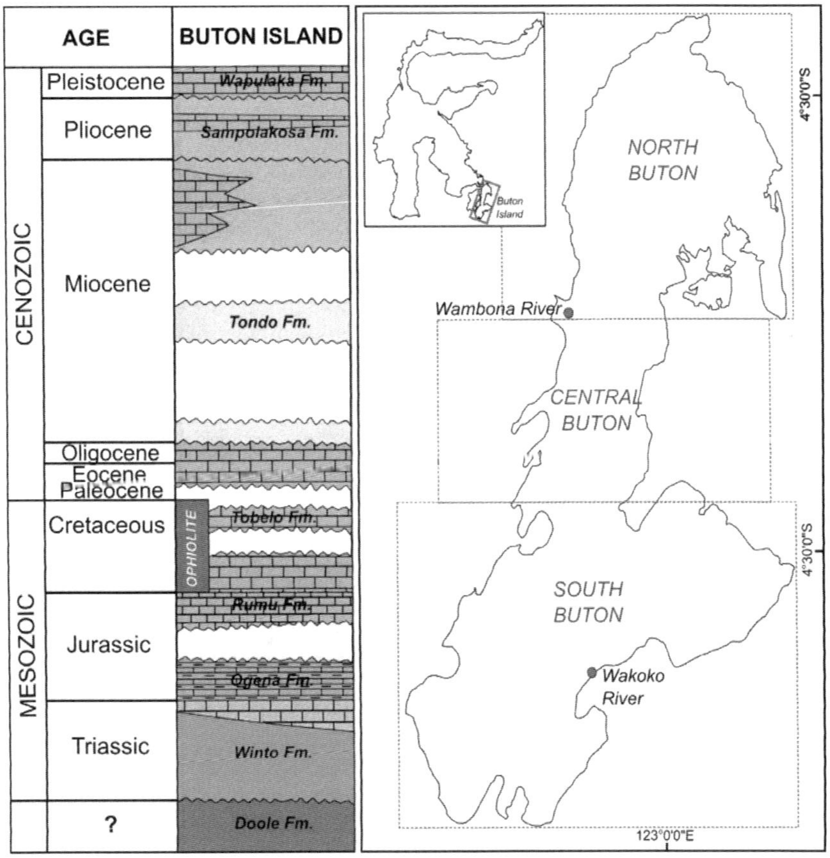

Figure 1. (A) Stratigraphy column of the Buton Island, showing the study area of Tondo Formation (modified from [3]); (B) Study area (red dots) in the Buton Island, Southeast Sulawesi.

IOP Conf. Series: Earth and Environmental Science **1525** (2025) 012011 doi:10.1088/1755-1315/1525/1/012011

2. Stratigraphy of Tondo Formation

The Tondo Formation is classified into three facies, consisting of limestone facies, coarse clastic facies, and fine grains clastic facies. The Tondo clastic facies are interpreted as deep marine turbidite fan, composing of older pre-Neogene and ophiolitic rocks with common presence of sedimentary structures such as fining upward sequences, cross bedding, graded bedding, and liquefaction structures. While, the Tondo fine clastic facies are interpreted as distal turbidites, consisting of mudstones, claystones, siltstones, and sandstones, with abundant of planktonic foraminifera, indicating from outer neritic to upper bathyal environment [3].

The thickness of the Tondo Formation varies. In South Buton, the thickness can reach ±800 m in Lasalimu area, while in the Central Buton region, the thickness increases from 1 to 1.3 km [8]. Previous study suggested the age of Tondo Formation as late Middle Miocene (N14) with depositional environment as deltaic [9].

The Tondo Formation in North Buton (figure 2) region consists of a lower limestone facies and an upper clastic division, consisting of coarse and fine clastic facies [10]. In addition, the Tondo Formation in South Buton region (figure 2) consists of dominantly clastic assemblage of rocks, wherein limestone units are developed locally on the flanks [10].

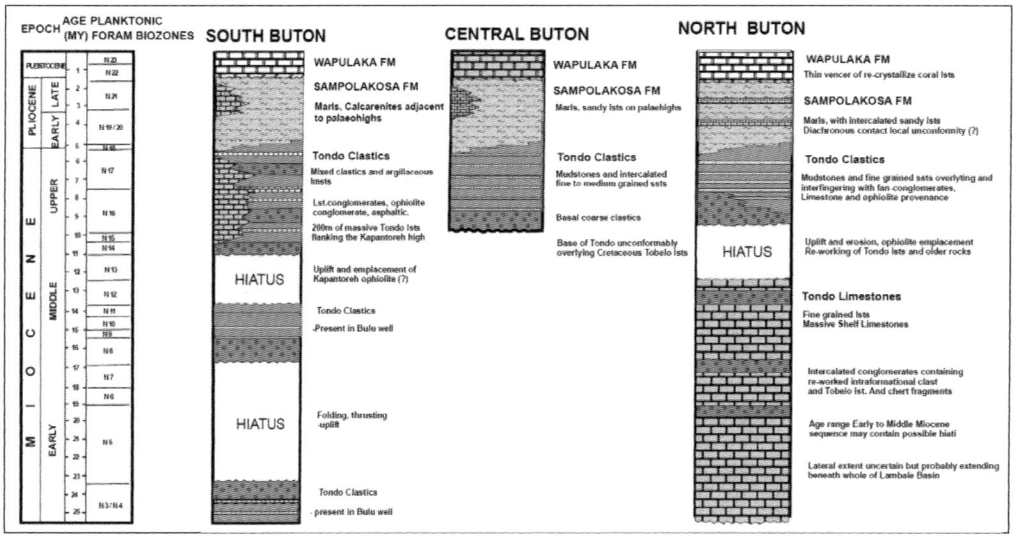

Figure 2. Distribution of the Tondo clastics in the Buton Island, consisting of South Buton, Central Buton, and North Buton regions [10].

3. Material and Methods

3.1. Sample Collection and Analysis

This study investigates the distribution of Tondo clastics in two areas in the Buton Island: the Wakoko River in South Buton region and the Wambona River in North Buton region.

The rock samples were collected from the Wambona River (Maligano District, North Buton) and the Wakoko River (Pasarwajo District, South Buton) (figure 1B). The outcrops were selected to represent the Tondo clastic rocks from North and South Buton regions. Sampling was conducted in last September 2024, and ten samples of Tondo clastics were collected from North and South Buton. The rock samples were collected only fresh rock samples. The sampling locations were recorded by Avenza Maps Apps.

In this study, we identify ten samples of Tondo clastics for several analyses, such as: microfossils analysis to provide age ranges as well as information of depositional environment, provenance study to identify the sediment source of Tondo clastic, and to constrain the evolution of Tondo Formation in the Buton Island.

4. Result and Discussion

4.1. Wakoko River

The Wakoko River is located in Pasarwajo of South Buton region (figure 1B). The coordinate of this area is 5°27'51.3"S and 122°51'34.7"E (table 1), lying near the main road of Lapodi Village. This area is composed of clastics of Tondo Formation, consisting of conglomerates which are well developed in the Wakoko River.

We identified eight stations in the Wakoko River (table 1) with thickness from 50 cm (figure 3D) as observed in station seven (ST7-WKK) up to 200 cm in the station six (ST6-WKK) (figure 3C). All stations consist of conglomerates with clasts are dominated by sedimentary rock (e.g., sandstone, siltstone), indicating of monomictic conglomerate. The size of clasts is varied from pebbles (4-64 mm) to boulders (>256 mm), while the matrix of conglomerate is fine to coarse sandstone and also calcareous. The grain shape of fragments is from angular to rounded. There is no structures or imbrications are observed in conglomerates, showing disorganized of clast fragments (figures 3A to 3D).

4.2. Wambona River

The Wambona River is located in the Maligano of North Buton region (figure 1B). The coordinate of this area is 4°51'29.4"S and 122°50'16.8"E (table 1). This river is exactly lying near the main road of Maligano-Pure, which be able to reach at approximately 90 minutes from the Labuan Ferry Port of North Buton. The area is also composed of clastics of Tondo Formation, consisting of conglomerate and sandstone beds, which are well developed in the Wambona River.

Two stations were identified in the Wambona River, i.e., ST1-WBO and ST2-WBO (table 1; figures 3E and 3F), showing interbedding of conglomerate and sandstone (figures 3C and 3D). The characteristics of conglomerate are greyish, matrix is calcareous sandstone, and clasts from angular to rounded. The size of clasts from granules (2-4 mm) to boulders (>256 mm). The fragments are dominated of sedimentary rocks, such as sandstone, indicating of monomictic conglomerate (figures 3E and 3F). The fragments distribution is disorganized and no indication of imbrications. Contact between conglomerate and sandstone is well identified in station one (ST1-WBO), showing of bedding plane is N285°E/7° (figure 3.E.1).

Sandstones in the Wambona River is quite interesting, which primary structures such as parallel lamination (figure 3.E.1) are well observed in station one (ST1-WBO) and two (ST2-WBO). A bedding position is identified at N285°E/7° in station one (ST1-WBO) and N320°E/7° in station two (ST2-WBO). The presence of microfossils is identified in sandstone of ST1-WBO (figure 3.E.1) and ST2-WBO (figure 3.F.1), suggesting a marine environment. The grain sizes of sandstone are from medium to coarse.

IOP Conf. Series: Earth and Environmental Science **1525** (2025) 012011 doi:10.1088/1755-1315/1525/1/012011

4.3. Distribution of Tondo Clastics in North and South Buton

This study recognized a different distribution of Tondo clastics between in North Buton and South Buton regions. The Tondo clastics in the Wakoko River is almost dominated by conglomerate rocks which mostly composed of sedimentary rock fragments. While in the Wambona River, the lithology is consisting of conglomerate and sandstone bedding. The characteristics of conglomerate rocks in the Wambona River is relatively equal with the conglomerate rocks in the Wakoko River. The presence of sandstone in the Wambona River suggests a different distribution of Tondo clastics between in north and south Buton.

Figure 3. (A–D) Outcrops of Tondo clastics in the Wakoko River (South Buton); (E–F) Outcrops of Tondo clastics in the Wambona River (North Buton). Notes: A–Station 3 (ST3-WKK); B–Station 5 (ST5-WKK); C–Station 6 (ST6-WKK); D–Station 7 (ST7-WKK); E–Station 1 (ST1-WBO); and F–Station 2 (ST2-WBO).

Tabel 1. General information of outcrop stations in the Wakoko and Wambona Rivers.

Location	Station	Coordinates	Area	Lithology	Formation
Wakoko River	ST1-WKK	5°27'51.3"S / 122°51'34.7"E	Pasarwajo (South Buton)	Conglomerate	Tondo
	ST2-WKK	5°27'51.3"S / 122°51'34.7"E	Pasarwajo (South Buton)	Conglomerate	Tondo
	ST3-WKK (fig. 3A)	5°27'51.3"S / 122°51'34.7"E	Pasarwajo (South Buton)	Conglomerate	Tondo
	ST4-WKK	5°27'51.3"S / 122°51'34.7"E	Pasarwajo (South Buton)	Conglomerate	Tondo
	ST5-WKK (fig. 3B)	5°27'51.3"S / 122°51'34.7"E	Pasarwajo (South Buton)	Conglomerate	Tondo
	ST6-WKK (fig. 3C)	5°27'51.3"S / 122°51'34.7"E	Pasarwajo (South Buton)	Conglomerate	Tondo
	ST7-WKK (fig. 3D)	5°27'51.3"S / 122°51'34.7"E	Pasarwajo (South Buton)	Conglomerate	Tondo
	ST8-WKK	5°27'51.3"S / 122°51'34.7"E	Pasarwajo (South Buton)	Conglomerate	Tondo
Wambona River	ST1-WBO (fig. 3E)	4°51'29.4"S / 122°50'16.8"E	Maligano (North Buton)	Conglomerates and sandstones beds	Tondo
	ST2-WBO (fig. 3F)	4°51'29.4"S / 122°50'16.8"E	Maligano (North Buton)	Conglomerate and sandstone beds	Tondo

5. Conclusion

Clastics of Tondo Formation are very well observed in the Wakoko River (North Buton) and the Wambona River (South Buton), composing of conglomerate and sandstone. The strata of Tondo clastics is different between in the Wakoko and Wambona Rivers, wherein in the Wakoko River only shows the conglomerate rocks, while in Wambona River shows the bedding of conglomerates and sandstones. We suggest that a different distribution of clastics of Tondo Formation occurred in North and South Buton.

References

[1] Satyana AH, Irawan C, Kurniawan W. Revisit Geology and Geochemistry of Buton Asphalt Deposits, SE Sulawesi: Implications for Petroleum Exploration of Buton Area. In: *Indonesian Petroleum Association, 37th Annual Convention & Exhibition*. 2013, pp. 1–18.

[2] Satyana AH, Purwaningsih ME. Collision of micro-continents with eastern Sulawesi: records from uplifted reef terraces and proven-potential petroleum plays. In: *Proc. Indon Petrol. Assoc., 35th Ann. Conv.* Jakarta: Indonesian Petroleum Association (IPA). Epub ahead of print 2011. DOI: 10.29118/IPA.1376.11.G.219.

[3] Davidson JW. The Geology and Prospectivity of Buton Island, S.E. Sulawesi, Indonesia. In: *Proc. Indon Petrol. Assoc., 20th Ann. Conv.* Jakarta: Indonesian Petroleum Association (IPA). Epub ahead of print 1991. DOI: 10.29118/IPA.2026.209.233.

[4] Katili JA. Past and Present Geotectonic Position of Sulawesi, Indonesia. *Proc Indones Pet*

Assoc 6th Annu Conv. Epub ahead of print March 1977. DOI: 10.1016/0040-1951(78)90166-X.

[5] Smith RB, Silver EA. Geology of a Miocene Collision Complex, Buton, eastern Indonesia. *Geol Soc Am Bull* 1991; 103: 660–678.

[6] Hamilton W. Tectonics of the Indonesian Region. *US Geol Surv*. Epub ahead of print 30 July 1979. DOI: 10.7186/bgsm06197301.

[7] Tanjung H, Sukarno N, Yuskar Y, et al. Field observation of southern Buton: an overview of hydrocarbon manifestation and its geological setting. In: *Proc. Indon Petrol. Assoc., 32nd Ann. Conv.* Jakarta: Indonesian Petroleum Association (IPA). Epub ahead of print 2018. DOI: 10.29118/IPA.2247.08.SG.074.

[8] Van Bemmelen RW. *The Geology of Indonesia*. Volume IA. 1949.

[9] Rochmanto B, Darwin. Depositional environment of the Tondo formation at Wakoko river, Pasar Wajo area, District of Buton Southeast Sulawesi. In: *Proc. Indon Petrol. Assoc., 31st Ann. Conv.* Jakarta: Indonesian Petroleum Association (IPA). Epub ahead of print 2007. DOI: 10.29118/IPA.320.07.G.101.

[10] Pertamina, ITB. *Hydrocarbon Potential Studies of Indonesian Basins, Microcontinental Sliver: Buton – Banggai-Sula – Seram*. 2003.

IOP Conf. Series: Earth and Environmental Science **1525** (2025) 012012 doi:10.1088/1755-1315/1525/1/012012

Analysis of Seismic Characteristics and Dynamic Behavior of Bili-Bili Dam using HVSR Method and Dynamic Model

Imelda Fitra Nugrani[1], Herman Parung[2], Miswar Tumpu[1], Abd Razak[3] Jusman[3]

[1] Disaster Management Study Program, The Graduate School, Hasanuddin University, Makassar, Indonesia
[2]Department of Civil Engineering, Hasanuddin University, Makassar, Indonesia
[3] Bili-Bili Dam Management Unit, BBWS Pompengan Jeneberang , Makasssar, Indonesia

*imeldafitra646@gmail.com

Abstract. The Bili-Bili Dam in Gowa Regency, South Sulawesi, is a critical infrastructure with a high seismic risk due to its location in a seismically active region. This study utilizes the HVSR [Horizontal-to-Vertical Spectral Ratio] method and dynamic analysis to evaluate the seismic characteristics and the dynamic response of the dam structure to earthquakes. The results reveal significant variations in the dominant soil frequencies, with the highest value at BLI 04 [2.62 Hz] and the lowest at BLI 09 [1.29 Hz], and the highest amplification at BLI 04 [A0 = 5.66] and the lowest at BLI 23 [A0 = 1.41]. Additionally, the dynamic analysis indicates that the first vibration mode is dominated by X-direction deformation with a mass participation of 97.24% and a natural frequency of 0.52 Hz. The correlation between the HVSR dominant frequencies and the mode shape frequencies reveals the potential for resonance in certain areas, such as at BLI 04, which requires specific mitigation measures. These findings provide valuable insights for enhancing the seismic resilience of the Bili-Bili Dam through a more integrated soil-structure interaction-based design.

1. Introduction

The Bili-Bili Dam, located in Bili-Bili Village, Bontomarannu District, Gowa Regency, South Sulawesi Province, is a critical infrastructure that has the potential to be affected by natural disasters, particularly earthquakes. Situated in Indonesia, a region known for its seismic vulnerability due to its geological position at the meeting point of three major tectonic plates—India-Australia, Eurasia, and the Pacific—the Bili-Bili Dam faces significant seismic risks. This study aims to analyze the seismic characteristics and dynamic behavior of the dam structure using the HVSR (Horizontal to Vertical Spectral Ratio) method and dynamic modeling to assess its earthquake resilience [1].

The HVSR method is used to analyze the seismic spectrum to identify the dominant soil frequencies that can influence the structural response of buildings to earthquakes [2][3]. Previous studies, such as those conducted by [1][4], have applied seismic methods to assess earthquake vulnerability in buildings in other seismically active regions, such as dams. Meanwhile, dynamic structural analysis is performed to evaluate the building's response to various seismic loads [5][6]. This study utilizes both methods to further explore the potential seismic impacts on the Bili-Bili Dam and provide deeper insights into its dynamic behavior, which has not been previously available for this location.

IOP Conf. Series: Earth and Environmental Science **1525** (2025) 012012 doi:10.1088/1755-1315/1525/1/012012

Previous research has mostly focused on seismic evaluation or dynamic behavior separately, without addressing both aspects simultaneously for large dam infrastructures like the Bili-Bili Dam. This gap presents an opportunity for more comprehensive research by integrating the HVSR method and dynamic analysis within a single study. The novelty of this research lies in the use of both approaches to provide a more accurate depiction of seismic impacts on the dam structure.

2. Research Method

Figure 1. Research Location [Bili-bili Dam, Gowa, South Sulawesi]

This study was conducted at the Bili-Bili Dam, in Bili-Bili Village, Bontomarannu District, Gowa Regency, South Sulawesi Province (see Figure 1). The research covers two main areas: above and below the dam structure, for the instrumentation setup, measurements were conducted with different orientations to ensure accurate data acquisition. On the left wing of the dam, the orthogonal direction of the site was aligned toward the west of the seismometer. Meanwhile, on the right wing, the orthogonal direction of the site coincided with the orthogonal orientation of the seismometer. This orientation strategy was implemented to standardize data collection and account for any potential directional influences on the recorded seismic signals. For seismic analysis, the HVSR [Horizontal-to-Vertical Spectral Ratio] method was employed, which is a passive geophysical technique used to assess the dynamic characteristics of the soil and local site effects. The recorded microtremor data were then analyzed to generate HVSR curves that show the spectral amplitude ratio between the horizontal and vertical components based on frequency. This ratio is formulated in the following equation: [2][3]:

$$HVSR = \frac{\sqrt{H_{orthogonal} + H_{Paraller}}}{V} \quad (1)$$

where $H_{Orthogonal}$ is the amplitude of the horizontal orthogonal component, $H_{Paraller}$is the amplitude of the horizontal east-west component, and V is the amplitude of the vertical component. This curve provides information about the dominant soil frequency, which is closely related to the thickness of the soil layers and the shear wave velocity of those layers. The dominant frequency (f_0) is calculated using the following equation [7]:

$$f_0 = \frac{V_s}{4H} \quad (2)$$

where Vs is the shear wave velocity in the soil layer, and H is the thickness of the soil layer above the bedrock. The spectral ratio analysis was performed separately for the areas above and below the dam to identify differences in ground response characteristics. In addition to spectral ratio analysis, a Rose Diagram Method was applied to evaluate the directional characteristics of seismic wave propagation at the study site

For the dynamic analysis, numerical modeling of the dam structure is performed using ETABS software, incorporating its geometry and material composition. The geometric parameters of the main dam are as follows: it has a height of 73 meters, a length of 750 meters, and a crest width of 10 meters. The crest elevation is recorded at 106.00 meters, while the total embankment volume is approximately 2,760,000 cubic meters. Modal analysis is conducted to determine the natural vibration period and calculate the natural frequency (f_n), which characterizes the dam's dynamic response to seismic loads. The f_n value is crucial for assessing the risk of resonance between the soil and the structure. If the natural frequency of the dam aligns with the dominant soil frequency, seismic wave amplification could lead to excessive deformation or structural damage [5] [6]. The f_n value from the modal analysis is compared with the dominant soil frequency from the HVSR analysis to assess the interaction between the dam structure and its underlying soil. This evaluation helps determine its impact on the dynamic response and seismic resilience of the dam.

3. Results and Discussion

3.1 Analysis of Horizontal to Vertical Spectral Ratio

The Horizontal-to-Vertical Spectral Ratio [HVSR] analysis conducted at the Bili-Bili Dam revealed significant variations in the seismic characteristics of the soil below [BLU] the dam structure (see Table 1).

Table 1. The HVSR results from the measurements below the dam

Measurement Point	f_0 [Hz]	A_0
BLU 04	1.96	2.68
BLU 09	3.09	1.61
BLU 23	2.59	1.83

Below the dam, the HVSR measurements highlight distinct seismic characteristics. The highest dominant frequency (f_0) was observed at point BLU 09, with a value of 3.09 Hz, reflecting stiffer material compared to other measurement points below the dam. The lowest dominant frequency below the dam was recorded at BLU 04, with 1.96 Hz, suggesting the presence of more flexible soil. Additionally, the highest amplification (A_0) below the dam was found at BLU 04, with an amplification factor of 2.68, indicating notable wave amplification at this location. In contrast, BLU 09 exhibited the lowest amplification factor of 1.61, corresponding to its higher stiffness. Variations in the values of f_0 and A_0 reflect differences in material characteristics at various locations both above and below the dam. Locations such as BLI 04 and BLU 09 show dynamic responses dominated by stiff material with mechanical properties capable of withstanding higher-frequency seismic waves. Conversely, locations with the lowest f0 values, such as BLI 09 and BLU 04, indicate the presence of materials with lower stiffness, which may amplify the resonance response due to an earthquake. The high amplification identified at BLI 04 and BLU 04 highlights the importance of considering the potential for wave amplification in these areas when conducting structural failure risk analysis.

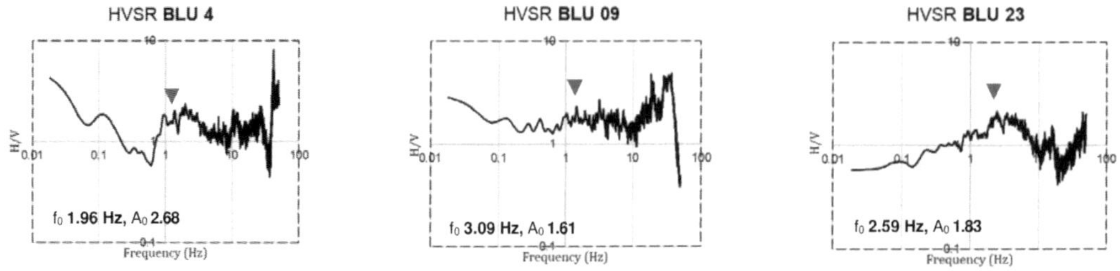

Figure 2. HVSR curve for each measurement point below the dam (BLU)

3.2 Microtremor Spectral Ratio of BLI and BLU site

Table 2. HVSR ratio of measurements above (BLI) and below the dam (BLU).

Measurement Point	f_0 (Hz)	A_0
BLI/BLU 04 [UD]	1.37	7.80
BLI/BLU 04 [Orthogonal]	1.09	8.21
BLI/BLU 04 [Paraller]]	3.52	5.22
BLI/BLU 09 [UD]	2.51	9.78
BLI/BLU 09 [Orthogonal]	2.72	5.08
BLI/BLU 09 [Paraller]	4.4	8.79
BLI/BLU 23 [UD]	2.68	4.38
BLI/BLU 23 [Orthogonal]	2.82	2.84
BLI/BLU 23 [Paraller]	5.63	1.43

The ratio between microtremor measurements above the dam (BLI) and below the dam (BLU) provides additional insights into the differences in stiffness and dynamic behavior between the structure and its underlying soil. The data show that the values of the dominant frequency (f_0) and amplification (A_0) vary significantly at different measurement points, depending on the orientation of the measurement components (Up-Down, Paraller, Orthogonal) The lowest f_0 ratio was recorded at BLI/BLU 04 *(*Orthogonal*)* with a value of 1.09 Hz, indicating similar stiffness between the structure and the underlying soil at this orientation. In contrast, the highest f_0 ratio was found at BLI/BLU 23 *(*Paraller*)* with a value of 5.63 Hz, indicating a significant difference in stiffness, where the structure is stiffer than the underlying soil. Regarding amplification, the highest A_0 value was recorded at BLI/BLU 09 *(*UD*)* with a value of 9.78, indicating significant wave amplification above the structure compared to the soil below it. Conversely, the lowest A_0 value was recorded at BLI/BLU 23 *(*Paraller*)* with a value of 1.43, reflecting minimal amplification response at this orientation.

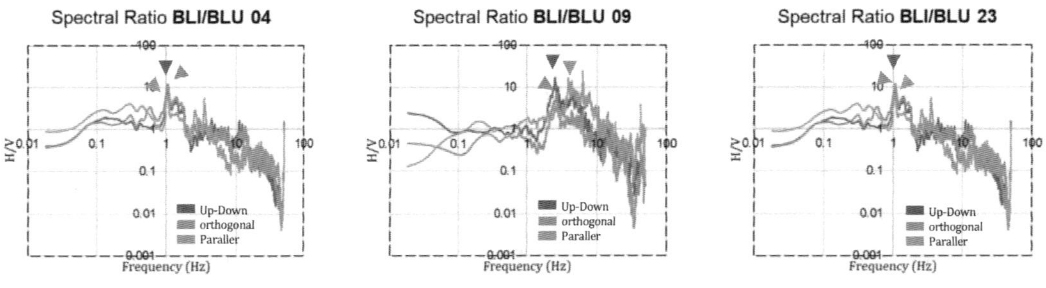

Figure 3. Spectral Ratio BLI/BLU (Fourier amplitude spectral (blue line shows vertical and orange and gray show two horizontal components, Paraller and Orthogonaln)

This phenomenon highlights the variation in dynamic response based on the measurement orientation. The Up-Down (UD) orientation shows the highest amplification values at most points, reflecting the dominance of vertical wave amplification due to the interaction between seismic waves and the concrete structure. In contrast, the orthogonal orientation at specific points, such as BLI/BLU 23, reflects a dominant response to horizontal waves, providing insights into the direction of seismic energy propagation that is more pronounced.

3.3 Diagram Rose Analysis

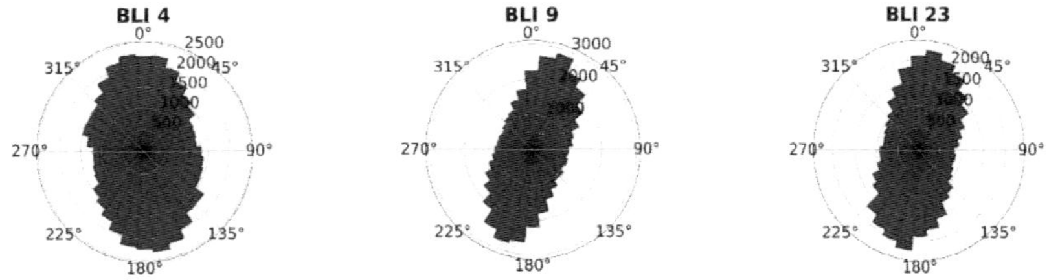

Figure 4. Rose diagram for each measurement point above the dam (BLI)

The rose diagram analysis identifies the dominant directions of motion of the dam, reflecting how seismic energy propagates in response to structural and geological influences (see Figure 4In BLI 04, the dominant directions are observed at 345° and 165° which are orthogonal to each other. Similarly, in BLI 09, the dominant directions at 25° and 205° which are orthogonal to each other show consistent energy propagation patterns in line with the local structural orientation. Similarly, in BLI 09, the dominant directions at 25° and 205° which are orthogonal show consistent energy propagation patterns in line with the local structural orientation. taken above the dam structure, indicate that the observed dominant directions are primarily influenced by perpendicular direction.

3.4 Analysis of Mode Shapes and Mass Participation

Table 3. Mode Shapes and Mass Participation in the X (Paraller) and Y (perpendicular) directions.

Mode	Period (second)	Natural frequency (f_n)	UX %	UY %	RZ %	SUM UX %	SUM UY %
1	1.91	0.52 HZ	97.24%	0.00%	0.00%	97.24%	0%
2	1.90	0.53 HZ	0.00%	0.00%	0.00%	97.24%	0%
3	1.88	0.53 HZ	2.41%	0.00%	0.00%	99.65%	0%
4	1.85	0.54 HZ	0.00%	0.00%	0.00%	99.65%	0%
5	1.82	0.55 HZ	0.26%	0.00%	0.00%	99.91%	0%
6	1.77	0.57 HZ	0.00%	0.00%	0.00%	99.91%	0%
7	1.71	0.58 HZ	0.06%	0.00%	0.00%	99.97%	0%
8	1.65	0.61 HZ	0.00%	0.00%	0.00%	99.97%	0%
9	1.58	0.63 HZ	0.02%	0.00%	0.00%	99.99%	0%
10	1.51	0.66 HZ	0.00%	0.00%	0.00%	99.99%	0%
11	1.43	0.70 HZ	0.01%	0.00%	0.00%	100.00%	0%
12	1.35	0.74 HZ	0.00%	0.00%	0.00%	100.00%	0%
13	1.28	0.78 HZ	0.00%	0.00%	0.00%	100.00%	0%
14	1.20	0.83 HZ	0.00%	0.00%	0.00%	100.00%	0%
15	1.13	0.89 HZ	0.00%	0.00%	0.00%	100.00%	0%
16	1.06	0.95 HZ	0.00%	0.00%	0.00%	100.00%	0%
17	1.04	0.96 HZ	0.00%	100.00%	0.00%	100.00%	100%
18	0.99	1.01 HZ	0.00%	0.00%	0.00%	100.00%	100%
19	0.93	1.08 HZ	0.00%	0.00%	0.00%	100.00%	100%
20	0.92	1.09 HZ	0.00%	0.00%	0.00%	100.00%	100%

The dynamic analysis of the Bili-Bili Dam structure (see Figure 4) shows a significant response to seismic loads, with the first mode shape dominating translation in the X direction, accounting for 97.24% mass participation, a vibration period of 1.91 seconds, and a natural frequency of 0.52 Hz. The third and 11th modes increase mass participation in the X direction to 99.65% and 100%, respectively. In the Y direction, full mass participation [100%] is achieved by the 17th mode, with additional contributions in modes 18 through 20. The 100% mass participation in the Y direction at mode 17 occurs because all effective modal contributions needed to fully represent the structure's response in that direction have been accumulated by this mode. The structure responds more dynamically to loads in the X direction compared to the Y direction.

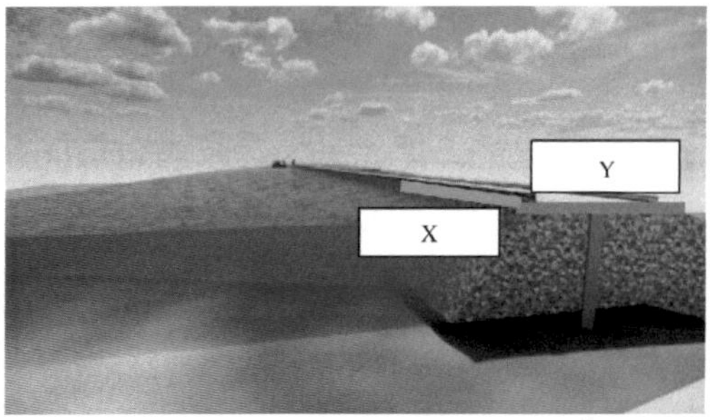

Figure 5. Rose diagram for each measurement point above the dam [BLI]

3.5 Correlation between Dominant Frequency and Mode Shape and Mass Participation

The natural frequency (f_0) identified at each measurement point of the Bili-Bili Dam is directly related to the first mode shape and mass participation in the X direction. At BLU 04, with f_0 of 1.96 Hz and mass participation reaching 97.24%, the dominant dynamic response in the X direction indicates a strong interaction between the structure and the underlying soil. In contrast, at BLU 09, the lower f_0 value (3.09 Hz) suggests softer material, altering the way the structure responds to seismic waves. The correlation between f_0 and the mode shape emphasizes the importance of considering local material characteristics in predicting the dynamic behavior of the dam structure.

4. Conclusion

This study reveals the relationship between the seismic characteristics of the soil, analyzed through HVSR, and the dynamic response of the Bili-Bili Dam structure. The variation in dominant frequency (f_0), ranging from 1.96 Hz to 3.09 Hz, reflects differences in material stiffness, with the highest value at BLU 09 and the lowest at BLU 04. The dominant frequency variation for the spectral ratio in the up-down direction ranges from 1.37 Hz to 2.68 Hz. In the parallel direction, it ranges from 3.52 Hz to 5.63 Hz, while in the orthogonal direction, it varies from 1.09 Hz to 2.82 Hz. The frequency alignment observed in this measurement indicates resonance at BLI 04 and the BLI/BLU 04 ratio, as well as at BLI 23 and the spectral ratio of BLI/BLU 23. These findings are consistent with the Rose diagram analysis, showing that the dominant direction in this study tends to be influenced by the perpendicular direction.

References

[1]. Syah C. Identifikasi Indeks Kerentanan Seismik Di Bendungan Bili-Bili Kabupaten Gowa Dengan Metode Horizontal To Vertical Spectral Ratio [Hvsr]. J Fis dan Ter. 2017;4[2]:160–7.

[2]. Mihaylov A, El Naggar H, Mihaylov D, Dineva S. Approximate analytical HVSR curve using multiple band-pass filters and potential applications. Soil Dyn Earthq Eng [Internet]. 2019;127[August]:105840. Available from: https://doi.org/10.1016/j.soildyn.2019.105840

[3]. Putti SP, Satyam N. Evaluation of Site Effects Using HVSR Microtremor Measurements in Vishakhapatnam [India]. Earth Syst Environ [Internet]. 2020;4[2]:439–54. Available from: https://doi.org/10.1007/s41748-020-00158-6

[4]. Rasyid AM, Rochman JPGN, Widodo A, Nabiha MR, Mikael Mario PP, Salam IF. Site classification Based on Shear Wave Velocity Inversion in the Jlantah Dam Construction Project using the HVSR [Horizontal to Vertical Spectral Ratio] Analysis Method. IOP Conf Ser Earth Environ Sci. 2024;1307[1].

[5] Harahap MF, Fauzan M. Perilaku Dinamik pada Struktur Apartemen Metro Galaxy Park terhadap Beban Gempa. J Tek Sipil Dan Lingkung [Internet]. 2019;04[03]:195–206. Available from: https://garuda.ristekbrin.go.id/documents/detail/1706256

[6] Amina TB, Mohamed B, Andre´ L, Abdelmalek B. Fluid-structure interaction of Brezina arch dam: 3D modal analysis. Eng Struct [Internet]. 2015;84:19–28. Available from: http://dx.doi.org/10.1016/j.engstruct.2014.11.011

[7] Baena-Rivera M, Arciniega-Ceballos A, Sa´nchez-Sesma FJ, Rosado-Fuentes A, Pardo-Dan˜ino JC. Directional HVSR at the Chalco lakebed zone of the Valley of Mexico: Analysis and interpretation. J Appl Geophys. 2024;228[June].

Investigation of Landslide Occurrence Using Microtremor: Landslide Case Study in BuloBulo Village, Barru, Indonesia

Almubdy Siraj Ramadhan[1*], Andi Muhamad Pramatadie[1], and Muhammad Altin Massinai[1]

[1] Department of Geophysics, Faculty of Mathematics and Natural Sciences, Hasanuddin University, Makassar 90245, Indonesia

*E-mail: almubdy35@gmail.com

Abstract. This research investigates the seismic resonance effect by utilizing the natural frequencies of the ground to study landslides in BuloBulo Village, Pujananting, Indonesia, with a focus on disaster mitigation. Microtremor surveys were conducted by mounting horizontal and vertical components at 15 points with a measurement distance of 11 m. The analysis involves HVSR correlation, microtremor particle directivity, from the analysis obtained, the frequency and amplification relationship which has a changing trend in the landslide area and away from the landslide area. Additionally, the polarization movement of microtremor direction has variation of vibration direction at each measurement point.

1. Introduction

Indonesia is one of the countries with high intensity of natural disasters [2]. According to a 2019 BNPB study, approximately 69,777 ha or 52.5% of the total area has a high vulnerability to landslides [1]. Pujananting sub-district, with a steep slope area of 9,640 ha, is one of the most prone areas to land movement, especially in BuloBulo village.

HVSR studies have provided important information on ground movement characteristics and are effective in mapping areas of ground instability due to their simplicity, speed, and low cost [4]. We continue this research with microtremor measurements at the landslide site of BuloBulo Village to identify seismic resonances and vibration polarization, hoping that the results can help mitigate landslide risk in Barru Regency.

2. Data and Method

2.1 Study Site

The research was conducted based on monitoring obtained at the BuloBulo landslide site, Pujananting, Barru, South Sulawesi, Indonesia. This landslide event resulted in the road access between villages being cut off, and stretched up to 20 m, while the height of the landslide cover material reached 3 m (Figure 1).

Figure 1. Documentation of landslides and measurement points.

Geologically, the area is located in the BantiMala Complex which is a tectonic assemblage of various lithologies ranging in age from Jurassic to Cretaceous. The main components of the Bantimala Complex are sandstone, conglomerate, shale, chert, silica shale, basalt, ultramafic rocks, schist and schist breccia [5] (Figure 2).

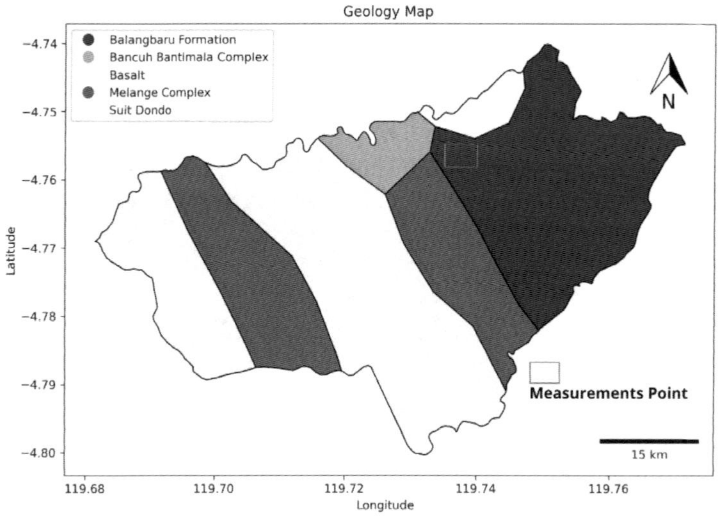

Figure 2. Geological Map of Research Area.

2.2 Measurement Site

Microtremor acquisition was conducted at 15 points around the landslide with a distance of 11 meters between stations. The coordinates of the location and measurement area are shown in (Figure 3). The data acquisition used Portable Digital Seismograph TDS-303 with short-period sensor type DS-4A.

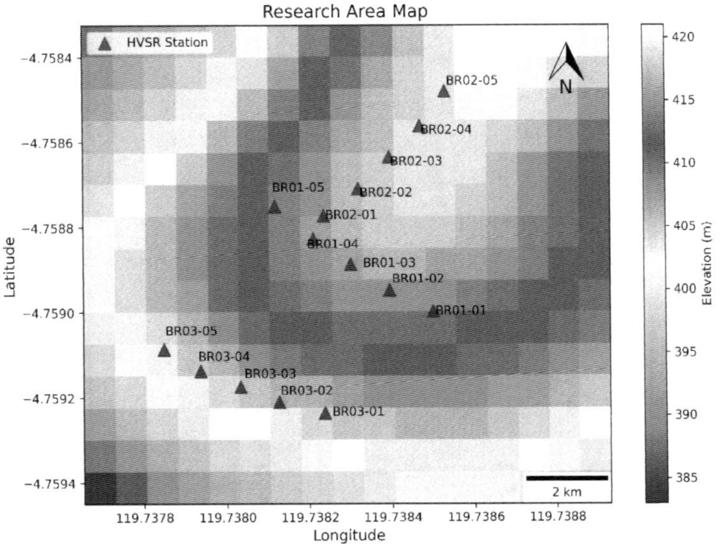

Figure 3. Map of research area and Microtremor Stations

Microtremor measurements for 15 minutes per point produced a good frequency spectrum. The time-domain data obtained consists of three components: vertical (Up-Down), and horizontal (East-West and North-South), as shown in Figure 4 (A-C). The data was then transformed to the frequency domain using the Fourier Transform, with the transformation results shown in Figure 4 (D-F).

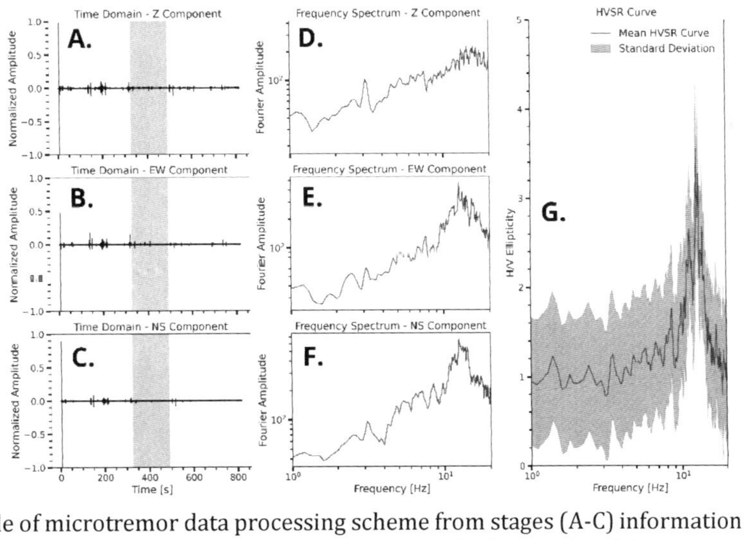

Figure 4. Example of microtremor data processing scheme from stages (A-C) information from 15 minutes of measurement for 3 recorded seismic components, (D-F) the result of fourier transformation into frequency spectrum, and (G) is the result of data processing with HVSR method. The shaded area is the selected window with a duration of 81.92 s for each window.

2.3 *HVSR (Horizontal Vertical Spectral Ratio)*

The HVSR method utilizes seismic response. In general, this method has been used to identify subsurface conditions such as landslides based on seismic wave resonance. The H/V component ratio is calculated by dividing the resultant Fourier spectrum of the horizontal component by the vertical component, as shown in equation (1):

$$\frac{H}{V} = \frac{\sqrt{H(EW)+H(NS)}}{V} \tag{1}$$

H(EW) and H(NS) are the Fourier amplitude spectra in the EW and NS directions, while V is the vertical spectrum. The H/V spectrum was calculated from an 81.92 s recording segment with a 0.1 Hz bandwidth Parzen window to smooth, minimize transient effects, and detect peak frequencies, as shown in figure 4(G) [6].

2.4 *Particle Microtremor and Root Mean Square (RMS)*

The polarity analysis of microtremor particle direction is performed by selecting the length of the time-domain window based on the HVSR curve, using the time-domain polarization method [3] to identify the dominant vibration distribution pattern. The input data in the polarization analysis is correlated with the window used in the determination of the HVSR curve, especially the particle motion in the North-South (NS) and East-West (EW) directions, as described in equation (2).

$$\theta = arctan\left(\frac{(y)}{(x)}\right) \tag{2}$$

In addition, we also analyzed the RMS amplitude for the three recording components, according to the window range selected at a duration of 81.92 s for each window based on the HVSR calculation.

3. Result and Discussion

The results of HVSR data processing show different wave frequency change patterns at each location. From the data processing results, the relationship between horizontal and vertical components is obtained. From the graph above, it can be seen that the high frequency value tends to be on the landslide field, while the low frequency is far from the landslide field (Figure 5).

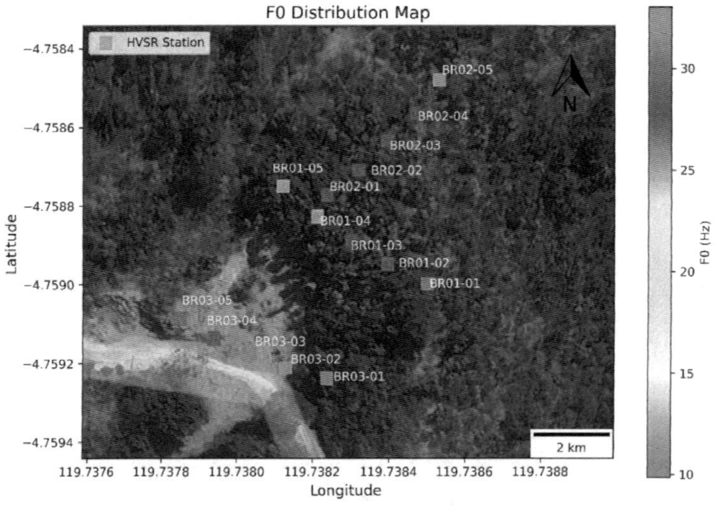

Figure 5. F_0 Distribution Map in research area.

As for the distribution of areas with large enough amplitude, it tends to have the same increasing trend in the area adjacent to the landslide slope (Figure 6). This is related to the degree of wave magnification associated with ground deformation or displacement.

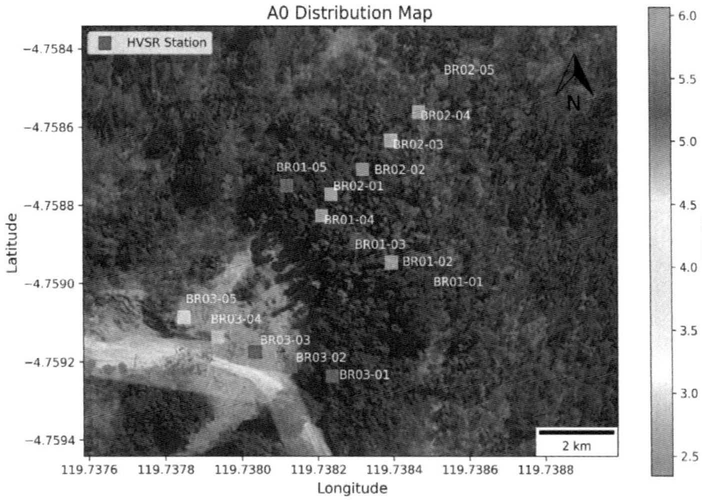

Figure 6. A0 Distribution Map in research area.

The microtremor particle analysis results show the distribution direction of the seismic component constituents in the horizontal component of the spectrum and the Root Mean Square (RMS) for the three components (Figure 7).

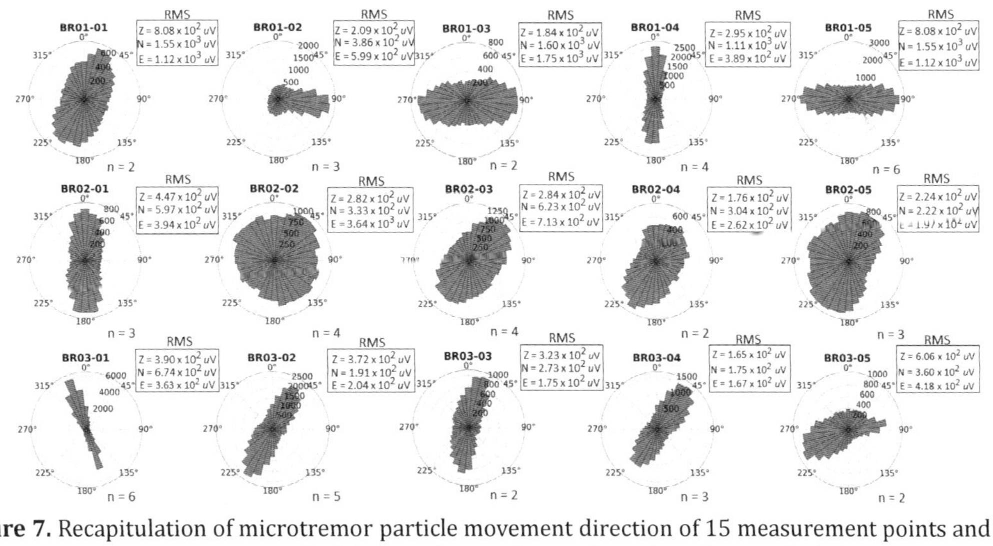

Figure 7. Recapitulation of microtremor particle movement direction of 15 measurement points and RMS Amplitude of three components. The label n indicates the number of windows included in the microtremor particle analysis, each n = 8192 data.

Based on the analysis of microtremor particles for two components and RMS amplitude for three components of the recordings. We found several indications in the study area. Some measurement points located in the area close to the landslide slope show varying orientations, as well as areas quite far from the landslide slope.

4. Conclusion

The results of the research conducted obtained that the trend of changes in frequency values and the increase in HVSR amplitude, has a significant relationship to the trend of rock deformation that occurs due to landslides. In addition, the trend of microtremor polarization movement shows variations in vibration direction at each measurement point.

Acknowledgments

I would like to express my sincere gratitude to the Regional Disaster Management Agency (BPBD) of Barru Regency and the government of Bulobulo Village for their support and assistance during the research process.

References

[1] Arsyad, U., & Hikmah, S. N. (2023). Analysis of Landslide Susceptibility Using Frequency Ratio Methods in the Lisu Watershed, Barru Regency. *IOP Conference Series: Earth and Environmental Science*, *1277*(1), 012020. https://doi.org/10.1088/1755-1315/1277/1/012020

[2] Djalante, R., Garschagen, M., Thomalla, F., & Shaw, R. (2017). Introduction: Disaster Risk Reduction in Indonesia: Progress, Challenges, and Issues. In R. Djalante, M. Garschagen, F. Thomalla, & R. Shaw (Eds.), *Disaster Risk Reduction in Indonesia* (pp. 1–17). Springer International Publishing. https://doi.org/10.1007/978-3-319-54466-3_1

[3] Jurkevics, A. (1988). Polarization analysis of three-component array data. Bulletin of the Seismological Society of America, 78(5), 1725–1743.

[4] Kolesnikov, Y. I., Fedin, K. V., & Ngomayezwe, L. (2019). Direct determination of resonant properties of near-surface sediments using microtremor. *Soil Dynamics and Earthquake Engineering*, *125*, 105739. https://doi.org/10.1016/j.soildyn.2019.105739

[5] Wakita, K., Sopaheluwakan, J., Miyazaki, K., Zulkarnain, I., & Munasri. (1996). Tectonic evolution of the Bantimala Complex, South Sulawesi, Indonesia. *Geological Society, London, Special Publications*, *106*(1), 353–364. https://doi.org/10.1144/GSL.SP.1996.106.01.23

[6] You, Z., Xu, P., Qian, J., Cao, L., Du, Y., & Fu, Q. (2022). Frequency-Bessel Transform Based Microtremor Survey Method and Its Engineering Application. *International Journal of Environmental Research and Public Health*, *19* (20), 13484. https://doi.org/10.3390/ijerph192013484

IOP Conf. Series: Earth and Environmental Science **1525** (2025) 012014 doi:10.1088/1755-1315/1525/1/012014

Application of Geoelectric Resistivity Method to Determine the Location and Depth of Groundwater in Maccini Baji Village, Lau District, Maros Regency

Syamsuddin[1*], Refsinawati M Nur[1], Makhrani[1], Bambang Harimei[1]

[1]Department of Geophysics, Faculty Mathemathics and Natural Science, University of Hasanuddin

*E-mail: syamsuddin@fmipa.unhas.ac.id, refsinwatimuhnur@gmail.com,

Abstract. As increase of the need for water, it is necessary to conduct research to identify the presence of groundwater. Groundwater is a very potential water source to meet human daily needs. Most of the people in Maccini Baji Village use springs and wells as water sources to meet their daily needs. The existence of groundwater can be known using the Wenner Schlumberger configuration geoelectric method. In this study, the acquisition of 2 passes with a space of 5 meters per electrode was carried out. From the results of the study, 3 layers were obtained that constituted the study area, namely slate material (19 Ωm - 89.7 Ωm), alluvium deposits (2.43 Ωm - 19 Ωm) and limestone (20 Ωm - 50 Ωm). The aquifer as a place where groundwater is found in the second layer (alluvium deposits) and accumulates in the northeastern part of the study area at a depth varying from 0-18 meters. Types of aquifers found in the research area.

1. Introduction

Groundwater is one of the good water sources for clean water and drinking water needs compared to other water sources. The need for water is always increasing along with the increase in population. The need for water, which is always increasing every year, makes people forget about the carrying capacity of nature which has a limit in meeting water needs. For example, during the rainy season, the water content in the aquifer will be abundant, but if it enters the dry season, the water discharge in the aquifer will decrease.

To find out the estimation of subsurface formations, it is necessary to conduct an investigation in order to find out whether or not there is a groundwater carrier layer commonly called an aquifer. According to Manrulu, 2018 [1] several methods of soil surface investigation can be carried out, including gravity methods, seismic methods, and geoelectric methods. From these methods, the geoelectric method is a method that is widely used and is often used in the identification of groundwater and the results obtained are quite good.

The tool used in the geoelectric method is called a Resistivity meter. There are several rules in the arrangement of the four electrodes, in geophysics known as electrode configuration. According to Wahyuni, in 2018 [2] the Wenner and Schlumberger electrode configurations were used in the implementation in the field which is not too difficult (quite flat and wide) and the penetration is not too deep.

The community in Maccini Baji Village uses springs and wells as the main source of water to meet their daily needs. In the dry season, water sources will experience a reduction in spring water discharge due to changes between the rainy season and the dry season. This situation is a problem that usually occurs every year due to the depletion of discharge so that the community's clean water needs are not met properly.

2. Basic Theory

2.1 Regional Geology

The regional geology of Maros Regency consists of the Tonasa Formation, the Camba Formation, the Camba Formation Volcanic, the Baturape-Cindako Volcanic, the Propylite Volcanic Rocks, the Diorite Breakthrough and Aluvium Deposits [3].

The research area is located in the western part of Maros Regency where the area is Aluvium Deposits, Swamps and Beaches which are illustrated in blue in Figure 1.

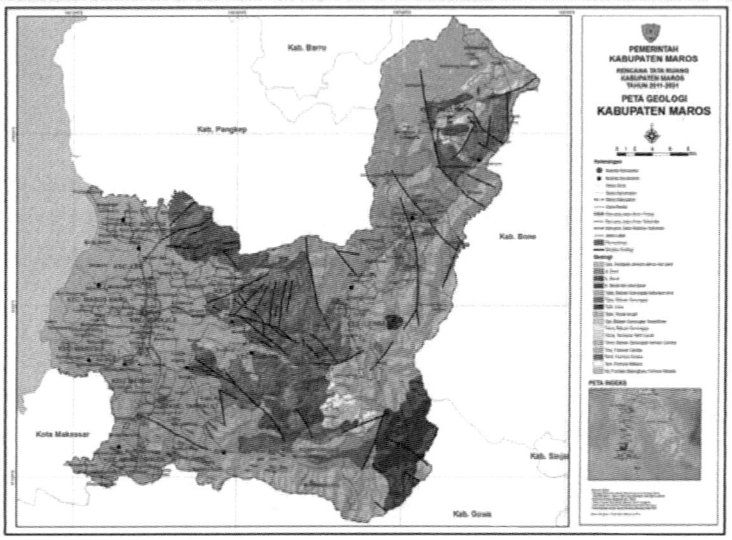

Figure 1. Geological map of Maros Regency [4].

2.2 Groundwater

Groundwater is a very broad aspect in relation to the hydrological cycle. Rainwater that falls to the earth's surface, part of it will be absorbed by plants, and the rest will flow in the soil as a surface run-off [5]. Surface water will flow through the river into a river discharge (stream flow) and some of it enters the ground through the infiltration process and some flows through the groundwater flow. At certain locations, the groundwater flow will come out as a spring and join the surface flow. Water that is infiltrated into the ground can undergo a percolation process into groundwater.

The properties of rocks to groundwater can be divided into four, namely Aquifer, Aquitard, Aquiclude and Aquifuge [6]. Based on the impermeability of the rocks that surround the aquifer, aquifer can be divide to confined aquifers, and semi-confined aquifers [7].

2.3 Porosity and Permeability

Porosity is the comparison between the empty space and the entire volume of rock or sediment expressed in percent (%). Porosity determines how much water can be contained in a rock.

IOP Conf. Series: Earth and Environmental Science **1525** (2025) 012014 doi:10.1088/1755-1315/1525/1/012014

Porosity is influenced by the size and shape of the mineral grains that make up the rock, the arrangement of the grains and the size of the pores.

Permeability also greatly affects the flow and amount of groundwater. Permeability is the ability of rocks or soil to pass or pass water through a porous medium. Permeability depends on factors such as the size of the cavities and the degree of interconnection between the cavities. Rocks with low porosity generally have low permeability and rocks with high porosity are not necessarily high permeability, because the magnitude of the relationship between cavities is very decisive [1].

2.4 Geoelectric method

Geoelectricity method is one of the geophysical methods that using concept of electricity to earth problems [6]. The purpose is to estimate the electrical properties of a medium or sub-surface rock formation, especially its ability to conduct or inhibit electricity (conductivity or resistivity).

The basic principle of the resistivity method is Ohm's Law which states that the potential difference that arises at the ends of a medium is directly proportional to the electric current flowing in the medium. In addition, Ohm's Law also states that the resistance of electricity is directly proportional to the length of the medium and inversely proportional to its cross-sectional area. The formulation of Ohm's two statements above, can be written as follows [8]:

$$V \propto I \text{ or } V = I \cdot R \quad (2.1)$$

The Wenner-Schlumberger configuration is a combination method between the Wenner configuration and the Schlumberger configuration. This configuration has a configuration with a constant spacing rule system with a comparative factor note "n" for this configuration is the comparison of the distance between the AM electrode and the distance between MN as shown in Figure 2 [9].

The formula for calculating the geometric factor based on Figure 2 [10] is:

$$K = \pi n\,(n + 1)\,a \quad (2.2)$$

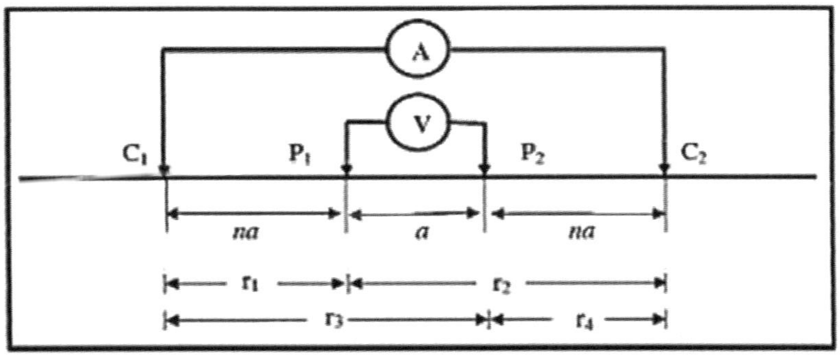

Figure 2. Electrode arrangement of the Wenner Schlumberger configuration [10].

3. Research Methods

The data collection process was carried out on October 21, 2021 in Maccini Baji Village, Lau District, Maros Regency. The data collection process was carried out using 2 tracks, each track has a span length of 235 meters. The first linestretches from coordinates (119°34'18.12"E, 4°57'57.66"S) to (119°34'13.98"E, 4°58'3.90"S) and the second linestretches from coordinates

(119°34'20.40"E, 4°57'59.40"S) to (119°34'12.84"E, 4°58'0.66"S). Data collection uses 48 electrodes with a space of 5 meters between the electrodes. Track 1 has a span direction from southwest to northeast and track 2 has a span direction from northeast to southwest. The position of the stretch intersects each other / cross (Figure 3). The data obtained from data acquisition is in the form of pseudo-resistivity values and is supported by Maros Regency geological data and sondir data to facilitate the interpretation process.

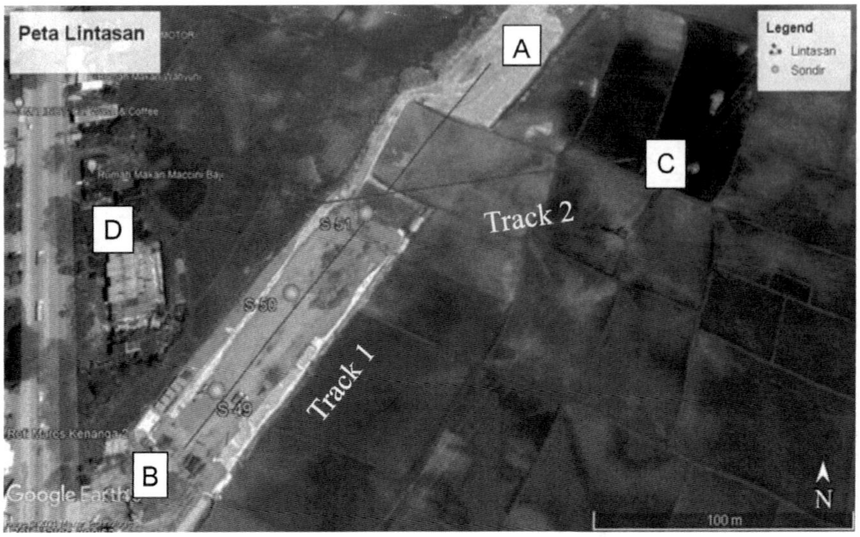

Figure 3. Data acquisition linemap.

4. Results and Discussion

4.1 Track 1

Based on the relationship between the results of the 2D cross-sectional inversion and the sondir data information, it can be interpreted as shown in Figure 4.

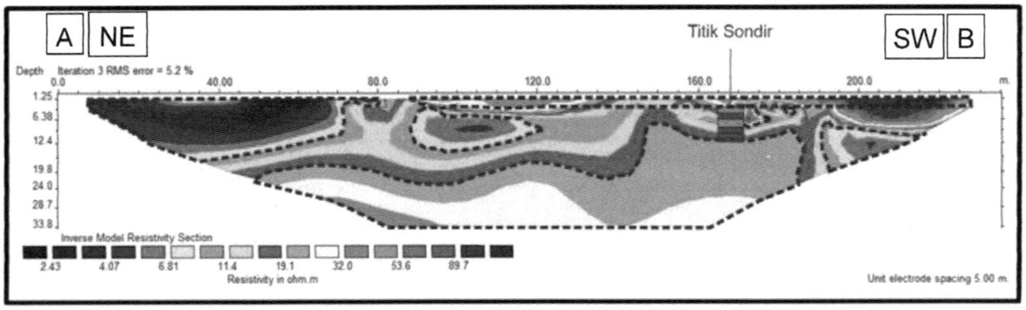

Figure 4. Combination of the result of the 2D resistivity cross-section inversion of track 1 with sondir data.

Based on the resistivity value in the cross-section above, it can be divided into three layers based on the resistivity value in the research area. The first layer with a resistivity value range (19

Ωm - 89.7 Ωm) is imaged in dark green to dark purple. The layer is interpreted as a slate material found at a distance of 85 meters to 235 meters with a depth of 0 - 2.5 meters. The second layer with a resistivity value range (2.43 Ωm - 19 Ωm) is imaged in dark blue to dark green. The layer is interpreted as an alluvial deposit based on a resistivity value table (Table 2.1), the geological information is then tied with sondir data showing the same layer information. The alluvium layer is at a depth varying from 0-23 meters. The third layer with a resistivity value range (20 Ωm - 45 Ωm) is imaged in moss green to brown color. The layer is interpreted as limestone found at depths varying from 13 to ≥33.8 meters.

Based on the resistivity table, geological information and sondir data with a resistivity value range (2.43 Ωm - 10 Ωm) are imaged with dark blue to light blue colors found in the second layer indicating the presence of groundwater. The material that acts as an aquifer is suspected to be a sandy silt where the information is tied to sondir data which provides the same information as the resistivity value contained at that depth. Sandy silt has good porosity and permeability so that it can store and pass water.

The distribution and types of aquifers found in the second layer of track 1:

- The distance of 0 - 80 meters at a depth of 0-17 meters is a shallow aquifer with a type of aquifer, namely a unconfined aquifer where the upper layer is a layer that has high permeability (not waterproof).
- The distance of 85 – 120 meters at a depth of 4 – 14 meters is a shallow aquifer with a type of aquifer, namely a unconfined aquifer because the upper part is a permeable sediment material.
- A distance of 165-180 meters at a depth of 4-8 meters is a shallow aquifer with a type of aquifer, namely a unconfined aquifer because the upper part is a permeable sedimentary material.
- A distance of 190-220 meters at a depth of 6-20 meters is a shallow aquifer type of aquifer, namely a unconfined aquifer because the upper part is a permeable sedimentary material.

4.2 Track 2

Based on the combined interpretation of the inversion results and penetration data contained in track 1, it can be connected with the interpretation contained in track 2 as seen in Figure 5.

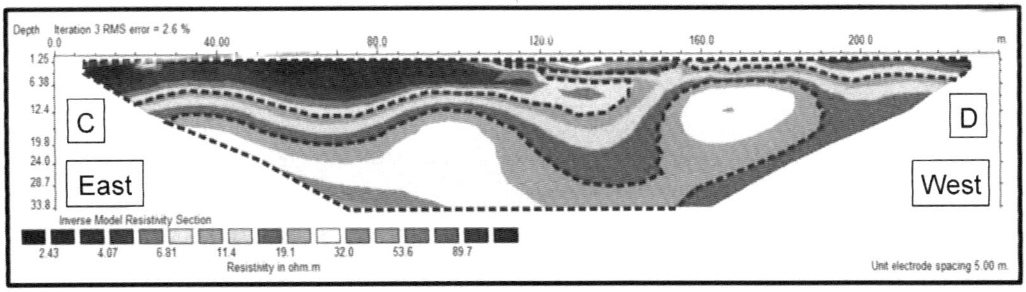

Figure 5. Interpretation of lineresistivity inversion results 2.

Based on the resistivity value in the cross-section above, it can be divided into three parts based on the resistivity value in the research area. The first layer with a resistivity value range (19 Ωm - 45 Ωm) is imaged in dark green to brown. The layer is interpreted as a slate material found at a distance of 110 meters to 155 meters with a depth of 0 - 2.5 meters. The second layer with a resistivity value range (2.43 Ωm - 19 Ωm) is imaged in dark blue to dark green. The layer is interpreted as an alluvial deposit based on the results of line1 interpretation. The alluvium layer is found at depths varying from 0-33.8 meters. The third layer with a resistivity value range (20 Ωm - 50 Ωm) is imaged with moss green to orange color. The layer is interpreted as limestone found at a depth varying from 6 to ≥33.8 meters. In the second layer with a resistivity value range (2.43 Ωm - 10 Ωm) which is imaged in dark blue to light blue, it is indicated that there is groundwater. The material that acts as an aquifer is suspected to be a sandy silt where the information is linked to the results of the interpretation of line1. Sandy silt has good porosity and permeability so that it can store and pass water.

The distribution and type of aquifers found in the second layer of track 2:

- The distance of 0-145 meters at a depth of 0-15 meters is a shallow aquifer with a type of aquifer, namely a unconfined aquifer, where the upper layer is a layer that has high permeability (not waterproof).
- A distance of 155 – 235 meters at a depth of 0–6 meters is a shallow aquifer with a type of aquifer, namely a unconfined aquifer, where the upper layer is a layer that has high permeability (not waterproof).

4.3 Pseudo 3D

Figures 6 and 7 show the combined results of the two cross-sections which are then processed to produce pseudo 3D. It can be seen that track 1 and track 2 intersect each other and show the overall subsurface shape in the research location area.

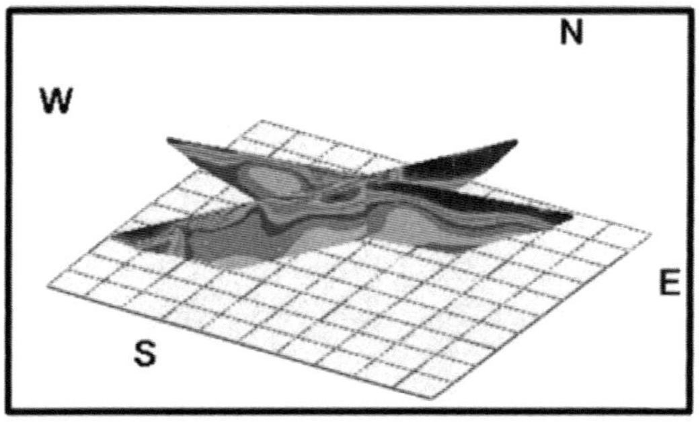

Figure 6. Pseudo 3D from the south and east.

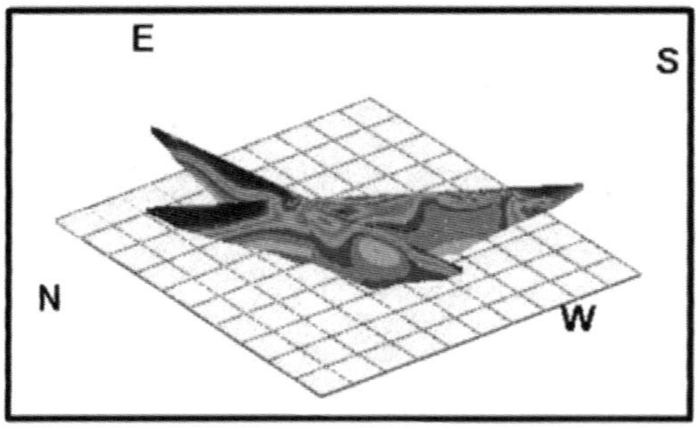

Figure 7. Pseudo 3D from north and west.

Based on the resistivity value in the cross-section above, it can be divided into three parts based on the resistivity value in the research area. The first layer with a resistivity value range (19 Ωm - 45 Ωm) is imaged in dark green to brown. The layer is interpreted as a slate material found at a distance of 110 meters to 155 meters with a depth of 0 - 2.5 meters. The second layer with a resistivity value range (2.43 Ωm - 19 Ωm) is imaged in dark blue to dark green. The layer is interpreted as an alluvial deposit based on the results of line1 interpretation. The alluvium layer is found at depths varying from 0-33.8 meters. The third layer with a resistivity value range (20 Ωm - 50 Ωm) is imaged with moss green to orange color. The layer is interpreted as limestone found at a depth varying from 6 to ≥33.8 meters. In the second layer with a resistivity value range (2.43 Ωm - 10 Ωm) which is imaged in dark blue to light blue, it is indicated that there is groundwater. The material that acts as an aquifer is suspected to be a sandy silt where the information is linked to the results of the interpretation of line1. Sandy silt has good porosity and permeability so that it can store and pass water.

5. Conclusion

Based on the results of the resistivity cross-sectional analysis, the research area consists of three types of layers, including the first layer with a resistivity value range (19 Ωm - 89.7 Ωm) interpreted as a sedimentary material, the second layer with a resistivity value range (2.43 Ωm - 19 Ωm) is interpreted as an alluvial deposit and the third layer with a resistivity value range (20 Ωm - 50 Ωm) is interpreted as limestone.

The aquifer is scattered in the second layer on the alluvial deposits and accumulates in the northeastern part of the study area at depths varying from 0 to 18 meters. The type of aquifer found in the study area is a unconfined aquifer because the cover layer has good permeability and porosity.

References

[1] Manrulu, R. H., Nurfalaq, A., & Hamid, I. D. (2018). Dugaan Sebaran Air Tanah Menggunakan Metode Geolistrik Resistivitas Konfigurasi Wenner dan Schlumberger di Kampus 2 Universitas Cokrominoto Palopo. Jurnal Fisika FLUX, 15(1).

[2] Wahyuni, D., Musa, M., & Sandra. (2018). Identifikasi Sebaran Akuifer Menggunakan Metode Geolistrik Hambatan Jenis di Wilayah Kecamatan Ampibado Kabupaten Parigi Moutong. Journal of Science and Technology

[3] Taslim, I. (2014). Perilaku Hidrologi Kawasan Karst Maros: Studi Kasus Gua Saleh Daerah Patunuangasue Kecamatan Simbang. Makassar: Universitas Hasanuddin.

[4] KLHS Kab. Maros. (2011). Laporan Akhir Kajian Lingkungan Hidup Strategis (KLHS) Kab. Maros. Maros: Pemerintah Kabupaten Maros.

[5] Aryanto, D. E., & Hardiman, G. (2017, Oktober). Kajian Multi Varian Faktor yang Berpengaruh Terhadap Infiltrasi Air Tanah Sebagai Dasar Penentuan Daerah Potensi Resapan Air. Proceeding Biology Education Conference, 14(1), 252-257.

[6] Nasution, E. M. (2013). Penyelidikan Zona Akuifer dengan Survei Pendugaan Geolistrik Metode Schlumberger Studi Kasus Daerah Kecamatan Kaliwungu dan Sekitarnya, Kabupaten Kendal, Jawa Tengah. Semarang: Universitas Diponegoro.

[7] Utaya, S. (1990). Pengantar Hidrologi. Malang: Aditya Media.

[8] Syamsuddin. (2007). Penentuan Struktur Bawah Permukaan Bumi Dangkal dengan Menggunakan Metode Geolistrik Tahanan Jenis 2D (Studi Kasus Potensi Tanah Longsor di Panawangan, iamis. Bandung: Institut Teknologi Bandung.

[9] Arsyadi, A. Q. (2017). Identifikasi Persebaran Air Lindi di TPA Ngipik Kabupaten Gresik dengan Menggunakan Metode Resistivitas 2D Konfigurasi Wenner Schlumberger. Surabaya: Institut Teknologi Sepuluh Nopember.

[10] Loke, M. H. (2004). Rapid 2D Resistivitas & IP Inversion Using the Last-Square Method, Geotomo Software. Malaysia.

IOP Conf. Series: Earth and Environmental Science **1525** (2025) 012015 doi:10.1088/1755-1315/1525/1/012015

Microtremor Measurements on Landslide Site at Pattiro Village, Gowa Regency

Rezki Agus[1,2*], Andi Muhamad Pramatadie[1]

[1] Department of Geophysics, Hasanuddin University, Makassar, Indonesia
[2] Agency for Meteorology, Climatology, and Geophysics (BMKG) Region IV, Makassar

*E-mail: noviana.rezki@gmail.com

Abstract. Landslide on Pattiro Village, Gowa Regency occurred at the end of May 2022. We measured microtremors at 14 locations within the landslide area. The horizontal-vertical spectral ratio (HVSR) method was then used to process and analyze these data in order to determine the resonance frequency (f_0) and H/V amplification factor (A_0). We divided the measurement location into 3 segments. First, we measured 7 points of site on the landslide site. The second segment was measured on 4 points, distanced 5 meters to the north from the first segment location which was non-landslide segment. On the third location we measured 2 points of site, located to the northwest about 120 meters from the first and second segments. The predominant frequency (f_0) values range from 5.07 Hz to 6.78 Hz, and the amplification factor (A_0) for these three segments ranges from 2.28 to 7.45. The results consistent with Kanai Classification, that predominant frequency (f_0) value from 4 Hz to 10 Hz as described as surface sediment thickness are from 5 meters to 10 meters which is categorized into medium level. Moreover, the H/V amplification factor (A_0) values in this study are categorized into 'medium risk' level.

1. Introduction

Gowa Regency is one of the areas most susceptible to landslides because 72.26% of its area has steep topography. Pattiro Village, which is one of the areas in Gowa Regency, is located about 62 km from the city of Gowa Regency. On January 24, 2019, the village was reported to have been hit by a large-scale landslide which resulted at least several houses being buried under the ground and several people were declared missing. Agency for Meteorology Climatology and Geophysics (BMKG) stated that the extreme weather that had occurred since January 22, 2019 was caused by the growth of cumulonimbus clouds in Makassar and spread wide range over the city which resulted a long duration of heavy rain and strong winds in the South Sulawesi region for several days.

At the end of May 2022, a landslide also occurred at Pattiro Village after being hit by heavy rains for several days according to local residents. In this study, we selected landslide area as a research site at an altitude of 1130 masl, precisely near the Bulubaria climbing registration post. The research area is about 23,000 square meters. This area is shown in Figure 1. Landslide is a rapid displacement of a mass of rock, remaining soil, or sediment adjacent to a slope in which the moving mass's center of gravity moves downward and outward [1].

IOP Conf. Series: Earth and Environmental Science **1525** (2025) 012015 doi:10.1088/1755-1315/1525/1/012015

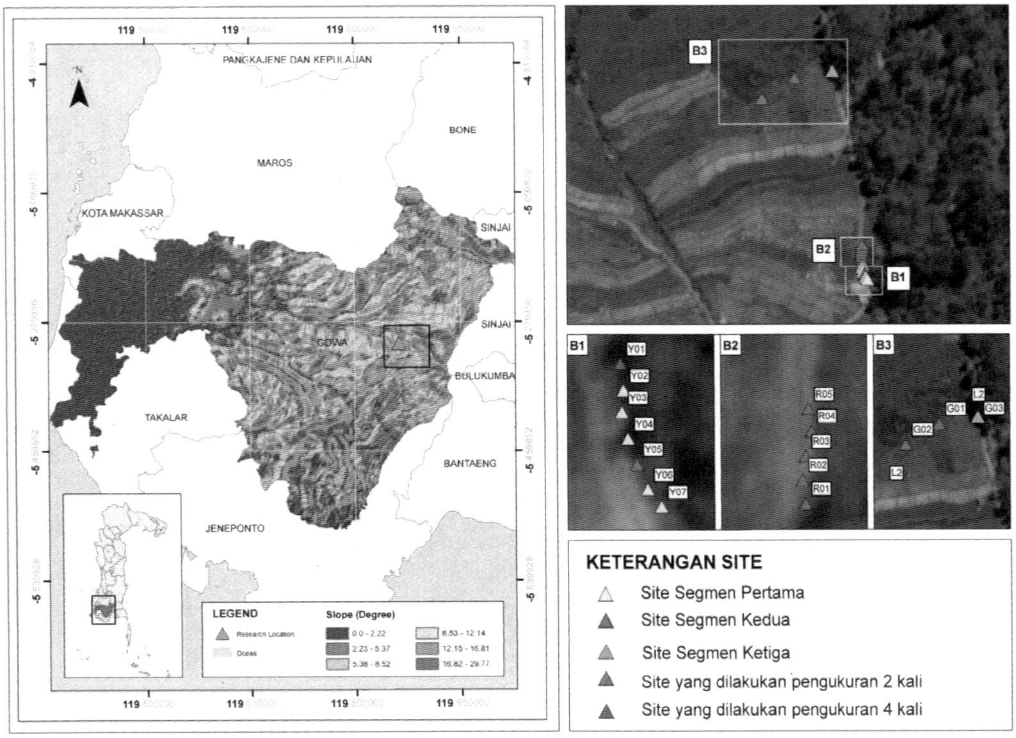

Figure 1. Research location at Pattiro Village, Gowa Regency. **a.** The green triangle in the box represents the research location. **b.** The triangle in the research area points the location points of microtremor measurements.

Several methods can be used to examine the landslide phenomenon, one of them is the Horizontal to Vertical Spectral Ratio (HVSR) method. In 1970, Nogoshi and Igarashi [2] introduced this method. But in 1989, Nakamura popularized it [3]. The HVSR method records ambient noise and examines the ratio of the seismometer's horizontal and vertical Fourier amplitude spectra [3,4,5,6,7]. Microtremor HVSR has been used to estimate the fundamental site frequency at locations with a significant subsurface impedance contrast [8]. This method has proven and widely used to estimate the fundamental period of soil deposits [5,9]. In this study, we applied HVSR method to estimate the resonance frequency (f_0) and H/V amplification factor (A_0).

2. Methodology

We deployed microtremor HVSR measurement equipment at observation points as shown in Figure 1b. The data collected during 3 times measurements (on June 21st 2022, on July 31st 2022, and also on September 4th 2022) followed the SESAME *H/V* User Guidelines [10]. The seismometers were installed in surficial soil-oriented East-West and North-South horizontally and levelled vertically. Using the STA/LTA anti-triggering algorithm, microtremors were measured for at least 10 minutes at each location. The data were filtered with a 0.2 – 25 Hz bandpass filter after being sampled at a rate of 100 sps [8,10]. Geopsy software was used to process the records [11]. The microtremor record example is shown in Figure 2.

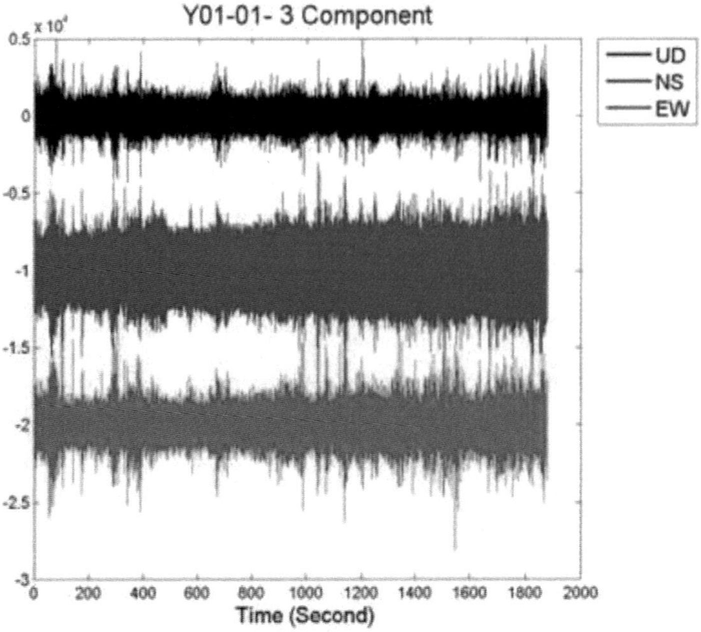

Figure 2. This figure shows three-component (up-down, east-west, and north-south) 9 minutes time-series data recorded at station point Y01.

The distribution of predominant frequency (f_0) values according to Kanai Classification is shown in Table 1 [12]. The peak of the *H/V* curve, which indicates the natural frequency of the soil deposits, can be used to determine the predominant frequency [13]. In 1989, Nakamura [3,5] postulated that the horizontal to vertical ratios in the frequency domain of seismic microtremors can be used to predict site response using the formula in equation:

$$HVSR\ (\omega) = \frac{\sqrt{H^2{}_{NS}\,(\omega) + H^2{}_{EW}\,(\omega)}}{V_{UD}\,(\omega)}$$

where HNS, HEW, and VUD are the north–south, east–west and vertical components, respectively, recorded at the site.

Meanwhile, the amplification values (A_0) are affected by S-wave velocity (V_s), density (ρ), P-wave velocity (V_s), and shear wave attenuation (Q_s) [13]. If the rock has been deformed (impacted by e.g. weathering, folding, or enlarged) the amplification value increases which alters its physical properties [12]. Additionally, the degree of rock deformation affects the amplification level [14].

IOP Conf. Series: Earth and Environmental Science **1525** (2025) 012015 doi:10.1088/1755-1315/1525/1/012015

Table 1. Soil classification based on predominant frequency values by Kanai [12].

Classification		Predominant Frequency (Hz)	Kanai Classification	Description
Type	Variety			
Type IV	Variety I	6.667 - 20	Rocks that are older than the tertiary consisting of sand, gravel, and other hard rocks	Hard rock dominates the surface sediment's thickness.
	Variety II	4 -10	Sand-gravel, sandy-hard clay, loam, and other types of alluvial rocks with a thickness of 5 meters.	Surface sediment's thicknesses between 5 and 10 meters, categorized into the medium category.
Type III	Variety III	2.5 - 4	Alluvial rocks with a thickness greater than 5 meters, such as loam, hard muds, and sand-gravel muds.	Surface sediment's thickness between 10 – 30 meters, categorized into the thick category.
Type II Type I	Variety IV	<2.5	Material known as alluvial material, made up of top soil, muds, etc., formed by delta sedimentation.	The surface sediment is very thick.

3. Results And Discussion

The result of frequency predominant values and *H/V* amplification factor are shown in Table 2. The predominant frequency (f_0) values range from 5.07 Hz to 6.78 Hz, as shown in Table 1. For these three segments, the amplification factor (A_0) ranges from 2.28 to 7.45.

Table 2. Results of frequency resonance (f_0) and H/V amplification factor (A_0).

Point	*Lat*	*Long*	f_0	A_0
Y01	-5.31047	119.8632	6.11	7.45
Y02	-5.31049	119.8632	6.17	4.66
Y03	-5.3105	119.8632	6.78	2.28
Y04	-5.31052	119.8632	5.70	5.90
Y05	-5.31053	119.8632	5.42	4.16
Y06	-5.31055	119.8632	5.23	6.10
Y07	-5.31056	119.8632	5.50	4.00
R01	-5.31044	119.8632	5.86	4.92
R02	-5.31043	119.8632	5.29	3.93
R03	-5.31041	119.8632	5.25	4.12
R04	-5.3104	119.8632	5.34	4.76
R05	-5.31038	119.8632	5.24	4.81
G01	-5.30953	119.8629	5.07	3.63
G02	-5.30964	119.8627	6.23	3.63

The results consistent with Kanai Classification [12] that predominant frequency values from 4 Hz to 10 Hz as described as surface sediment thickness grouped into the medium category 5–10 meters. Moreover, the *H/V* amplification factor values in this study are categorized into 'medium risk' level [14]. These results also match with the visual observation on the research area.

4. Conclusion

In this study, the predominant frequency values range from 5.07 Hz to 6.78 Hz, while the amplification factor varies from 2.28 to 7.45 for these three segments. The results consistent with Kanai Classification that predominant frequency value from 4 Hz to 10 Hz as described as surface sediment thickness are from 5 meters to 10 meters categorized into Type IV that consists of sandy-gravel, sandy hard clay, loam, etc. Moreover, the *H/V* amplification factor values in this study are categorized into 'medium risk' level.

References

[1] Arnone E, Noto L V, Lepore C and Bras R L 2011 Physically-based and distributed approach to analyze rainfall-triggered landslides at watershed scale *Geomorphology* 133(3-4), 121-131
[2] Nogoshi M and Igarashi T 1970 On the propagation characteristics of microtremors *J. Seism. Soc. Japan* **23** 264-280
[3] Nakamura Y 1989 A method for dynamic characteristics estimation of subsurface using microtremor on the ground surface *Railway Technical Research Institute* Quarterly Reports 30(1)
[4] Amitrano D, Gaffet S, Malet J P and Maquaire O 2007 Understanding mudslides through micro-seismic monitoring: the Super-Sauze (South-East French Alps) case study *Bulletin de la Société Géologique de France* 178(2) 149-157
[5] Gupta R K, Agrawal M, Pal S K, Kumar R and Srivastava S 2019 Site characterization through combined analysis of seismic and electrical resistivity data at a site of Dhanbad, Jharkhand, India *Environmental Earth Sciences* 78(6) 1-13
[6] Imposa S, Grassi S, Fazio F, Rannisi G and Cino P 2017 Geophysical surveys to study a landslide body (north-eastern Sicily) *Natural Hazards* 86(2) 327-343
[7] Whiteley J S, Chambers J E, Uhlemann S, Wilkinson P B and Kendall J M 2019 Geophysical monitoring of moisture-induced landslides: a review *Reviews of Geophysics* 57(1) 106-145
[8] Molnar S, Sirohey A, Assaf J, Bard P Y, Castellaro S, Cornou C, ... and Yong A A 2022 Review of the microtremor horizontal-to-vertical spectral ratio (MHVSR) method *Journal of Seismology* 1-33
[9] Acerra C, Aguacil G, Anastasiadis A, Atakan K, Azzara R, Bard P Y, ... and Zacharopoulos S 2004 Guidelines for the implementation of the H/V spectral ratio technique on ambient vibrations measurements, processing and interpretation *No. European Commission–EVG1-CT-2000-00026 SESAME* Brussels Belgium: European Commission
[10] Abdelrahman K, Al-Otaibi N, Ibrahim E and Binsadoon A 2021 Landslide susceptibility assessment and their disastrous impact on Makkah Al-Mukarramah urban expansion, Saudi Arabia, using microtremor measurements *Journal of King Saud University-Science* 33(5) 101450
[11] Wathelet M, Chatelain J L, Cornou C, Giulio G D, Guillier B, Ohrnberger M and Savvaidis A 2020 Geopsy: A user-friendly open-source tool set for ambient vibration processing *Seismological Research Letters* 91(3) 1878-1889
[12] Rasimeng S, Yogi I B S and Rananda E 2020 Analysis and Zonation of Land Vulnerability Areas in Pekon Karangrejo Ulubelu Tanggamus Using Microzonation Method *Jurnal Geofisika HAGI* 18(01) 14-18
[13] Syamsuddin E and Assegaf M A H 2021 Dasar-dasar Akuisisi Data MASW dan Mikrotremor *Unhas Press*
[14] Marjiyono 2010 Estimasi Karakteristik Dinamika tanah dari Data Mikrotremor Wilayah Bandung Tesis *Institut Teknologi Bandung*

IOP Conf. Series: Earth and Environmental Science **1525** (2025) 012016 doi:10.1088/1755-1315/1525/1/012016

Low Shear Wave Velocity Zones as Evidence for Liquefaction Susceptibility: Case Study of Building X in Mamuju

Hasnan Sutadi[1*], Andri Moh. Wahyu Laode[1]

[1] Departement of Geophysics, Hasanuddin University, Makassar, Indonesia

*E-mail: sutadih18h@student.unhas.ac.id

Abstract. Recurrent earthquakes in Mamuju Regency in 2021-2022 triggered a co-disaster in the form of liquefaction, which had a significant impact on urban infrastructure. One case occurred in a building located near a refueling station, which experienced subsidence due to liquefaction. Post-liquefaction studies generally rely on conventional geotechnical methods, which are relatively time-consuming and costly. This study aims to evaluate the potential impact of liquefaction on buildings using a geophysical approach, specifically through cross-sectional analysis of shear wave velocity (Vs) using the Multichannel Analysis of Surface Waves (MASW) method. Data were collected using the Common Mid Point Cross-Correlation (CMP-CC) technique to obtain a better Vs cross-section resolution than the conventional MASW method. The results of the Vs cross-section interpretation allow the identification of subsurface zones that are more susceptible to liquefaction, which are indicated in areas referred to as Low Velocity Zone. This information can be used as a basis for determining areas that require special attention in geotechnical repair and mitigation efforts.

1. Introduction

Liquefaction is a major concern in geotechnical engineering. Because the impact can be very destructive, including the collapse of building structures, land shifts, and sudden sinking of infrastructure. Liquefaction is one of the impacts caused by recurrent earthquakes in Mamuju Regency which occurred on January 14, 2021, January 15, 2021, and June 8, 2022 with a magnitude of 5.9 Mw, 6.2 Mw, and 5.8 Mw respectively [1,2]. One of the affected buildings is a building X that near a gasoline filling station in Mamuju with an indication of the appearance of deformation and sand boil between the paving blocks around the building. Previous research using the NSPT method to analyze the condition of the area around the building after the liquefaction showed that there was ground settlement in the building [3].

Meanwhile, conventional geotechnical investigations such as Standard Penetration Test (SPT) are invasive, relatively time-consuming and expensive, and limited to point-based measurements, which may not capture lateral variability in heterogeneous subsurface conditions. In contrast, geophysical methods-particularly Multichannel Analysis of Surface Waves (MASW)-offer a non-invasive, rapid, and laterally continuous alternative to characterize subsurface shear wave velocity (Vs). This allows for more comprehensive subsurface mapping compared to discrete SPT data points.

IOP Conf. Series: Earth and Environmental Science **1525** (2025) 012016 doi:10.1088/1755-1315/1525/1/012016

A key outcome of MASW interpretation in the case of liquefaction potential areas is the identification of Low Velocity Zones (LVZ) - subsurface layers characterized by much lower shear wave velocities compared to the surrounding material. Research by Wu (2022) [4] analyzing recurrent liquefaction areas in Taiwan due to the Jiasian and Meinong earthquakes found that LVZ areas are strongly correlated with the appearance of surface deformation due to liquefaction. Areas identified as LVZ often correlate with soft, loose, water-saturated sediments that are highly susceptible to deformation during seismic shaking, and are thus considered important indicators of liquefaction susceptibility. Detecting the LVZ allows delineation of zones that require geotechnical reinforcement or further investigation.

Figure 1. a) Building X affected by liquefaction, b) and c) Deformation of paving blocks, and d) sand boil appearance of paving blocks.

2. Method

2.1 Data Acquisition

The research was conducted around Building X in Mamuju Regency, West Sulawesi Province on March 13, 2023 using the Multichannel Analysis of Surface Wave method with Roll A Long Configuration. Data collection was carried out using 24 4.5 Hz geophones with a spacing of 1 meter, shooting from east to west. The total shooting was carried out 13 times which adjusted to the conditions in the field. We also use Bor log data as data taken on December 20-25, 2021 as secondary data in this study.

Figure 2. Measurement map

2.2 Data Processing

Processing was performed using the Common Mid Point Cross Correlation (CMP CC) method to cross-correlate geophone pairs that have the same geometry. Hayashi & Suzuki (2004) [5] have shown that lateral resolution using CMP CC processing is better than conventional processing methods. The process of obtaining geophone pairs from CMP CC uses the Seisimager v7.0.0.0 application.

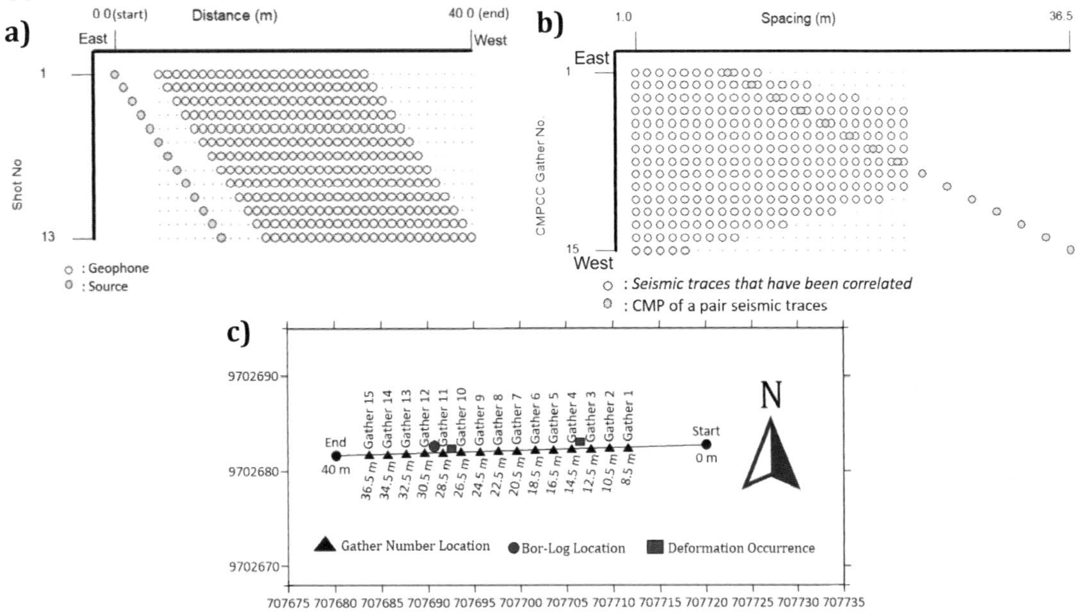

Figure 3. a) Field measurement geometry, b) Data processing geometry with CMP CC, c) Trajectory geometry view

The number of gather numbers is calculated by selecting the center point (CMP) and then creating a new mid point based on the bin size setting. The bin size length is 2 times the geophone spacing length, so a bin size of 2 m is used (Figure 3). Gather number will stop when the CMP CC no longer has enough seismic traces that can produce a good dispersion image. Thus, 15 gather numbers were obtained.

Furthermore, the processing to obtain the dispersion image until the inversion is processed using Geopsy v3.4.2. The advantage of Geopsy v3.4.2 is that it has a feature to display dispersion images on a logarithmic scale and has a Neighborhood Algorithm (NA) inversion feature. Logarithmic resolution allows for better picking because the model layers in the forward dispersion curve model can be very different with relatively small differences in picking. Meanwhile, NA inversion is one of the non-linear inversions that can provide the best solution when inversing dispersion curves when compared to linear inversion. This advantage can accommodate the shortcomings of linear inversion which is often trapped in a local minimum solution and far from the actual state, especially the inversion of dispersion curves that are indicated to have LVZ areas. Meanwhile, Sambridge (1999) [6] also showed the superiority of NA inversion over non-linear inversion such as Genetic Algorithm (GA) in exploring solutions with the lowest misfit.

Neighborhood Algorithm is a direct search method used to solve non-linear inversion problems. NA inversion will produce random samples of pairs of physical and geometric parameters in the parameter space, so that each sample has a set of parameters corresponding to one base model by calculating dispersion curves (forward models) on all models. The forward

IOP Conf. Series: Earth and Environmental Science **1525** (2025) 012016 doi:10.1088/1755-1315/1525/1/012016

modeling of the dispersion curves obtained for each parameter pair is then used as the calculation model. The calculation model obtained from the random parameter pair is then evaluated using the misfit value.

$$Misfit = \sqrt{\sum_{i=1}^{n} \frac{(y_{di} - y_{ci})^2}{{\sigma_i}^2 n_f}} \quad (1)$$

Where y_{di} is the observed phase velocity at each picking frequency f_i, y_{ci} is the calculated phase velocity at each frequency f_i, n_f is the amount of phase velocity data at each frequency f_i, and σ_i is the uncertainty associated with the observation data at frequency f_i. Based on equation (1), a misfit value of less than 1 indicates that the calculated dispersion curve has a fit or has a standard deviation of ±1 to the observed dispersion curve data. Whereas a misfit of more than 1 indicates a poor fit to the observed dispersion curve. However, there is currently no concept in calculating the misfit of dispersion curves that is universally accepted [7].

To run NA inversion on the dispersion curve, it is necessary to set physical parameters as the search space and tuning parameters as the solution exploration space. The physical parameters

Table 1. The physical parameters.

	v_S (m/s)	v_P (m/s)	μ (Poisson Ratio's)	ρ (*Density*) kg/m^3	***d*** (*depth*) (m)
Layer 1	100-350	500-3500	0,2-0,5	1200-2500	1-5
Layer 2	50-250	200-3500	0,2-0,5	1200-2500	1-30
Layer 3	80-250	200-3500	0,2-0,5	1200-2500	1-30
Layer 4	200-250	500-3500	0,2-0,5	1200-2500	*Half-space*

and tuning parameters in this study are based on the observation of picking dispersion curves, the physical parameter space proposed by Foti. Meanwhile, the tuning parameters use the defaults of the Dinver feature -part of the Geopsy application- which searches for 10000 models. The physical parameters used are presented in Table 1. The model with the best misfit is selected. The data processing process in geopsy to obtain the v_S profile is shown in Figure 4.

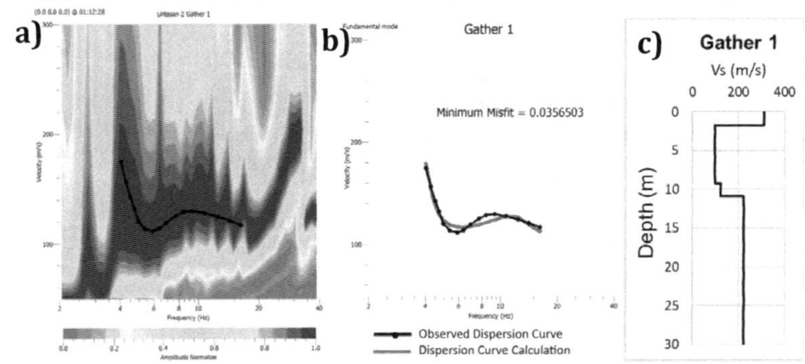

Figure 4. Examples of data processing in Gather; a) picking dispersion curves, b) inversion of dispersion curves, and c) inversion results of v_S profile.

IOP Conf. Series: Earth and Environmental Science **1525** (2025) 012016 doi:10.1088/1755-1315/1525/1/012016

3. Results and Discussion

Each gather or trace pair will be processed to obtain a dispersion curve. The dispersion curve of each gather on the survey line shows the distribution of picking from a frequency of 4-30.4 Hz at a phase velocity of 89.5-196.1 m/s (Figure 5). These dispersion curves tend to vary, indicating the dynamic state of the subsurface at each midpoint. The presence of an increase in phase velocity at higher frequencies also indicates the presence of a Low Velocity Zone (LVZ) vertically as shown from Wathelet (2005) in his research to identified hard first layer using forward modell or subsurface model where the upper layer is harder than the layer below. This situation in the v_S profile is a consideration for model parameters in the inversion process on table 1. Using properties parameter in table 1 and parameter tuning to search 10000 models to run inversion, the profile model with the lowest misfit is obtained as shown in Figure 6. Each vs profile was then interpolated to obtain a lateral vs subsurface view (Figure 7).

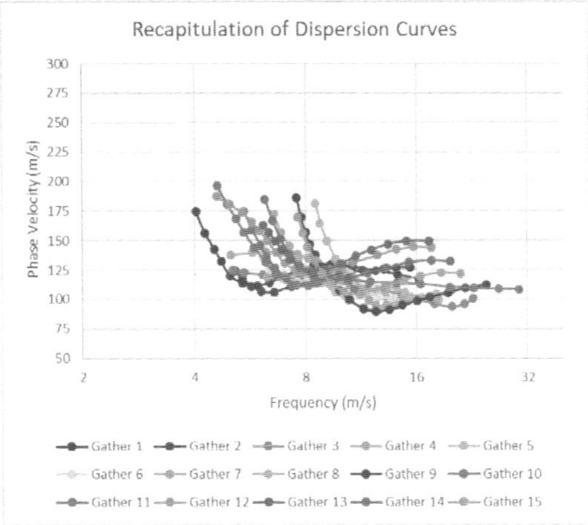

Figure 5. Recapitulation of the observed dispersion curve.

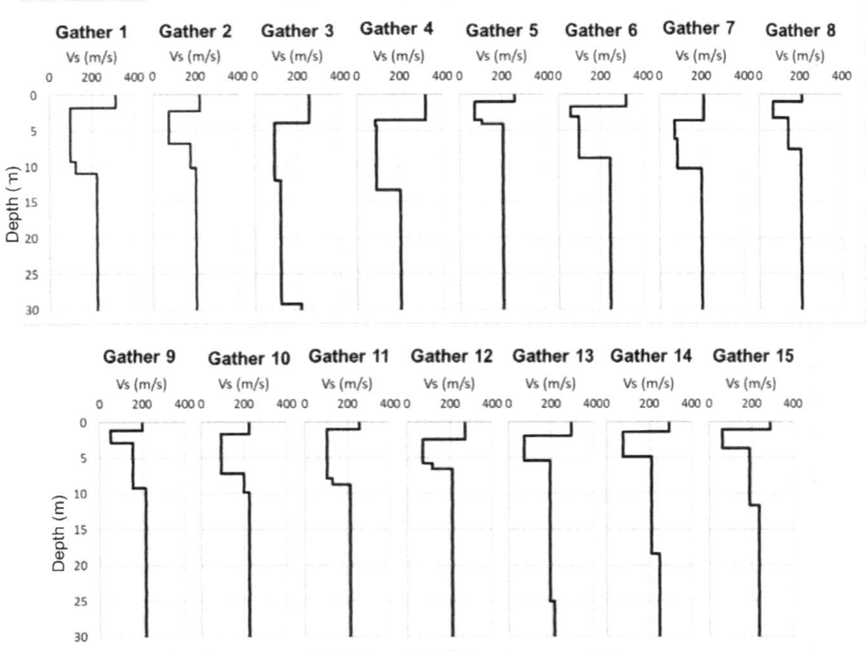

Figure 6. Recapitulation of the v_S profiles

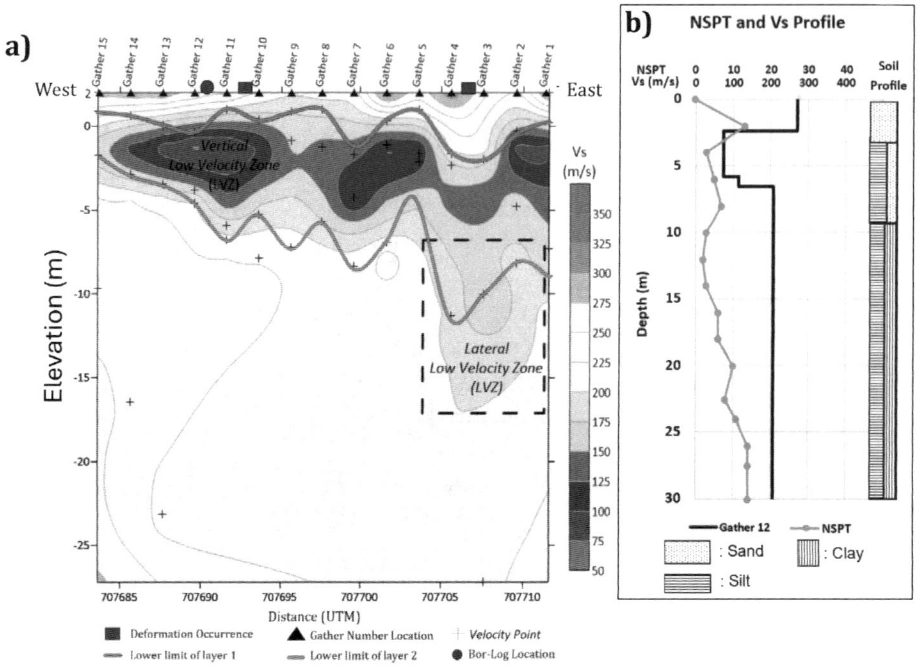

Figure 7. a) Recapitulation of the v_S profiles, b) Side by side between NSPT, v_S profile, and soil profile

The v_S cross section can be interpreted in three layers (Figure 7). The 1st layer section appears harder at elevations 2 m to -2 m with v_S in the range of 202-320 m/s, then weakens in the 2nd layer from elevations 1 to -2 m in the west and deepens in the east reaching elevation -11 m with lower v_S in the range of 52-203 m/s, then hardens again with v_S in the range of 119-248 m/s continuously to the subsurface. The presence of a layer with softer physical parameters is also shown in the drill data which found lower NSPT data at depths of 4-6 meters (elevations -2 to -4 m). The high v_S of layer 1 compared to layer 2 in both trajectories is also due to the presence of building embankments and also because the measurements were taken over paving blocks (Figure 1).

The 2nd layer that has low v_S values of 52-203 m/s at elevations -2 to -11 m can be categorized as vertical low velocity zones (LVZs). Vertical LVZs can also be seen from the shape of the dispersion curve which has an increase in phase velocity at higher frequencies. The range of v_S values in the 2nd layer corresponds to the range for geomaterials categorized as soft clay and loose sand. This 2nd layer correlates with the presence of silty sand in the drill data located at depths of 3 to 9.5 m (elevation -1 to -7.5 m). Silty sand layers tend to have good porosity and therefore easily store water.

Silt content can reduce the compaction efficiency of sand grains and can cause pore water pressure accumulation during saturation. A layer saturated with water will increase the density of the material and result in a low v_S value. As a result, the water-saturated sand layer will easily lose its shear strength as the pore pressure increases dramatically. This condition, when triggered by earthquake shaking, can lead to liquefaction [8]. The investigation of liquefaction events reported by Chang et al. (2011) [9] due to a 6.4 Mw earthquake in Taiwan also reported that liquefaction events at a depth of 5-6 m occurred in fine silty sand which is also still within the depth range of identification of the 2nd layer in this study.

The occurrence of deformation on track 2 could be a representation of the area being close to the lateral weak shear velocity zones (LVZs) which are marked with dashed black lines from elevation -7 m to -17 m as shown in Figure 24. This situation was also found in the study of Wu (2022) who analyzed recurrent liquefaction events in Taiwan. Lateral LVZs are characterized by v_S changes that increase in depth but do not experience significant lateral variations. The appearance of Lateral LVZs is caused by the low v_S profile in the 2nd layer of the gather 3 with a thickness of about 8 m and then compaction at the elevation of -27 m. Similar to the 2nd layer, the low v_S value in the Lateral LVZs compared to the surrounding area can be caused by the filling of material pores by water. In this vs. cross section, the Lateral LVZs are found on the east side of the track which is supported by the appearance of deformation in the range of gathers 3-4. In addition, along gathers 1-4 have areas that are weaker than the surrounding area. This is in line with the results of a study conducted by Wu (2022) who found that the appearance of many surface liquefaction bursts correlates with the identification of subsurface LVZs zones located at depths of 5-18m. These lateral LVZs have the potential as water channels that can trigger surface liquefaction events.

4. Conclusion

The results of the velocity vs. subsurface cross section identification can be categorized into three layers, with the first layer tending to be harder than the second layer. This situation indicates the existence of Vertical Low Velocity Zones which are also visible in the dispersion curve observation data. In addition, based on the interpolation of vs values, the appearance of Low Velocity Zones laterally is obtained. These lateral LVZs are characterized by v_S changes that increase in depth but do not experience significant lateral variations. The LVZ in this study corresponds to the presence of silty clay obtained from the drill. This zone has the potential to trigger liquefaction events with the characteristics of the material that easily stores water and is suspected to be the flow direction of liquefaction bursts that cause deformation at the surface.

Acknowledgment

The authors would like to thank HAGI Sulselbar for funding this research. Also to Ahmad Fauzy Arief, Wirawan Saleh, and Syaifullah who have helped in the data acquisition process.

References

[1] Supartoyo, Suwargana, H., & Karim, A. (2022). Dampak Gempa Bumi di Sulawesi Barat dan Upaya Mitigasi. 7(2), 104–118

[2] Yuwana, D. A., Kurniah, & Buana, T. W. (2021). Manifestasi Likuefaksi dan Gerakan Tanah Akibat Gempa Bumi 14 Dan 15 Januari 2021 Di Majene, Sulawesi Barat. Geo Spatial Proceeding, 106–113. [in Indonesian]

[3] Arsyad, A., Amaliyah, A. A. N., Paerong, S., & Djamaluddin, A. R. (2022). Liquefaction Potential Assessment for the City of Mamuju Sulawesi by using N-SPT based methods. *Indonesian Geotechnical Journal*, *1*(3), 37-55.

[4] Wu, Y. (2022). Case History Changes in Soil Conditions Before and After Earthquakes at A Repetitive Soil Liquefaction Site in Taiwan. *Geophysics*, *87*(3), 179–191.

[5] Hayashi, K., & Suzuki, H. (2004). *CMP cross-correlation analysis of multi-channel surface-wave data*. *57*(1), 7–13.

[6] Sambridge, M. (1999). Geophysical inversion with a neighbourhood algorithm—I. Searching a parameter space. *Geophysical journal international*, *138*(2), 479-494.

[7] Wathelet, M. (2005). Array recordings of ambient vibrations: surface-wave inversion. [Disertasi Doktoral, Universit´e de Li`ege].
[8] Bao, X., Jin, Z., Cui, H., Chen, X., & Xie, X. (2019). Soil liquefaction mitigation in geotechnical engineering: An overview of recently developed methods. *Soil Dynamics and Earthquake Engineering*, *120*, 273-291.
[9] Chang, W. J., Ni, S. H., Huang, A. B., Huang, Y. H., & Yang, Y. Z. (2011). Geotechnical reconnaissance and liquefaction analyses of a liquefaction site with silty fine sand in Southern Taiwan. *Engineering geology*, *123*(3), 235-245.

Spatial Autocorrelation (SPAC) Method for Subsurface Shear Wave Velocity Profiling: A Non-invasive Approach to Site Characterization

Erfan Syamsuddin[1*], Ahmad Fauzy Arif[1], Makhrani[1], Ardy Arsyad[2] and Rustan Effendi[3]

[1] Department of Geophysics, Hasanuddin University, Makassar, Indonesia
[2] Department of Civil Engineering, Hasanuddin University, Makassar, Indonesia
[3] Geophysics Engineering Department, Tadulako University, Palu, Indonesia

*E-mail: erfan.syamsuddin@unhas.ac.id

Abstract. The Spatial Autocorrelation (SPAC) method is considered a robust method for site characterization and seismic hazard assessment, being a non-destructive but credible method for determining subsurface shear wave velocity profiles. This research uses the SPAC method to obtain the 1D profile that focuses mainly on theory and the direct fitting of spatial autocorrelation coefficients, which leads to a higher accuracy with the employment of Rayleigh wave modes for both fundamental and higher modes. The field setup consists of seven vertical-component geophones arranged in the nested triangular configuration with radii of 9.24m, 4.62m, 2.31m, and 1.15m. The comprehensive dispersion curve was obtained using the SPAC method at frequencies from 7 to 50 Hz, showing a slight depth increase. At shallower depths, the results were from 200 m/s to 400 m/s, indicating the presence of loose sediments and clay-rich deposits. In contrast, deeper layers revealed values of up to 1000 m/s, possibly related to more compacted formations. The dispersion curves derived from different radii display a stable trend of the phase velocity with depth, and the Bessel function fitting process disperses a high correlation. The research findings demonstrate the SPAC technique's efficiency in recording the correct velocity profiles and presenting clear and reliable results for other geophysical data processing methods, such as envelope detection, which significantly increases the reliability of the results. The non-invasive feature of SPAC is an advantage in that it reduces the logistic problems associated with active-source surveys, and, therefore, it is very useful in long-term monitoring projects and in geologic settings that are different from each other.

1. Introduction

The Spatial Autocorrelation (SPAC) method, one of the most sought-after methods, is a non-invasive technique that can be used to determine the profiles of shear wave velocity of the subsurface materials, which are crucial for site characterization and seismic hazard analysis. It is the method that uses the ambient noises in the seismic system and such noises in the microtremors to read the spatial autocorrelation coefficients and then link them to the dispersion characteristics of surface waves, which are predominantly Rayleigh waves that dominate isotropic

IOP Conf. Series: Earth and Environmental Science **1525** (2025) 012017   doi:10.1088/1755-1315/1525/1/012017

noise fields [1], [2]. These Vs profiles are vital for the in-depth investigation of the site effects, such as the seismic amplification that are the most important reasons to justify seismic response studies and, thus, the structural design. SPAC seems to answer the cost-effectiveness problem in low-level equipment and passive data collection, among others. However, the main factors such as array effects, anisotropic ambient noise, and the inability to resolve deeper layers remain to be overcome [3], [4]. Due to the recommendation point of view on item effectiveness, SPAC was used to perform an introductory analysis which helped subtle meanings to be transferred related to the subsurface of the study area.

This study determined the 1D shear wave velocity profile, which has been employed as a basic tool for site characterization and seismic hazard assessment through the SPAC method. Furthermore, this study expressed the theoretical way that the SPAC method confers, allowing for direct fitting of spatial autocorrelation coefficients to increase the accuracy while integrating the fundamental and the higher Rayleigh wave modes for a deeper structural resolution [5], [6]. This methodology provides a unique opportunity for a thorough examination of site effects. The presence of array geometries and the anisotropy fault produced by ambient noise is also a constraint; these will be addressed. The Vs profile obtained in this study will serve as a critical diagnostic tool in advancing seismic response analyses and guiding future geotechnical studies.

2. Methodology

The spatial autocorrelation (SPAC) approach used a seismic survey to get the shear-wave velocity profile of the subsurface. The array used for the field configuration comprises seven vertically aligned geophones placed in various nested triangle arrangements. The geophones are arranged in an array to create various radii of 9.24 m, 4.62 m, 2.31 m, and 1.15 m, respectively, to achieve multi-scale resolution in the frequency domain (Hayashi et al., 2022). The survey location was meticulously chosen to minimize human noise interference and optimize the quality of the microtremor recordings.

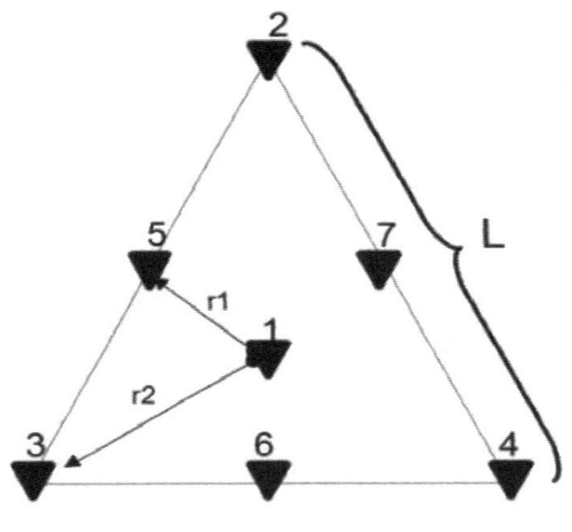

Figure 1. Geophone array configuration used in the SPAC survey

The Spatial Autocorrelation (SPAC) method is a passive seismic technique that can investigate the depths of elastic properties in the subsurface by analyzing the ambient seismic noise recorded at sensor arrays. This method is grounded in computing SPAC coefficients $\rho(f, r)$, that cohere seismic wavefields among station pairs as a function of interstation distance r and

IOP Conf. Series: Earth and Environmental Science **1525** (2025) 012017
doi:10.1088/1755-1315/1525/1/012017

frequency f at its core. These are derived from the zeroth-order Bessel function of the first kind, typically represented by

$$\rho(f,r) = J_0\left(\frac{2\pi f r}{c(f)}\right)$$

The symbol $c(f)$ which signifies the timing that the surface waves reach point r is the key to the Rayleigh wave's phase velocity [1], [7]. The SPAC method allows the determination of shear wave velocity (V_s) profiles from a set of surface waves recorded by the microtremor method. It does this by obtaining the phase velocity data from those measurements before inverting them to a V_s deep profile [1], [4]. The observed SPAC coefficients were computed using the normalized cross-correlation function of the seismic noise signals between the two stations as follows:

$$\rho(f,r) = \frac{\int u_i(t)u_j(t+\tau)\ dt}{\sqrt{\int u_i^2(t)\ dt \cdot \int u_j^2(t)\ dt}}$$

where $u_i(t)$ and $u_j(t)$ represent the recorded seismic displacement and velocity at stations i and j, respectively [8].

First, surface wave dispersion curves are obtained from SPAC coefficients,whichs use iterative forward modeling methods like least-squares optimization [3], [4] allows one to generate V_s profiles from SPAC coefficients. Alternatively, recent advancements have introduced direct inversion methods that bypass the intermediate dispersion curve step by directly fitting the observed SPAC coefficients to the theoretical models [4], [5]. These approaches improve computational efficiency and reduce inversion uncertainties, particularly when incorporating higher-mode Rayleigh waves to enhance depth resolution [5], [9].

Table 1. SPAC Survey Acquisition Parameters

Parameter	Gradient (%)
Number of geophones	7
Natural frequency	4.5 Hz
Array configuration	Nested triangle
Radii (m)	9.24, 4.62, 2.31, 1.15
Sampling interval	0.0001 s
Sampling frequency	1000 Hz
Recording duration	21.28 min

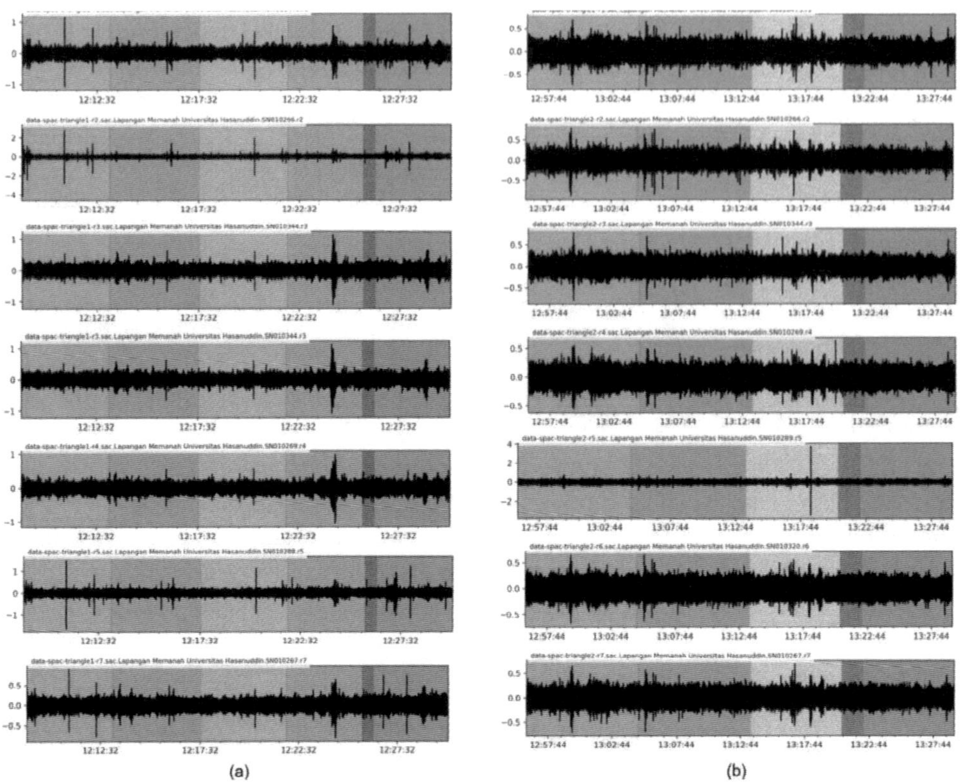

Figure 2. SPAC seismogram data for (a) Triangle 1 and (b) Triangle 2. Each shaded color represents the window division

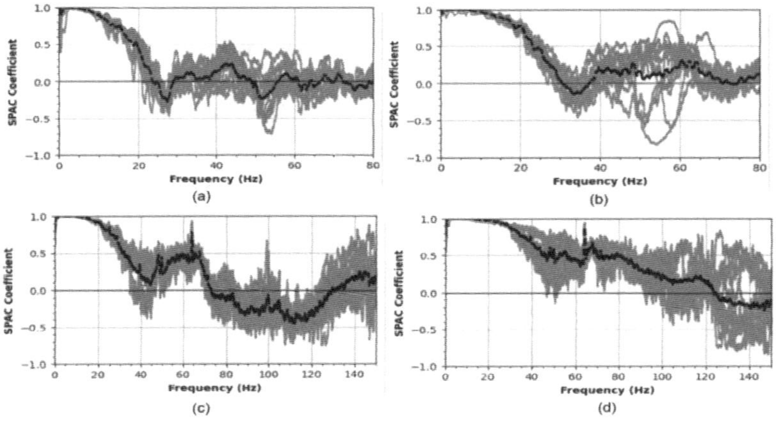

Figure 3. The SPAC coefficient for (a) SPAC-1 and (b) SPAC-2, (c) SPAC-3, and (d) SPAC-4. The gray curves indicate the SPAC coefficients for each receiver pair with the same radius, whereas the black curve represents the average.

IOP Conf. Series: Earth and Environmental Science **1525** (2025) 012017 doi:10.1088/1755-1315/1525/1/012017

3. Results and Discussion

SPAC analysis provided a comprehensive dispersion curve spanning a frequency range of 7–45 Hz. The observed dispersion characteristics indicated a gradual increase in the shear wave velocity with depth (Figure 4). The fundamental mode of Rayleigh wave dispersion was extracted from the recorded seismic data, and the inversion results successfully delineated the subsurface velocity structure. At shallower depths, the Vs values ranged from approximately 200 to 400 m/s, suggesting the presence of loose sediments and clay-rich deposits. As the depth increased, beyond 10 m, the Vs values steadily rose, reaching 1000 m/s at greater depths, indicative of more compacted layers, such as weathered rock or stiff clay formations.

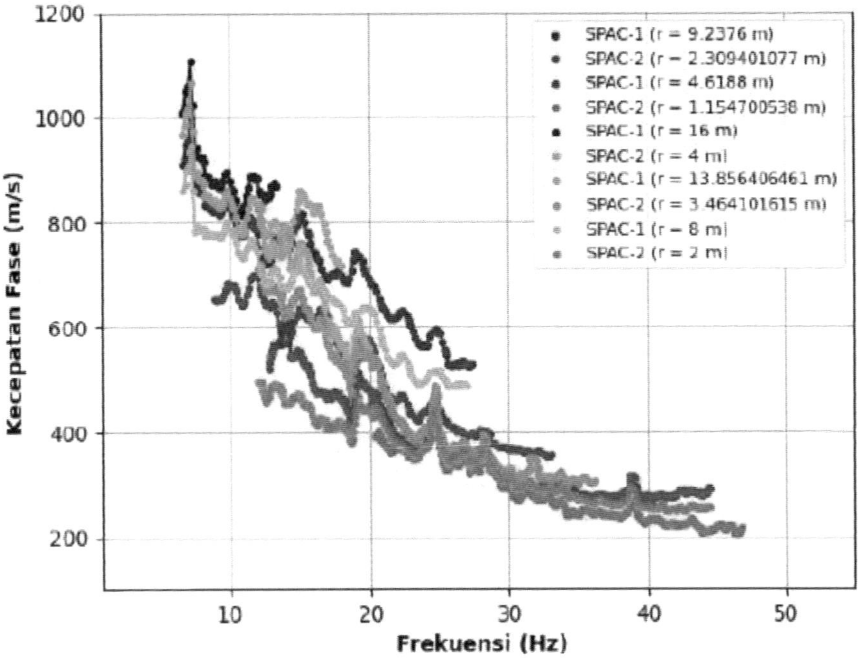

Figure 4. Observed dispersion curve or phase velocity variation from microtremor records using SPAC method.

The extracted dispersion curves (Figure 4) for different radii highlight a consistent trend of increasing phase velocity with depth. The coherence of the SPAC coefficient over various distances reinforced the reliability of the obtained dispersion data. The Bessel function fitting process also demonstrated a high degree of correlation, ensuring that the extracted Vs profile accurately represented subsurface characteristics.

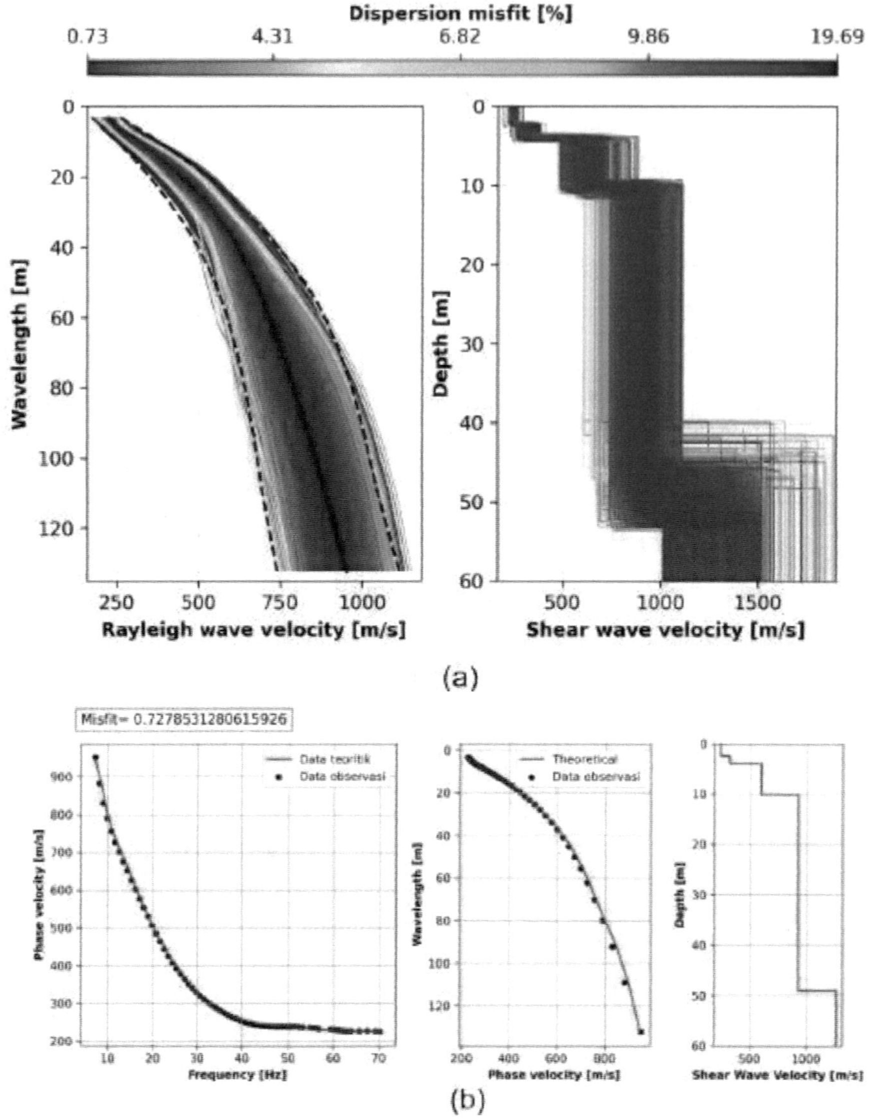

Figure 5. a) Inverted Vs profile, b) comparison between the observed and the theoretical phase velocity from the lowest misfit model.

Primarily, the findings suggest that the SPAC technique grants the ability to acquire the velocities of the subsurface with exceptional accuracy. The coming endeavors would have the option to add some other ones like passive seismic methods, for example, the Horizontal-to-Vertical Spectral Ratio (HVSR), for a more detailed creation of the velocity, the interpretation itself, and further depth-dependent interpretation.

4. Conclusion

This study's outcomes show that the Spatial Autocorrelation (SPAC) approach successfully identifies the deeper subsurface layers compared to conventional active seismic methods. The ability to use ambient noise in seismic wave analyses for passive surface work unlocks the potential of that tool for investigating earthquake shaking and damage, especially in urban areas and other uneconomic or forbidden spots for the application of active sources. As a result of extracting the fundamental information from seismic surface waves, SPAC yields sufficient details on the matter, providing the first self-standing information on the amount of energy loss in the layers due to wave propagation. In addition, there are many differences between surface waves.

The interpretation of surface-wave data is a key aspect of site characterization and earthquake (seismic) hazard assessment. SPAC is well-positioned to resolve the depth-to-the-rock interface and the phase velocity dispersion across the whole frequency spectrum, allowing for a more detailed and continuous representation of the subsurface stratigraphy. Similarly, this work mode is powerful for engineers and geotechnical specialists to distinguish between compaction layers and unconsolidated sediments based on the data collected. Moreover, the approach provides a basis for a further and directly associated development of borehole lithological features and SPAC-derived information. The strong agreement between the generated profiles and borehole lithological data further confirmed the SPAC method's stability.

Moreover, the non-invasive SPAC method helps to do away with the logistic constraints that come with active-source surveys and, in the process, reduce environmental disturbances, simultaneous with being direct about the subsurface model. The strengths of this technology make it especially appropriate for monitoring, which includes seismic microzonation studies, infrastructure development, and hydrogeological assessments. This is followed by integrating SPAC with other passive seismic methods, for example, the Horizontal-to-Vertical Spectral Ratio (HVSR), which will permit more precise subsurface characterization and better modeling accuracy.

By integrating more complex arrays and the most recent inversion techniques, which will raise the subsurface velocity model resolution, researchers can utilize the SPAC platform in many geological environments in the future. The ongoing utilization of machine learning algorithms as tools for automated SPAC analysis could also lead to more reliable and faster data processing, making the method even more usable in geophysical investigations. Generally, SPAC is very much in demand for developing geophysical tools and procedures to provide the necessary information for identifying the situation on the site, reducing seismic risks, and planning infrastructure.

References

[1] H. Okada, "Theory of Efficient Array Observations of Microtremors with Special Reference to the SPAC Method," Exploration Geophysics, vol. 37, pp. 73–85, 2006, doi: 10.1071/EG06073.

[2] I. Cho, T. Tada, and Y. Shinozaki, "A generic formulation for microtremor exploration methods using three-component records from a circular array," Geophys J Int, vol. 165, pp. 236–258, 2006, doi: 10.1111/J.1365-246X.2006.02880.X.

[3] M. Asten, "Site Shear Velocity Profile Interpretation From Microtremor Array Data by Direct Fitting Of Spac Curves," pp. 1069–1080, 2006.

[4] M. Wathelet, D. Jongmans, and M. Ohrnberger, "Direct Inversion of Spatial Autocorrelation Curves with the Neighborhood Algorithm," Bulletin of the Seismological Society of America, vol. 95, pp. 1787–1800, 2005, doi: 10.1785/0120040220.

[5] Q. Qin, L. Wang, Y. Luo, C. Chen, A. Bian, and X. Song, "Direct Inversion of Multi-mode Rayleigh-wave Spatial Autocorrelation Coefficients from High-Frequency Ambient Noise," GEOPHYSICS, 2024, doi: 10.1190/geo2023-0704.1.

[6] I. Cho, T. Tada, and Y. Shinozaki, "Assessing the applicability of the spatial autocorrelation method: A theoretical approach," J Geophys Res, vol. 113, 2008, doi: 10.1029/2007JB005245.

[7] M. Asten, "On bias and noise in passive seismic data from finite circular array data processed using SPAC methods," Geophysics, vol. 71, 2006, doi: 10.1190/1.2345054.

[8] H. Nakahara, "Formulation of the spatial autocorrelation (SPAC) method in dissipative media," Geophys J Int, vol. 190, pp. 1777–1783, 2012, doi: 10.1111/J.1365-246X.2012.05591.X.

[9] A. Kocaog˘lu and K. Fırtana, "Estimation of shear wave velocity profiles by the inversion of spatial autocorrelation coefficients," J Seismol, vol. 15, pp. 613–624, 2009, doi: 10.1061/41095(365)88.

Geochemical Trends of CaO and MgO in Marine Carbonates: Insights from XRF Analysis

Erfan Syamsuddin[1,3], Adi Maulana[2,3], Ulva Ria Irfan[2,3*], and Muhammad Alimuddin Hamzah Assegaf[1]

[1] Department of Geophysics, Hasanuddin University, Makassar, Indonesia
[2] Department of Geological Engineering, Hasanuddin University, Makassar, Indonesia
[3] Study Program of Earth and Environmental Technology, Faculty of Engineering, Hasanuddin University, Makassar, Indonesia

[*] E-mail: ulvairfan@unhas.ac.id

Abstract. The analysis of calcium oxide (CaO) and magnesium oxide (MgO) concentrations in marine carbonates using X-ray fluorescence (XRF) offers important new perspectives on depositional settings and diagenetic processes. The geochemical patterns of CaO and MgO in limestones and dolostones were investigated using XRF spectrometry to determine the primary depositional settings and establish linkages between the oxide ratios and sedimentological features. XRF tests reveal a negative connection between CaO and MgO values, implying dolomitization over time from stratigraphic and geographical changes in samples obtained from a marine sedimentary basin in Southeast Sulawesi, Indonesia. Dolomitized samples exhibit lower CaO values (30.0–42.5%), and rising MgO concentrations (4.5–15.5%), whereas limestone samples show high CaO concentrations (45.0–54.5%). Stratigraphic changes in the CaO-MgO ratios indicate either limited dolomitization in lower units, suggestive of open sea deposition, or elevated dolomitization in higher intervals, presumably arising from Mg-rich diagenetic fluids. Analyzing CaO-MgO ratios concerning other oxides (SiO_2, Al_2O_3) helps to better understand depositional habitats; larger Ca/Mg ratios imply shallow marine settings and lower ratios indicate confined environments with major siliciclastic input. Utilizing stratigraphic and facies analyses combined with XRF-derived geochemical proxies, knowledge of carbonate deposition and subsequent post-depositional processes is enhanced, thus underscoring the interaction between main marine deposition and secondary diagenetic modification in the control of CaO-MgO distribution within carbonate successions.

Keywords: *CaO, MgO, Marine carbonates, XRF, Dolomitization, Depositional environments, Diagenesis*

1. Introduction

Reconstructing the depositional and diagenetic histories of the ocean requires limestones and dolostones. Found in numerous marine habitats ranging from shallow to deep seas, limestones—are mostly composed of calcium carbonate ($CaCO_3$). The diagenetic replacement of calcium carbonate by magnesium ions generates dolostones with a composition mostly of magnesium carbonate ($MgCO_3$). The geochemical composition of these carbonates, especially the calcium oxide (CaO) to magnesium oxide (MgO) ratio, is a crucial indicator for interpreting depositional conditions, seawater chemistry, and diagenetic processes. A consistent method for measuring main and trace elements is X-ray fluorescence (XRF) spectroscopy, hence improving geological interpretations [1].

Understanding marine carbonates largely depends on the CaO/MgO ratio, which provides essential details on the original depositional conditions and later diagenetic changes. Main deposition in low-energy basins, open lagoons, low magnesium input, or other marine ecosystems marked by calcium carbonate precipitation frequently falls at high CaO/MgO ratios [2]. On the other hand, high Mg-rich fluids passing over carbonate deposits during diagenesis enable partial or total dolomitization—a process seen by elevated MgO levels [3]. Petrographic and mineralogical studies have helped to clarify the variations between main depositional factors, including seawater Mg/Ca ratios and biological inputs, and secondary processes, such as fluid-rock interactions and recrystallization, so improving the analysis of CaO and MgO concentrations in sedimentary environments [4].

Even with the progress of geochemical methods, linking CaO/MgO ratios to depositional processes offers many challenges. Much earlier research has focused on dolomitization and diagenetic alterations without regard for the starting conditions of seawater chemistry or mechanisms encouraging carbonate precipitation [5], [6]. For example, while studies have shown that XRF spectrometry is useful in establishing important oxide compositions such as CaO and MgO [7], the degree to which geochemical trends precisely recreate original depositional contexts remains underexplored. In mixed carbonate-siliciclastic depositional systems especially, concealed geochemical signals include terrigenous inputs and detrital pollutants [4], [8]].

This study aimed to examine the geochemical patterns of CaO and MgO in marine carbonates, emphasizing using XRF analysis to identify depositional environments and clarify the correlations between oxide ratios and sedimentological factors. This study utilizes XRF data to close an important knowledge gap on the impact of seawater chemistry, depositional energy, and diagenetic processes on the geochemical fingerprint of marine carbonates. It offers a complete framework to explain the relevance of the CaO/MgO ratio in marine sedimentary systems of the study area.

2. Methods

Using X-ray fluorescence spectroscopy (XRF), the geochemical composition of marine carbonate samples was investigated with a particular focus on CaO and MgO concentrations. This method derived diagenetic processes, including dolomitization, by describing the oxide variation of Ca and Mg elements with depth. Sixty-three carbonate samples were collected from well-defined carbonate layers in the study area, as shown in (Figure 1). The collection spots were delineated and displayed on a geological map, showing a systematic grid of sample sites inside a research area in Southeast Sulawesi, Indonesia. The sample sites, identified by alphanumeric codes (e.g., GE-272 and GE-730), were systematically distributed to provide comprehensive geographic coverage of the limestone lithologies of the region.

The sampling method sought to include notable spatial and stratigraphic changes. Samples were selected based on field observations to reduce the impact of weathering and diagenesis, concentrating primarily on typical limestones and partially or fully dolomitized carbonate rocks. The samples were carefully prepared to ensure homogeneity and minimize contamination before XRF analysis. Before undergoing geochemical analysis, the samples were prepared using standard protocols. These processes included creating fusion beads rather than pellet pressing and drying the materials at 1005°C to remove moisture. Fusing the material involved mixing one gram with nine grams of a 50:50 mix of lithium metaborate (LiM) and lithium tetraborate (LiT).

For the XRF tests, they melted the material at 1005°C to make a uniform glass bead. This approach reduced the influence of extraneous elements, ensuring steady XRF data. They configured the XRF equipment using verified carbonate reference materials to make it run more precisely and consistently.

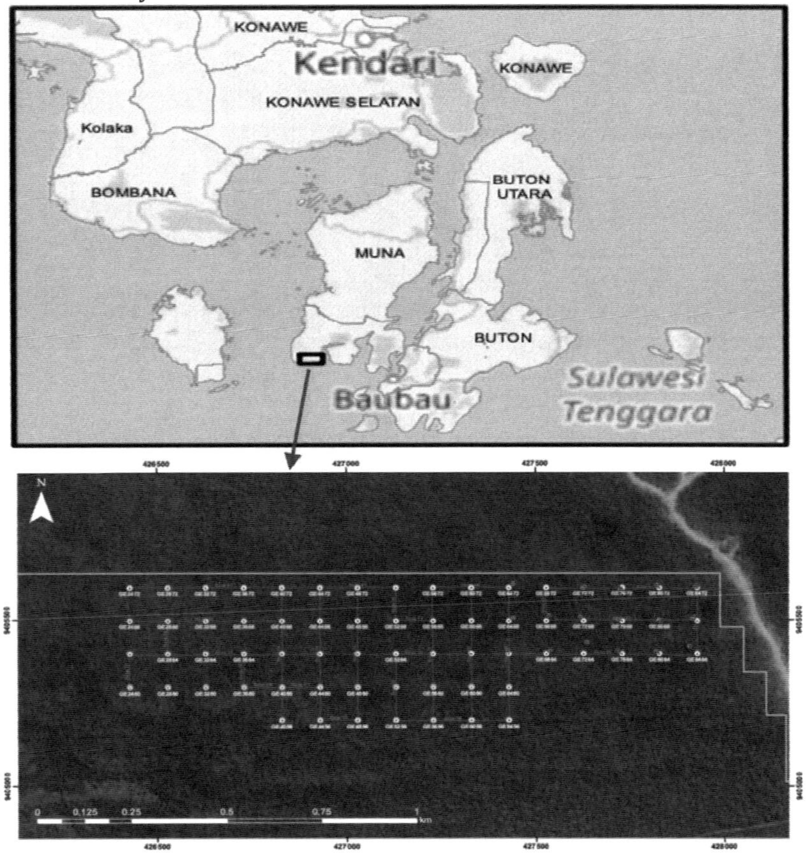

Figure 1. Sample collection locations in Southeast Sulawesi, Indonesia

Operating with a high-resolution silicon drift detector for exact elemental measurement, an Epsilon 4 XRF spectrometer (Malvern Panalytical) performed XRF experiments. Carbonate matrices' instrument parameters were changed to suitably record changes in CaO and MgO concentrations by different excitation levels. The quality control measures used on the obtained geochemical data were duplicate analysis and reference standard comparisons. Data processing with baseline adjustments and normalization strategies comes first, considering possible matrix effects.

Geochemical patterns were interpreted using statistical analysis after data collection. The negative correlation between the CaO and MgO concentrations was investigated using correlation analyses, verifying the dolomitization tendency in earlier work. Histograms and depth-based variation graphs were created to further show the oxide distributions and pinpoint important geochemical transitions. Whereas the frequency distributions of these oxides were determined using histograms, the inverse connection between CaO and MgO was investigated using scatter plots. Examined to provide further understanding of the diagenetic processes were depth-dependent fluctuations in CaO and MgO. These studies helped to identify many carbonate facies and offered an understanding of depositional conditions and diagenetic overprinting. Combining XRF-derived geochemical data with sedimentological interpretations helps us to better grasp dolomitization mechanisms in the research region and marine carbonate development.

3. Results and Discussion

The scatter plot of CaO vs MgO (Figure 2) in marine carbonates demonstrated a robust negative correlation (R^2=0.94), indicating a clear inverse link between the two oxides. As CaO diminishes and MgO escalates, this trend exemplifies the dolomitization process, indicating a progressive calcium substitution by magnesium within the carbonate matrix. The equation of the fitted trend line (y=-0.98x+54.48) supports this substitution, by established ideas of diagenetic alteration in carbonate environments. The results derived from XRF analysis allowed us to ascertain likely depositional settings based on the detected geochemical variations. Elevated CaO concentrations above 44% imply the existence of pure limestone, whereas heightened MgO values beyond 14% indicate partial to full dolomitization, representing various phases of marine diagenesis.

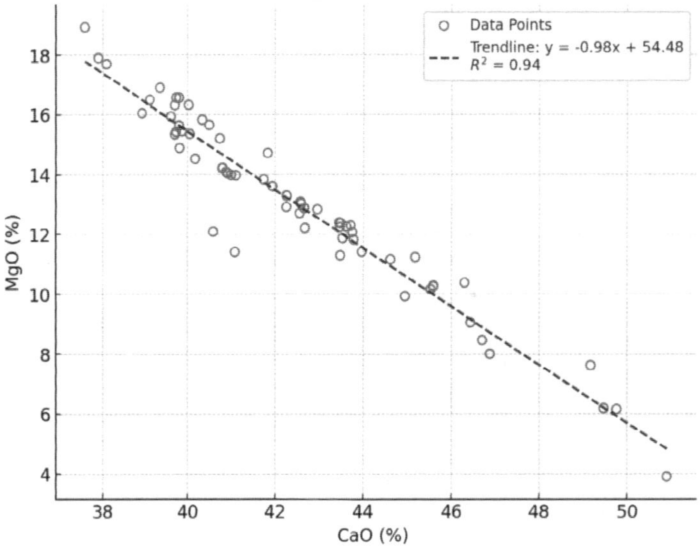

Figure 2. Inverse correlation between CaO and MgO in marine carbonates

The distribution histograms of CaO and MgO further reinforce these findings by illustrating the frequency of the oxide concentrations across the dataset. The CaO histogram shows a bimodal distribution, with peaks around 40-44% and a tail extending toward higher values (>48%), suggesting two dominant carbonate lithologies, possibly limestone, with varying degrees of dolomitization. Conversely, the MgO histogram exhibited a more unimodal distribution, with values clustering between 10-15%, typical of mixed limestone-dolomite facies. The observed trends imply that the carbonate sediments underwent differential diagenetic processes, where Mg-rich fluids selectively replaced Ca, forming dolomitic zones in specific depositional settings.

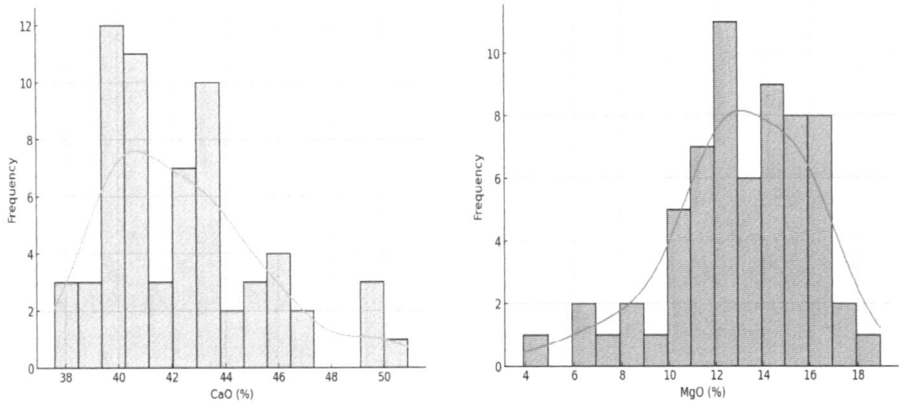

Figure 3. Frequency distribution of CaO and MgO in marine carbonates

This study's geochemical patterns provide significant insights into the sedimentological influences on carbonate deposition and diagenesis. The regular fluctuation of CaO and MgO with depth indicates a gradual tendency of dolomitization in deeper carbonate strata, perhaps associated with fluid-rock interactions and changes in depositional energy. These patterns correspond with existing geochemical theories of marine carbonates, whereby elevated MgO at higher depths is ascribed to extended exposure to magnesium-rich brines or early diagenetic modification. The results underscore the importance of geochemical proxies in recreating historical environmental conditions, validating the use of XRF analysis as a reliable approach for discerning carbonate diagenetic patterns. Subsequent research should include mineralogical evaluations and petrographic analysis to further substantiate these geochemical conclusions.

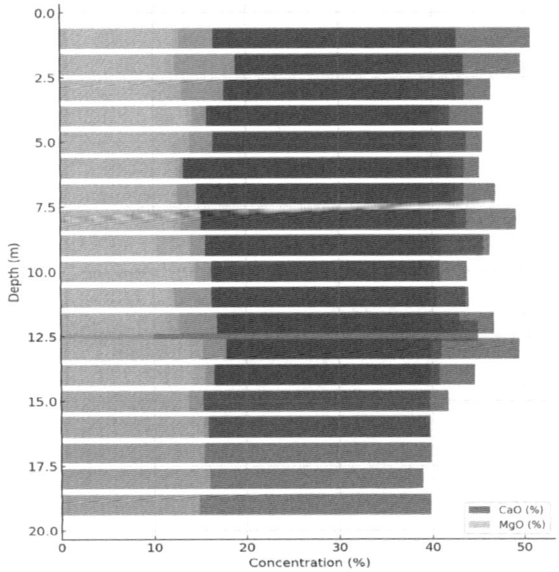

Figure 4. Depth-dependent variation of CaO and MgO concentrations in marine carbonates

The geochemical distribution adheres to traditional dolomitization models, whereby an increased MgO with depth signifies a more profound diagenetic change. Shallower carbonate strata with high CaO remain mostly unmodified, but deeper regions with increased MgO indicate enhanced dolomitization resulting from extended exposure to Mg-rich brines. The conversion from limestone to dolomite underscores the influence of diagenetic fluid interactions on marine carbonates. The top stratigraphic intervals exhibit gradual dolomitization, with Ca/Mg ratios below 10 associated with magnesium enrichment, presumably affected by limited marine environments or hypersaline brines. Comparable patterns were noted in the Darnah Formation [3] and Fulra Limestone [9].

Integrating depth-based geochemical trends with surface-level oxide distributions (Figures 2 and 3) provides a comprehensive understanding of these carbonates' depositional and post-depositional histories. Combining XRF-based geochemical profiling and statistical analysis enables a more detailed reconstruction of past environmental conditions, depositional settings, and diagenetic overprinting. Future research should incorporate petrographic microscopy and stable isotope analysis to further delineate the diagenetic pathways responsible for these geochemical trends. Such an approach would refine the interpretations of carbonate reservoir quality and provide insights into broader paleoenvironmental reconstructions.

The geochemical patterns observed in this study highlight the interplay between primary marine deposition and secondary diagenetic alteration that controls the CaO-MgO distribution within carbonate succession. Primary depositional factors, including seawater Mg/Ca ratios and depositional energy, influence the initial mineralogy (e.g., aragonite vs. calcite). In contrast, secondary dolomitization, evident from increased MgO concentrations in specific stratigraphic intervals, reflects fluid-rock interactions, likely associated with burial diagenesis or exposure to Mg-rich hypersaline conditions.

These results correspond with prior research highlighting basin-specific influences on dolomitization, including fluid movement paths and changes in sedimentary facies [2], [5]. Integrating CaO-MgO ratios with stratigraphic and facies analysis improves our capacity to recreate marine depositional environments and post-depositional histories, providing a more nuanced knowledge of carbonate diagenesis.

4. Conclusion

The XRF results provide significant quantitative evidence for dolomitization in marine carbonates, with robust CaO-MgO correlations reinforcing theories of incremental magnesium absorption. Linking these geochemical patterns to prediction models and examining anomalies or deviations in further detail will provide more comprehensive insights into depositional and diagenetic processes. Integrating isotopic, petrographic, and facies-derived information might enhance environmental and tectonic interpretations of Buton's carbonate systems.

References

[1] M. Wilson and D. Bosence, "Platform-top and ramp deposits of the Tonasa Carbonate Platform, Sulawesi, Indonesia," *Geological Society, London, Special Publications*, vol. 126, pp. 247–279, 1997, doi: 10.1144/GSL.SP.1997.126.01.16.

[2] B. Singh, V. Srivastava, and S. Kanhaiya, "Sedimentological and geochemical characteristics of the late middle Eocene dolostone succession, Kachchh, western India," 2018. doi: 10.1002/gj.3378.

[3] A. M. Elfeituri, I. M. A. El Leil, and S. S. El-Elkhfifi, "Mineralogical and Geochemistry Study of Darnah Formation in Wadi Al Kharsha, Northeast Libya," 2021. doi: 10.36811/ljbs.2021.110071.

[4] N. Samin, "Mineralogy and Geochemistry of Qulqula Limestone in Shenrui, Halabja Governorate, Northeastern Iraq," 2021. doi: 10.33899/earth.2021.170382.

[5] T. Taha and E. Abdullah, "Mineralogy and geochemistry of Mishrif Formation from selected oilfield, south east of Iraq," 2020. doi: 10.25130/j.v25i1.939.

[6] R. A. Agliullina, Z. V Akhmetzyanova, G. V Ivanov, V. D. Sitdikov, A. Nikolaev, and A. V Malinin, "Features of Elemental Analysis of Dolomites of the Lower Permian Deposits By X-Ray Fluorescence Spectroscopy," 2024. doi: 10.17122/ngdelo-2024-5-139-148.
[7] Jamaluddin, A. Darwis, and M. A. Massinai, "X-Ray Fluorescence (XRF) to identify chemical analysis of minerals in Buton island, SE Sulawesi, Indonesia," *IOP Conf Ser Earth Environ Sci*, vol. 118, 2018, doi: 10.1088/1755-1315/118/1/012070.
[8] P. R. Ikhane, O. O. Oyebolu, and A. Alaka, "X-Ray Fluorescence and Stable Isotopes Analyses of Ikpeshi Marble, South Southern Nigeria," vol. 10, pp. 102–113, 2020, doi: 10.46881/ajsn.v10i0.183.
[9] V. K. Srivastava and B. P. Singh, "Depositional environments and sources for the middle Eocene Fulra Limestone Formation, Kachchh Basin, western India: Evidences from facies analysis, mineralogy, and geochemistry," 2019. doi: 10.1002/gj.3154.

IOP Conf. Series: Earth and Environmental Science **1525** (2025) 012019 doi:10.1088/1755-1315/1525/1/012019

Modeling The Tsunami Potential of Mt. Awu Pyroclastic Density Currents

Sri Wahyunia[1*], Amiruddin[1] and Saaduddin[1]

[1] Department Geophysics, Hasanuddin University, Makassar, Indonesia

*E-mail: sriwahyunia2000@gmail.com

Abstract. Mount Awu has a history of tsunami generation due to its pyroclastic flows. This fact reveals that it could regenerate the tsunami in the future. Pyroclastic flows comprise dilute and dense pyroclastic density currents (PDC). Dense PDC flows are considered to have properties similar to those of landslides. Therefore, the modeling applied a similar principle to landslide-induced tsunami simulations. Simulations aimed to model the potential run-up and the run-out distance of tsunami waves generated by the dense PDC Mount Awu. The modeling applied shallow water equation using Cornell Multi-grid Coupled Tsunami Grid (COMCOT). The simulation resulted in the tsunami height reaching 0.08–4.01 meters. Further tsunami waves reach the west coast of Sangihe Besar Island and several small islands around Mt. Awu. This study provides significant benefits in disaster mitigation for those who have authority over Mount Awu and its vicinity.

1. Introduction

Indonesia is located in the subduction zone among 3 tectonic plates, namely Eurasia, Pacific, and Indo-Australia. Consequently, there are many active volcanoes, most of which are scattered on volcanic islands and have been active since ancient times [1]. When a volcano erupts, the material can cause following hazard, such as landslides, debris avalanches, and tsunamis [2]. The latest hazard is highly likely to occur in Indonesia because it is an archipelagic country. Volcanic tsunamis are generated by a variety of mechanisms as a result of volcanic activities, such as underwater explosions, caldera collapse, blast, subaerial and submarine landslide, and pyroclastic flows [3].

Mount Awu has a history of tsunami generation. It is an active stratovolcano in the northernmost part of Sangihe Besar Island, Sangihe Islands Regency, North Sulawesi Province (Figure 1). The formation of a subduction zone due to the interaction between the Maluku Sea Plate, Sangihe Arc, and Halmahera Arc affects the activity of Mount Awu [4]. In addition, the existing geological structures, such as faults, calderas, craters, and volcanic alignments, increase the Mount Awu activity [5].

The eruption of Mount Awu has been recorded in history since the 1640s, and its activity continues to be monitored until today [4]. From the historical records, two events resulted in tsunamis. The former tsunami is in 1856 due to pyroclastic flows [1] with a tsunami magnitude of 2 [6]. The latter is in 1892 due to a pyroclastic surge that produced a tsunami with a run-up of 0.75 meters [7] and a tsunami magnitude of 1 [6]. Based on this fact, this study was conducted to model the potential run-up and the run-out distance of tsunami waves generated by the dense pyroclastic density current of Mount Awu.

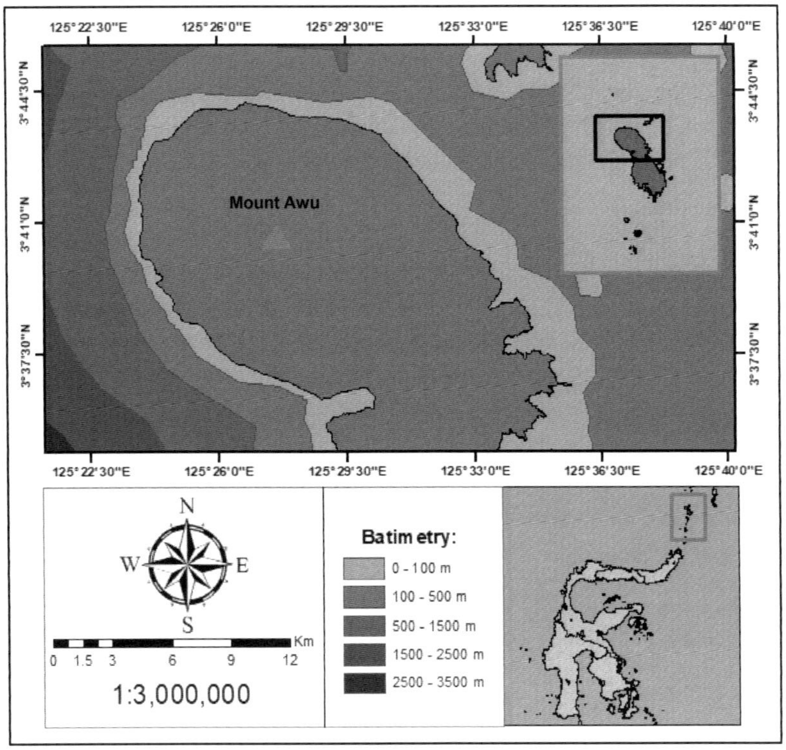

Figure 1. Location of Mount Awu

2. Methods

2.1 The Cornell Multi-grid Coupled Tsunami (COMCOT)

The Cornell Multi-grid Coupled Tsunami or COMCOT model can be used to solve simple linear and non-linear shallow water equations. The COMCOT model serves to describe the propagation and run-up of simulated tsunami waves [8]. The shallow water equations in COMCOT will be formulated in the form of free surface displacement and volume flux [9]. In addition, the COMCOT model for solving shallow water equations is applied using spherical and cartesian coordinates [8]. In this study, a non-linear shallow water equation that includes bottom friction effects with a spherical coordinate system is used to describe the tsunami wave movement process [10]. This is because the tsunami simulation will be conducted close to the coastal area. So, tsunami waves approaching the coastal area will have a larger amplitude with a smaller wavelength, referred to as the shoaling effect [11].

2.2 Geometry of Mount Awu for COMCOT Block Model

The selection of dense pyroclastic density current of Mount Awu as a tsunami source in this simulation because it has the same simulation principle of landslide-induced tsunami [12]. The Mount Awu block model in this simulation is created by the result of reviewing literatures including geological properties, eruption history, disaster-prone areas, and the surface morphology. Samodra (1994) found the pyroclastic flow deposits of Mount Awu in the coastal area of Mitung. Supported by historical eruption records showing that areas such as Mitung, Angges, and other coastal regions have previously been impacted by pyroclastic flows [13]. Additionally, disaster-prone area assessments indicate that Mitung and Angges have the potential to be affected by lahar flows, pyroclastic flows, lava flows, and incandescent lava falls (Figure 2a).

From the surface morphology, it is evident that a steep cliff southeast of Mount Awu's crater restricts the movement of pyroclastic flows (shown by the red line in Figure 2b), which will later redirect and flow southward, impacting areas like Mitung and Angges. Given these facts, the Mitung and Angges areas will be used as the simulated block model region. The block model requires length, width, and thickness parameters to represent the characteristics of PDC. The PDC volume used in this study refers to a range of approximately $\sim 10^4$ - 10^8 m^3, as reported in previous research [14]. The study by Shimizu et al. (2019) indicated that dense PDC, characterized by a high particle concentration, accounts for approximately 40–50% of the total PDC volume [15]. Additionally, the length and thickness of dense PDC range from 10^3 - 10^5 m [14] and 0,01–10 m [16]. For the simulation, a length of 4000 m was applied, based on the PDC flow path from Mount Awu to the coastal area, along with a thickness of 10 m. Considering the assumed total PDC volume of 10^8 m^3 and the proportion of dense PDC at 50%, the model width in the simulation was calculated as 1250 m using the block volume formula (Figure 2c).

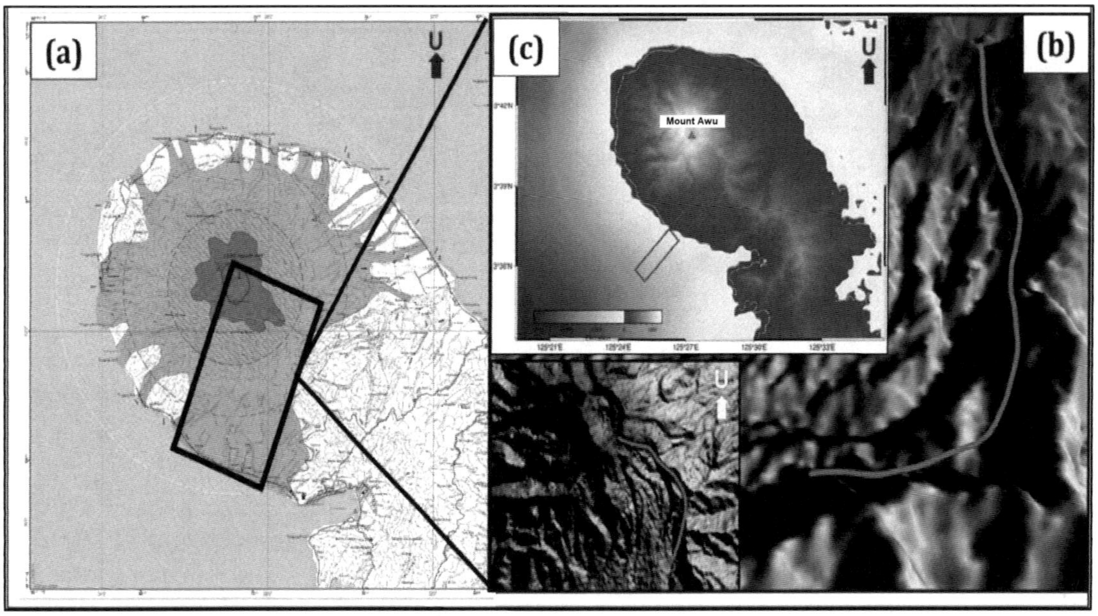

Figure 2. (a) Mount Awu Disaster Prone Area Map [17] , (b) Mount Awu Surface Morphology [18], and (c) COMCOT Block Model for Tsunami Simulation Due to Dense PDC Mount Awu

3. Results and Discussion

3.1 Location of Tide Stations (Ts) and Tsunami Travel Time of Mount Awu

The placement of tide stations (Ts) around Sangihe Besar Island is divided into three areas (Figure 3). First, five Ts in the black box are positioned on small islands south of Sangihe Besar Island. Second, nine Ts in the yellow box are distributed along the western coast of Sangihe Besar Island, with some located close to the tsunami source. Third, the red box contains five Ts placed along the northern to eastern coast and on a small island near Sangihe Besar Island.

Figure 3(b) shows the travel time of the first tsunami wave, influenced by the distance of the tide stations (Ts) from the tsunami source, ranging from 1.49 to 14.58 minutes. These values are derived from the Ts stations closest and farthest from the tsunami source, namely Ts-12 at Angges Coast and Ts-19 at Nusa Tabukan Dua II (Figures 4, 5, and 6).

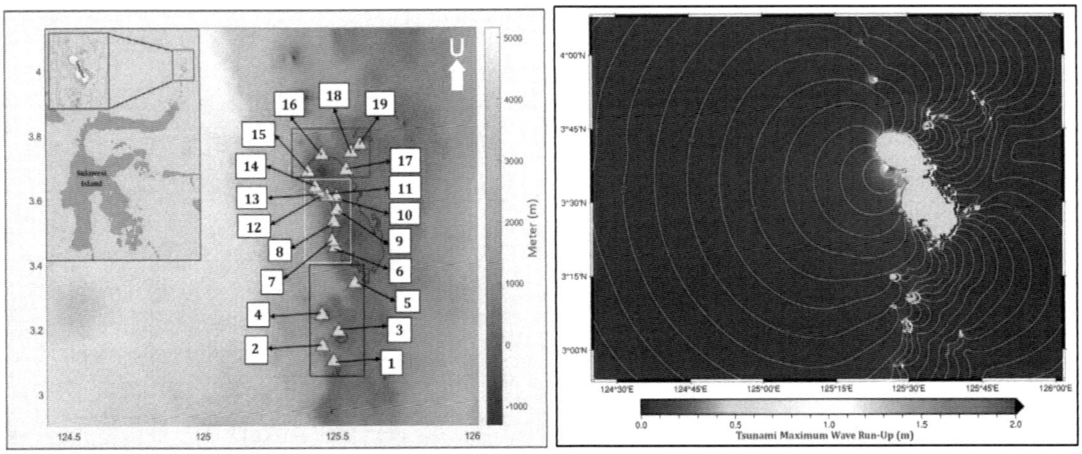

Figure 3. (a) Location of Tide Stations (Ts) of Mount Awu, and (b) Tsunami Travel Time of Mount Awu

3.2 Maximum Run-Up Of Tsunami Waves Recorded At Each Tide Station (Ts)

3.2.1 Tide Stations (Ts) in Black Box Area

Five tide gauge stations were placed at Ts-1 on Salingkere Island, Ts-2 on Mahengetang Island, Ts-3 on Kahakitang Island, Ts-4 on Kamala Island, and Ts-5 on Bebalang Island, as shown in the black box (Figure 3a). The recorded tsunami run-up potential at these stations ranged from 0.120 to 0.67 meters, with arrival times between 12.4 and 22.5 minutes (Figure 4). The highest tsunami wave was recorded at Ts-4 on Kamala Island.

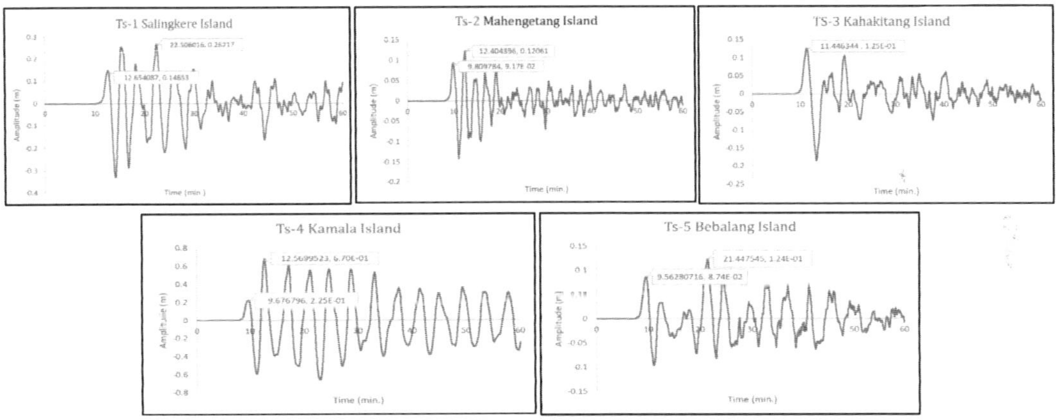

Figure 4. Tide Stations (Ts) in Black Box Area

3.2.2 Tide Stations (Ts) in Yellow Box Area

Nine tide stations (Ts) were placed along the western coast of Sangihe Besar Island, recording potential run-up of tsunami waves around 0.193 to 4.010 meters, with arrival times between 3.2 and 19.1 minutes. Several locations, such as Barangkalang, Angges, Mitung, Kolongan Akembawi, and Tahuna Bay, reported wave heights exceeding 1 meter, posing significant risks of casualties, infrastructure damage, and economic and environmental impacts. As the capital of the Sangihe Islands Regency, Tahuna, located near Tahuna Bay, is particularly vulnerable to severe damage in the event of a large tsunami. The coastal area from Angges to Kolongan Akembawi faces the highest risk due to its proximity to the tsunami source.

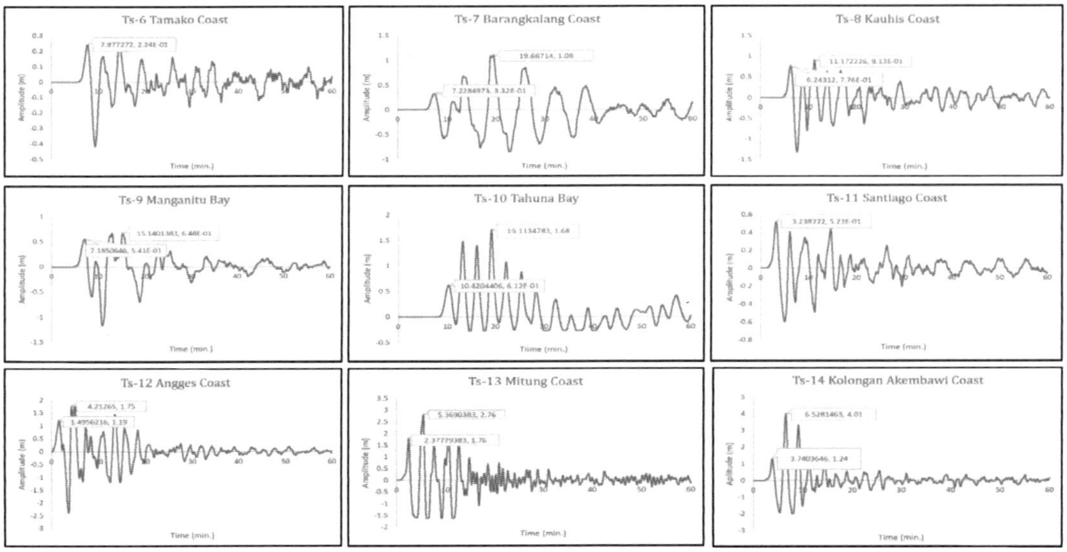

Figure 5. Tide Stations (Ts) in Yellow Box Area

3.2.3 Tide Stations (Ts) in Red Box Area

Five tide stations in Ts-15 Kendahe I Coast, Ts-16 Kendahe II Coast, Ts-17 Naha Airport, Ts-18 Nusa Tabukan I, and Ts-19 Nusa Tabukan II, positioned within the marked red box, recording potential run-up of tsunami waves around 0.080 to 0.593 meters, with arrival times between 7.7 and 31.5 minutes (see Figure 6). Simulation results indicate that only the Kendahe I Coast area is likely to be affected by a tsunami, while other areas are minimally impacted in the event of a tsunami triggered by Mount Awu's activity.

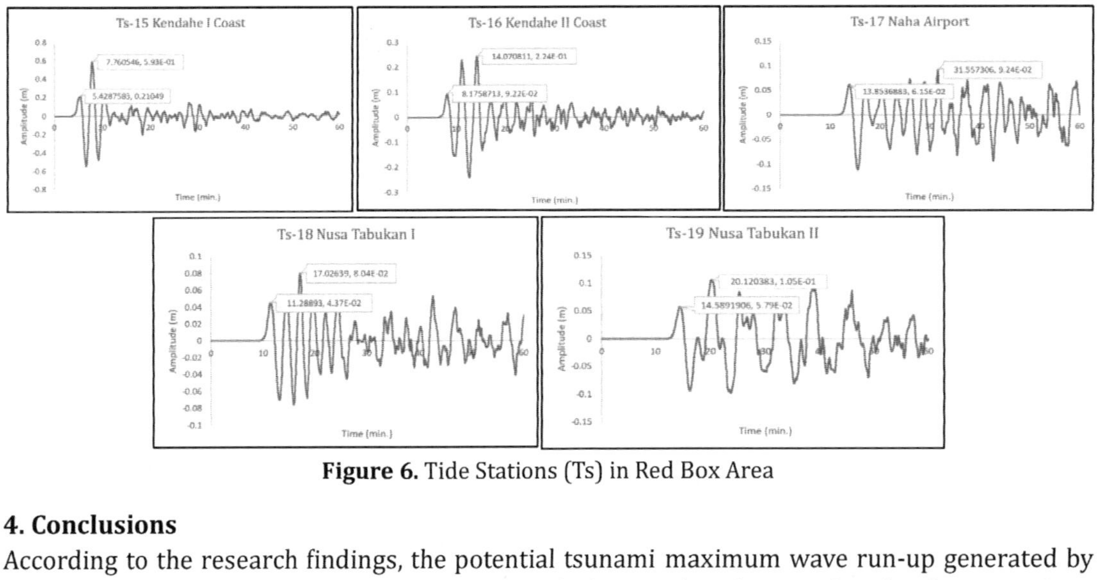

Figure 6. Tide Stations (Ts) in Red Box Area

4. Conclusions

According to the research findings, the potential tsunami maximum wave run-up generated by dense PDC Mount Awu recorded at tide stations (Ts) around in the Sangihe islands ranges from 0.08 to 4.01 meters (see Figure 3(b)). The lowest tsunami run-up of 0,08 meters was recorded at Ts-18 Nusa Tabukan I with an arrival time of 17 minutes. while the highest run-up of 4,01 meters with an arrival time of 6.5 minutes was recorded at Ts-14 Kolongan Akembawi, located close to the tsunami source. In addition, the resulting tsunami has the potential to reach several west and north coastal areas of Sangihe Besar Island, including Barangkalang, Angges, Mitung, Kolongan Akembawi, Tahuna Bay, and Kendahe I. and the small island nearby, Kamala Island.

References

[1] Hidayat A, Marfai M A and Hadmoko D S 2020 Eruption on Indonesia's volcanic islands: a review of potential hazards, fatalities, and management *IOP Conf. Ser. Earth Environ. Sci.* **485**

[2] Brown S K, Loughlin S C, Sparks R S J, Vye-Brown C, Barclay J, Calder E, Cottrell E, Jolly G, Komorowski J-C, Mandeville C, Newhall C G, Palma J L, Potter S and Valentine G 2015 Global volcanic hazard and risk *Global Volcanic Hazards and Risk* (Cambridge University Press) pp 81–172

[3] Paris R, Switzer A D, Belousova M, Belousov A, Ontowirjo B, Whelley P L and Ulvrova M 2014 Volcanic tsunami: A review of source mechanisms, past events and hazards in Southeast Asia (Indonesia, Philippines, Papua New Guinea) *Nat. Hazards* **70** 447–70

[4] Dewanto B G, Rachmat H, Kriswati E, Yanis M, Shomim A F and Julius A M 2023 Study of geothermal and volcanic activity at Mount Awu , the deadliest active volcano in North Sulawesi Province , Indonesia using Optical Satellite Imagery *J. Volcanol. Geotherm. Res.* **438** 1–14

[5] PVMBG 2014 *Gunung Awu, Pulau Sangir, Sulawesi Utara*

[6] Hamzah L, Puspito N and Imamura F 2000 Tsunami Catalog and Zone Indonesia *J. Nat. Disaster Sci.* **22** 25–43

[7] Mutaqin B W, Lavigne F, Hadmoko D S and Ngalawani M N 2019 Volcanic Eruption-Induced Tsunami in Indonesia: A Review *IOP Conf. Ser. Earth Environ. Sci.* **256**

[8] Harig S, Zamora N, Gubler A and Rakowsky N 2022 Systematic Comparison of Tsunami Simulations on the Chilean Coast Based on Different Numerical Approaches *GeoHazards* **3** 345–70

[9] Dresback K M, Kolar R L and Dietrich J C 2005 On the form of the momentum equation for shallow water models based on the generalized wave continuity equation *Adv. Water Resour.* **28** 345–58

[10] Liu P L F, Cho Y S, Yoon S B and Seo S N 1995 Numerical Simulations of the 1960 Chilean Tsunami Propagation and Inundation at Hilo, Hawaii *Tsunami: Progress in Prediction, Disaster Prevention and Warning* (Springer, Dordrecht) pp 99–115

[11] Wang X and Power W L 2011 *COMCOT : a tsunami generation propagation and run-up model* (GNS Science)

[12] Battershill L, Whittaker C N, Lane E M, Popinet S, White J D L, Power W L and Nomikou P 2021 Numerical Simulations of a Fluidized Granular Flow Entry Into Water: Insights Into Modeling Tsunami Generation by Pyroclastic Density Currents *J. Geophys. Res. Solid Earth* **126** 1–21

[13] Samodra H 1994 Peta Geologi Lembar Sangihe dan Siau, Sulawesi

[14] Roche O, Phillips J C and Kelfoun K 2013 Pyroclastic density currents *Modeling Volcanic Processes* (Cambridge University Press) pp 203–29

[15] Shimizu H A, Koyaguchi T and Suzuki Y J 2019 The run-out distance of large-scale pyroclastic density currents: A two-layer depth-averaged model *J. Volcanol. Geotherm. Res.* **381** 168–84

[16] Jones T J, Beckett F, Bernard B, Breard E C P, Dioguardi F, Dufek J, Engwell S and Eychenne J 2023 Physical properties of pyroclastic density currents: relevance, challenges and future directions *Front. Earth Sci.* **11** 1–21

[17] Kriswati E, Primulyana S and Dirasutisna S 2016 Volcanic Hazard Map Of Awu Volcano, North Sulawesi Province

[18] Badan Informasi Geospasial 2024 Demnas Gunung Awu https://tanahair.indonesia.go.id/portal-web/unduh/demnas

IOP Conf. Series: Earth and Environmental Science **1525** (2025) 012020 doi:10.1088/1755-1315/1525/1/012020

How could the Haya Slide in the Makassar Strait threaten the coastal population in its vicinity?

Saaduddin[1*], Husnul Mar'iah[1], and Sri Wahyunia[1]

[1] Department of Geophysics, Hasanuddin University, Makassar, Indonesia

*E-mail: saad@sci.unhas.id

Abstract. The Haya slide is identified as a mass-transport and complex (MTC) through seismic reflection surveys in the Makassar Strait, represents a significant geological hazard. The downslope movement of this mass deposit poses the potential to generate submarine landslide-induced tsunamis, which could threaten coastal populations along Sulawesi's west coast and Kalimantan's east coast. We modeled using the Cornell Multi-grid Coupled (COMCOT) to evaluate the tsunami potential, focusing on simulating wave heights near the source and in densely populated surrounding areas. The simulation results reveal that tsunami wave heights are 3.809 meters at 1 km from the landslide source. Along the western coast of Sulawesi, wave heights are projected to reach approximately 1.325 meters, while on Kalimantan's eastern coast, wave heights may reach 0.048 meters. These variations in wave height highlight the significant impact proximity and coastal orientation can have on tsunami risk. This preliminary study provides crucial insights into the potential tsunami hazards associated with the Haya slide, offering a foundational basis for disaster mitigation strategies to protect vulnerable coastal regions from future landslide-triggered tsunami events.

1. Introduction

Landslide-induced tsunamis are categorized into two primary mechanisms, including subaerial landslides occur due to the displacement of surface materials, such as caldera collapses and pyroclastic flows [1]. These events are influenced by several factors, including the volume of the material, initial acceleration, and maximum velocity (Figure 1a) [2]. Conversely, submarine landslides result from the movement of materials underwater along a consistent slope. The size of these landslides significantly affects the wave period and amplitude of the resulting tsunami (Figure 1b) [3].

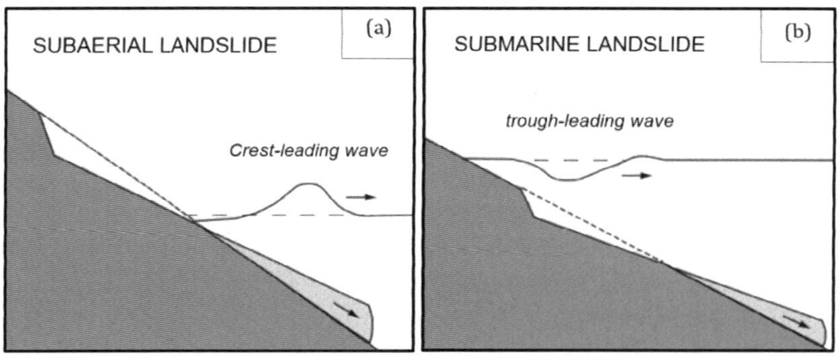

Figure 1. Tsunami Generation Mechanisms from (a) Subaerial and (b) Submarine Landslides [3]

The area with potential for tsunami generation due to submarine landslides is the Makassar Strait, located between Sulawesi Island and Borneo. This region consists of the northern and southern Makassar basins, which play a critical role in sedimentation processes within the area [1]. Previous studies using seismic refraction methods in the Makassar Strait have revealed the presence of complex material transport deposits (MTC), which have been named the Haya Slide (Figure 2) [4]. MTCs typically form in marine or freshwater environments, particularly along slopes where the continental crust transitions into deeper waters. In the southern Makassar Strait, MTCs are characterized by soft sediment layers that facilitate mass transport processes, potentially triggering submarine landslides [5]. These landslides can generate tsunamis, posing significant risks to human lives and infrastructure along the coastal areas near the landslide source [6]. Given the potential impact, this study aims to estimate tsunami hazards, focusing on wave height simulations near the source and in densely populated coastal areas.

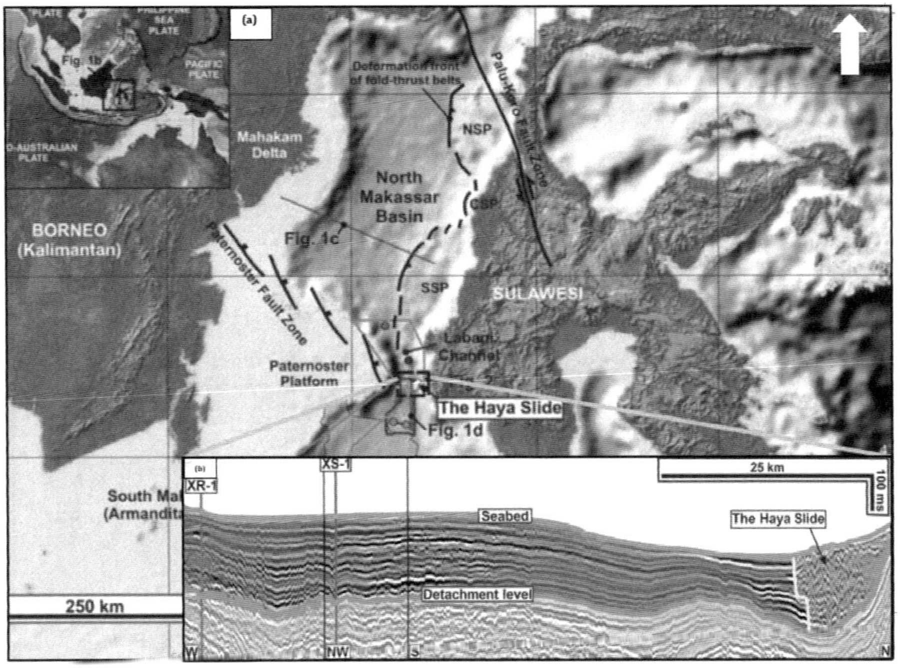

Figure 2. Endapan sedimen di daerah longsoran Haya [4]

2. Method

2.1 Modelling Using COMCOT

Advances in tsunami modeling, including both mathematical and numerical approaches, have significantly enhanced the understanding of tsunami hazards [7]. COMCOT (Cornell Multi-grid Coupled Tsunami model) is a numerical tsunami simulation method used to assess wave propagation, height, arrival time, and inundation [8]. This model proves to be highly effective in analyzing various tsunami-generating mechanisms, such as earthquake and submarine landslides [9]. Submarine landslides can generate tsunami waves, with their characteristics highly dependent on the properties of the surrounding water body [10]. The Makassar Strait exhibits a complex vertical current structure, where incoming water masses initially form surface currents before transitioning into strong subsurface flows. These subsurface currents undergo narrowing as they approach Labani Channel, situated between the islands of Kalimantan and Sulawesi. Given this, it can be concluded that the study area, known as Haya Slide, is located within the Labani Channel region [11]. The strong currents in the Makassar Strait play a crucial role in the

Indonesian Throughflow (ITF), a major water mass transport connecting the Pacific and Indian Oceans [12]. Considering the dynamics of ITF in this region, the submarine landslide mechanism in the study area is significantly influenced by the intense subsurface currents, which contribute to the instability of the Material Transport Complex (MTC), particularly the soft sediment layers found in this location. When strong subsurface currents induce instability in these sediment layers, mass movement occurs. In areas with significant elevation differences, this mass displacement can trigger a submarine landslide, potentially leading to a tsunami due to the sudden displacement of water volume.

The fundamental concept in COMCOT modeling on generating landslide tsunami is the Block Model. The potential submarine landslide in the Haya Slide region is modeled using a block representation, with thickness values derived from previous seismic refraction studies indicating variations between 140 – 200 m. A thickness of 140 m was selected as it represents the majority of the study area [4]. The landslide direction was assigned westward, based on the elevation differences surrounding the Haya Slide area, with a length of 1,050 m and a width of 5,425 m (Figure 3). Furthermore, a basal friction coefficient of 0.01 was applied to analyze the effects of shear resistance during the landslide, based on prior research investigating tsunami generation from submarine landslides [13].

Figure 3. Haya Slide Block Model

3. Result and Discussion

This study utilizes 18 tide stations (Ts) strategically positioned near the tsunami source, along the densely populated western coast of Sulawesi Island, as well as on Ambo Island and near Borneo Island (Figure 4). Three Ts are located to the west, and two Ts to the east of the tsunami source to observe the shape and propagation of the generated tsunami waves. Additionally, one Ts is placed on Ambo Island and one near Borneo Island to assess the tsunami's potential reach due to the Haya slide. Furthermore, 11 Ts are deployed along the western coast of Sulawesi, specifically in the densely populated West Sulawesi region to monitor wave height, arrival time, reach, and the associated risk posed by the Haya slide.

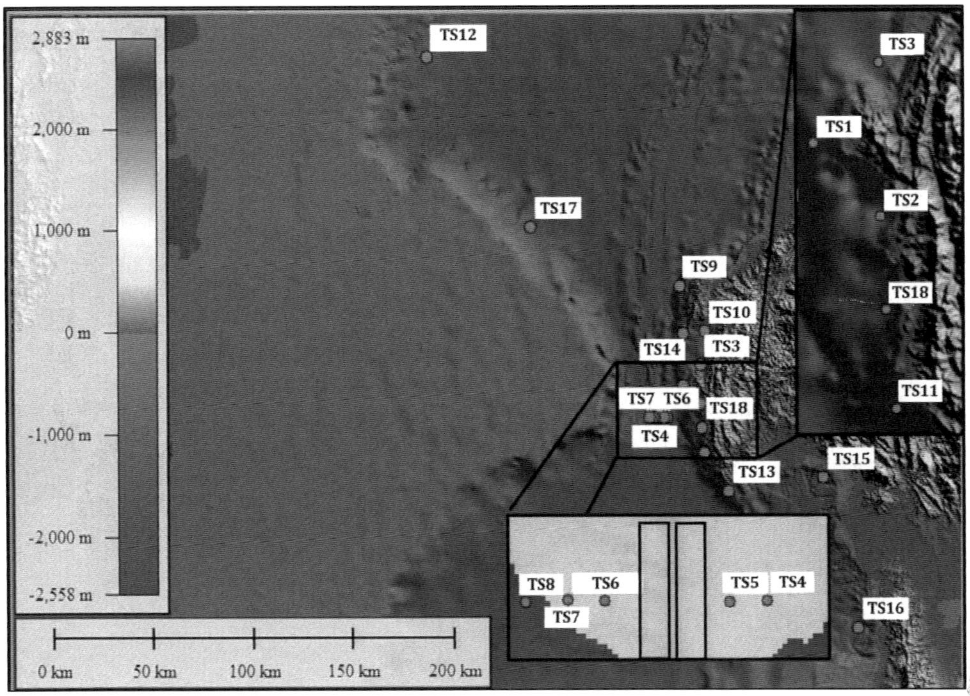

Figure 4. Distribution of Tide Stations (TS) from the Haya Slide

Tsunami modeling was conducted over a duration of 1 hour to illustrate wave propagation in all directions and the maximum wave amplitude (height). Figure 5 demonstrates that the yellow-colored area is highly susceptible to tsunami impacts, with a maximum wave height of approximately 3.809 meters. This area extends to the western coast of Sulawesi Island, located to the east of the tsunami source. The increase in wave amplitude near the coastal area occurs due to the phenomenon that tsunami waves tend to amplify as they approach shorelines.

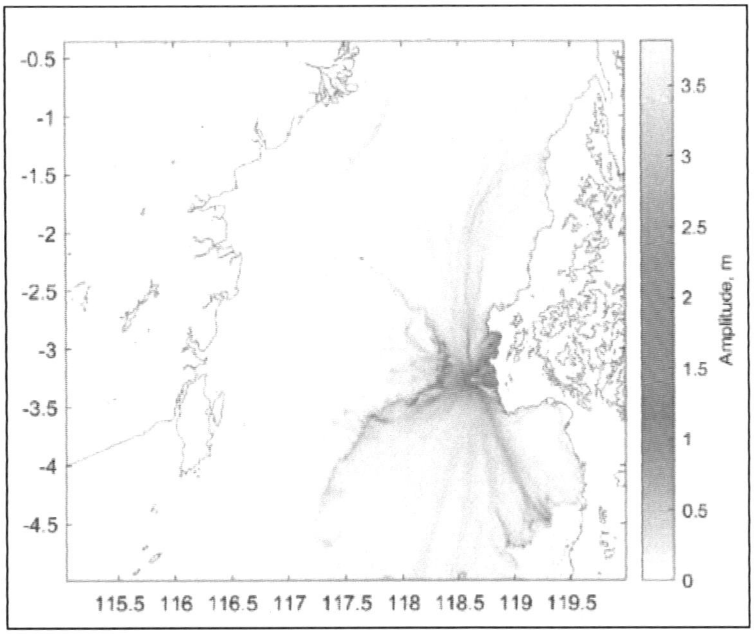

Figure 5. Maximum Height of the Tsunami Wave due to Haya Slide

IOP Conf. Series: Earth and Environmental Science **1525** (2025) 012020 doi:10.1088/1755-1315/1525/1/012020

Table 1 Potential Impact Area, Maximum Wave Height, and Tsunami Arrival Time Due to Haya Slide

No.	Coordinates (lat, long)	Ts	Wave Height (m)	Time (minutes)	Location of Tide Stations
1	118.6589730000, -3.2679320000	4	3.606	2.10	Near to the Source
2	118.6469340000, -3.2683270000	5	3.809	1.95	
3	118.6062750000, -3.2679320000	6	2.163	1.13	
4	118.5944320000, -3.2677350000	7	1.809	1.30	
5	118.5808130000, -3.2681290000	8	1.588	1.48	
6	117.5831993304, -1.6915686801	12	0.048	56.12	Ambo island and near the island of Borneo
7	118.0461546139, -2.4362247426	17	0.066	45.32	
8	118.7388855009, -3.1284158035	1	1.008	4.41	The western coastline of Sulawesi Island
9	118.8158305705, -3.2094621475	2	1.131	5.35	
10	118.8147966329, -3.0393221776	3	0.880	7.85	
11	118.7228643541, -2.6935188436	9	0.162	27.44	
12	118.8383841545, -2.8891148693	10	0.666	22.72	
13	118.8342663000, -3.4242170161	11	0.392	5.57	
14	118.9429279700, -3.5908315766	13	0.236	11.99	
15	118.7352634452, -2.9038483524	14	0.568	19.90	
16	119.3715378902, -3.5280492784	15	0.142	59.88	
17	119.5272862837, -4.1872634093	16	0.124	52.88	
18	118.8223923628, -3.3132577710	18	1.325	5.28	

Based on the table above, Ts-4, 5, 6, 7, and 8 are located very close to the landslide source area. The tsunami wave heights generated are relatively high, ranging from 1.588 m to 3.809 m, with a rapid wave arrival time of approximately 1.13 to 2.10 minutes. Ts-12 is located near Borneo Island, and Ts-17 near Ambo Island, exhibit much smaller wave heights around 0.048 m and 0.066 m, with significantly longer arrival times of 56.12 and 45.32 minutes. Meanwhile, the Tsunamis placed along the western coast of Sulawesi Island show varying wave heights, ranging from 0.124 m to 1.325 m, with arrival times ranging from 5.28 to 59.88 minutes, depending on the distance from the tsunami source. Additionally, of the 11 Ts placed along the western coast of Sulawesi, three Ts recorded relatively high tsunami waves: Ts-1 located at the northern Onang coast, Ts-2 at the Ulidang coast, and Ts-18 at the Sendana coast. These three areas are part of Majene Regency, an area with a dense population, making them highly vulnerable to the potential impacts of a tsunami, such as loss of life, infrastructure damage, and economic loss in the event of a tsunami caused by the Haya slide.

4. Conclusion

Based on the simulation, the tsunami potential resulting from Haya slide shows high waves near the source (Ts-4 to Ts-8) with heights ranging from 1.588 to 3.809 m and rapid arrival times (1.13 to 2.10 minutes). On the other hand, Ts-12 and Ts-17, located farther from the landslide source, recorded much smaller wave heights of 0.048 m and 0.066 m, with longer arrival times ranging from 45.32 to 56.12 minutes. Meanwhile, the densely populated areas along the western coast of Sulawesi, particularly Majene Regency (Ts-1, Ts-2, and Ts-18), recorded tsunami waves with heights ranging from 1.008 to 1.325 m. This indicates a serious threat to densely populated regions with the potential for loss of life, infrastructure damage, and economic losses, highlighting the critical importance of mitigation measures and early warning systems.

References

[1] Gumbira G, Wibowo M, Khoirunnisa H, Karima S and Kongko W 2021 Modelling of Potential Tsunami in Makassar Strait, Indonesia: Submarine Landslide-Induced Tsunami Simulations *Res. Sq.*

[2] Sabeti R and Heidarzadeh M 2022 Numerical Simulations of Tsunami Wave Generation by Submarine Landslides: Validation and Sensitivity Analysis to Landslide Parameters *J. Waterw. Port, Coastal, Ocean Eng.* **148** 1–13

[3] Paris R, Switzer A D, Belousova M, Belousov A, Ontowirjo B, Whelley P L and Ulvrova M 2014 Volcanic tsunami: A review of source mechanisms, past events and hazards in Southeast Asia (Indonesia, Philippines, Papua New Guinea) *Nat. Hazards* **70** 447–70

[4] Nugraha H D, Jackson C A, Johnson H D and David M 2020 offshore Indonesia

[5] Armandita C, Morley C K and Rowell P 2015 Origin, structural geometry, and development of a giant coherent slide: The South Makassar Strait mass transport complex *Geosphere* **11** 376–403

[6] Chen Z, Heller V and Briganti R 2024 Numerical modelling of tsunami propagation in idealised converging water body geometries *Coast. Eng.* **189** 104482

[7] Richard G L, Msheik K and Duran A 2023 A preliminary depth-integrated model for tsunamis propagation including water compressibility and seafloor elasticity *Eur. J. Mech. B/Fluids* **99** 84–97

[8] Tri Laksono F A, Aditama M R, Setijadi R and Ramadhan G 2020 Run-up Height and Flow Depth Simulation of the 2006 South Java Tsunami Using COMCOT on Widarapayung Beach *IOP Conf. Ser. Mater. Sci. Eng.* **982**

[9] Liu Y, Lyu Z, Wang L and Yu C 2024 A Simulation Method of HFSWR Tsunami Echo Based on COMCOT *2024 Photonics & Electromagnetics Research Symposium (PIERS)* (IEEE) pp 1–7

[10] Ramalho I and Ribeiro C 2010 Tsunami hazard induced by a newly discovered landslide in the Tagus delta off Lisbon (Portugal) *Nat. Hazards* 2010

[11] Mayer B and Damm P E 2012 The Makassar Strait throughflow and its jet *J. Geophys. Res. Ocean.* **117** 1–14

[12] Brackenridge R E, Nicholson U, Sapiie B, Stow D and Tappin D R 2020 Indonesian throughflow as a preconditioning mechanism for submarine landslides in the makassar strait *Geol. Soc. Spec. Publ.* **500** 195–217

[13] Han W, Xu X, Wang J, Xiao L, Zhou K and Guo X 2023 Safety Assessment of Coastal Bridge Superstructures with Box Girders under Potential Landslide Tsunamis *J. Mar. Sci. Eng.* **11**

IOP Conf. Series: Earth and Environmental Science **1525** (2025) 012021 doi:10.1088/1755-1315/1525/1/012021

Physical Characteristic Analysis of Sediments along Galesong Beach, Takalar Regency

Sakka[1*], Nurul Izzah[1], Amiruddin[1]

[1] Laboratory of Coastal and Ocean Dynamics, Geophysics Department, Faculty of Mathematics and Natural Sciences, Hasanuddin University, Makassar, Indonesia

*E-mail: sakka@sci.unhas.ac.id

Abstract. This study examines the physical characteristics of sediments along Galesong Beach, focusing on sedimentation issues that disrupt local fishing activities. The analysis includes sediment fall velocity, initiation of motion, and suspension rates, employing van Rijn and Shields equations for accurate evaluation. Sediment samples were collected from six locations in nearshore waters, providing a representative profile of the coastal sediment composition. Sieve analysis results reveal that the sediments are predominantly fine to very fine sand, indicating potential for easy resuspension and transport under certain hydrodynamic conditions. The findings show that sediment fall velocity range from 0.01 to 0.19 m/s, initiation of motion describing critical mobility numbers vary between 0.077 and 0.031, and suspension rates are between 0.084 and 0.070. These parameters illustrate the dynamics of sediment movement in the area and highlight the susceptibility of sediments to resuspension, which may contribute to sedimentation issues near the coast. The results offer valuable input for the Takalar Regency Government in planning and managing coastal zones, providing essential data for coastal management strategies aimed at mitigating sedimentation and improving conditions for fishing activities along Galesong Beach. This research thus supports sustainable coastal management and local economic activities.

1. Introduction

Coastal areas, as complex and dynamic environments, are prone to changes and disruptions in sediment balance. The process of sedimentation is driven by sediment transport and excessive sediment mobilization caused by hydro-oceanographic factors and human activities, such as fishing. The extent of shoreline morphological changes influenced by these factors significantly depends on the characteristics of the sediment particles that compose the beach [1–4].

Shoreline changes are caused by erosion resulting from the influence of waves and tidal currents. However, it is important to emphasize that the magnitude of these effects depends on the type of sediment material that constitutes the beach [5]. Therefore, this study investigates the physical characteristics of beach sediment particles, focusing on grain size parameters to assess sediment settling velocity, as well as the initiation of motion and sediment suspension. These factors are analyzed to understand sediment mobilization leading to sedimentation and deposition at several locations along the Galesong coastline in Takalar Regency.

2. Material and Methods

Geographically, the study area is located between -5º19'00'' to -5º20'30'' S and 119º19'30'' to 119º21'30'' E. It is bordered by the Takalar mainland to the east, Sanrobengi Island to the west, and rivers situated south of Bayowa Beach and north of Boddia Beach, as well as south of Boe Beach. The location of the study area is illustrated in Figure 1.

2.1 Data Collection

Sediment sampling from the seabed in coastal waters was conducted to analyze grain size distribution and sediment density. The sampling process utilized a bottom grab sampler, and

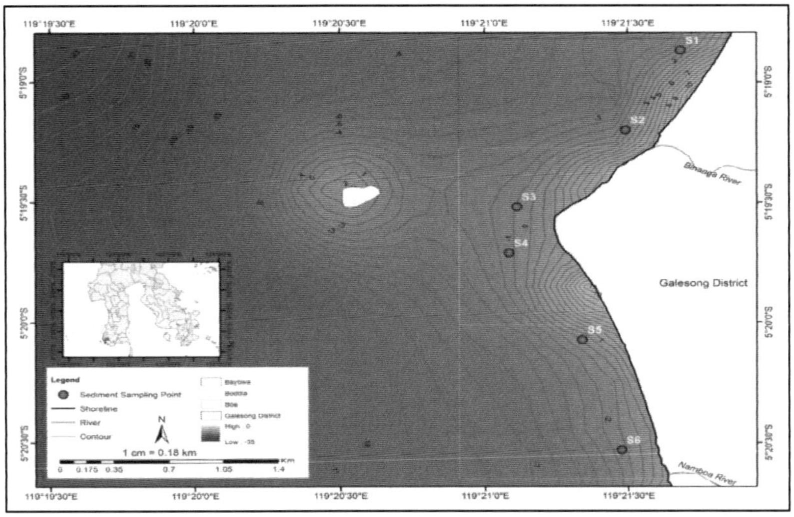

Figure 1. Research location

sample locations were determined using a purposive sampling method. Sediment sampling was carried out at six locations in May 2024, coinciding with the east monsoon season, during which wind predominantly blows from the east to southeast.

Additional data collection, including seawater temperature and density, was conducted using a thermometer and an aerometer. Data collection points were aligned with each sediment sampling location.

2.2 Data Analysis

The collected data were analysed using three approaches: analysis of sediment density and mean sediment particle diameter; determination of sediment particle settling velocity; and assessment of sediment particle initiation of motion and suspension. Laboratory analysis involved measuring sediment density and the median grain size (d_{50}). The analysis of mean sediment diameter (d_{50}) was conducted using sieves with diameters of 2 mm, 1 mm, 0.5 mm, 0.25 mm, 0.125 mm, 0.063 mm, and 0.031 mm. This process aimed to separate grain sizes based on sieve diameters and their classifications.

To determine the settling velocity of sediment particles, assuming that the particles are spherical, [6] derived the following equations based on Stokes' law:

for diameters between $0{,}000032 \text{ m} < d_{50} < 0{,}000062 \text{ m}$

$$w_{\text{s,stokes}} = \frac{(s-1)gd_{50}^{\;2}}{(18v)} \tag{1}$$

IOP Conf. Series: Earth and Environmental Science **1525** (2025) 012021 doi:10.1088/1755-1315/1525/1/012021

and for diameters between 0,000062 m $< d_{50} <$ 0,002 m

$$w_{s,stokes} = 10\left[\frac{v}{d_{50}}\right]\left[\left\{1 + 0{,}01(s-1)g\frac{d_{50}{}^{3}}{v^{2}}\right\}^{0{,}5} - 1\right] \quad (2)$$

where: $w_{s,stokes}$ is the settling velocity of the sediment particles (m/s); s is the relative density of the sediment particles; g is the acceleration due to gravity (9.81 m/s^2); v is the kinematic viscosity of water (m^2/s); and d_{50} is the median particle diameter (m).

Following that, to determine the initiation of sediment movement using the Shields parameter, the following equation is used:

$$\theta_{cr,shield} = \frac{0{,}25}{d_{50}} + 0{,}055\left(1 - exp(-0{,}02D_*)\right) \quad (3)$$

The study on critical shear stress (Critical Shear Stress for Suspension) to determine the sediment that is suspended, with the equation [7]:

$$\tau_{cr,susp,Van\ Rijn} = \theta_{cr,Van\ Rijn} \cdot (\rho_s - \rho_a)g \cdot d_{50} \quad (4)$$

where $\tau_{cr,susp,Van\ Rijn}$ is the critical shear stress for sediment suspension, according to Van Rijn (N/m^2). $\theta_{cr,Van\ Rijn} = \frac{0{,}3}{1+D_*} + 0{,}1\left(1 - Exp(-0{,}05D_*)\right)$ is the critical angle in Van Rijn's formulation.

3. Results and Discussion

Galesong Beach, Takalar Regency, has depths reaching up to 35 meters. However, the area near the shoreline is identified as a shallow coastal region. This is consistent with the findings of Karamma et al. (2020), the depth gradually decreases towards shallower zones. Sediment sampling was conducted approximately 400 meters from the shoreline, where the depth is only 1 meter, categorizing it as a shallow water area (Fig. 1).

3.1 Sediment Particle Sieve Analysis Results

Based on the grain size classification table according to the Wentworth scale, the sieve analysis results show the following (as illustrated in Figure 2): at sampling point S1, the sediment is predominantly fine sand (41.19%). At sampling point S2, the sediment is dominated by mud, consisting of a mixture of coarse mud (50.90%) and medium mud (22.76%). At sampling point S3, the sediment is dominated by very fine sand (50.93%). At sampling point S4, the sediment is predominantly fine sand (62.87%). At sampling point S5, the sediment is dominated by very fine sand (48.44%) and fine sand (37.33%). At sampling point S6, the sediment is dominated by fine sand (41.16%) and very fine sand (34.90%). The complete results of the sieve analysis can be seen in Table 1.

Table 1. The percentage of different sediment classifications (based on particle size) at each sampling point

Classification	Average (%)					
	S1	**S2**	**S3**	**S4**	**S5**	**S6**
Very coarse sand (2 mm)	0	0	0	0	0	0
Coarse sand (1 mm)	2.59	0.98	0.68	0.44	0.04	1.42
Medium sand (0.5 mm)	27.90	3.32	5.96	11.45	3.40	12.24
Fine sand (0.25 mm)	41.19	10.25	23.46	62.87	37.33	41.16
Very fine sand (0.125 mm)	17.56	10.63	50.93	24.32	48.44	34.90
Coarse mud (0.063 mm)	5.78	50.90	10.20	0.54	2.60	6.77
Medium mud (0.031 mm)	4.21	22.76	8.15	0.19	7.35	3.29

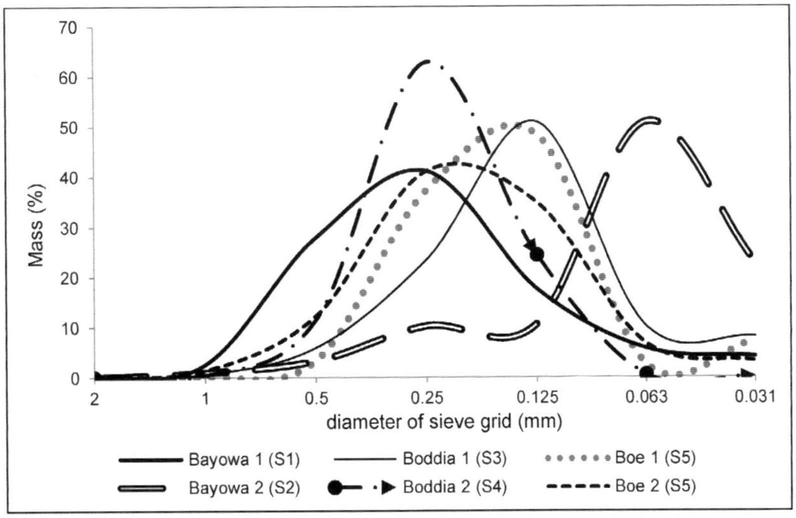

Figure 2. Grain Size Distribution Graph of Sediment

At sampling point S2, the sediment is predominantly mud (coarse and medium mud) since point S2 is directly in front of a river. This causes finer particles to be deposited as they move from the upstream to the river mouth, eventually reaching the coastal waters. Additionally, the sampling was conducted during the east monsoon season, when winds typically blow from the east, potentially causing sediment to move westward, resulting in sedimentation to the north or south of the beach. The influence of the beach's slope and the location of the Binanga River, which is surrounded by well-established vegetation (such as mangrove plants), helps drive the river currents, pushing sediment from upstream to the river mouth (toward the coastal waters). This causes mud-type sediments to settle at the river mouth, right in front of sampling point S2.

3.2 Fall velocity of particle sediment

The grain size analysis conducted shows that the average sediment diameter at Galesong Beach ranges from 0.088 to 0.360 mm. The φ value is obtained from the formula –log2d, where d is the sediment particle diameter [8]. The average value serves as a reference to calculate the fall velocity, which is then computed using the van Rijn equations in equations (1) and (2), resulting in the values shown in Table 2.

Table 2. Fall Velocity Values of Sediment Particles in the Waters of Galesong Beach

Grain Size of Sediment Particles		Fall velocity (m/s)
d_{50}	$phi(\varphi)$	
0.088	3.50	0.010
0.190	2.40	0.054
0.215	2.22	0.069
0.270	1.89	0.109
0.330	1.60	0.163
0.360	1.47	0.194

Based on the d_{50} values, the fall velocity in the waters of Galesong Beach ranges from 0.01 to 0.19 m/s. This value indicates that the grain size follows a logarithmic function with the fall velocity of sediment particles. As the grain size increases, the fall velocity also increases (Fig. 3).

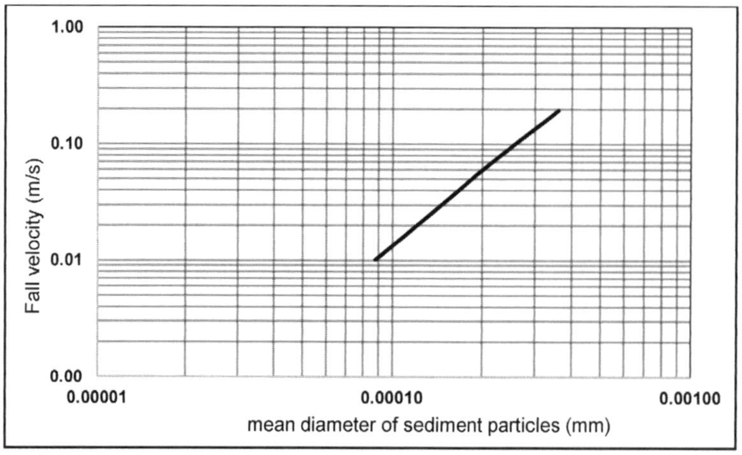

Figure 3. The relationship between sediment particle diameter and the fall velocity of sediment particles

3.3 Initiation of motion and suspension

The determination of sediment movement initiation is illustrated using the Shields parameter approach, while sediment suspension is represented through the Van Rijn approach. Based on the Shields parameter, sediment particles will begin to move within a range of values between 0.03 and 0.05, but may exceed these standard values depending on the particle characteristics and flow conditions in the water. On the other hand, for sediment suspension according to Van Rijn's criteria, sediment will be suspended within a range of 0.2 to 0.5. If the value falls between 0.1 and 0.2, the sediment will only move along the bed and will not truly be suspended (Fig. 4). When the value is less than 0.1, the sediment is considered to be stationary.

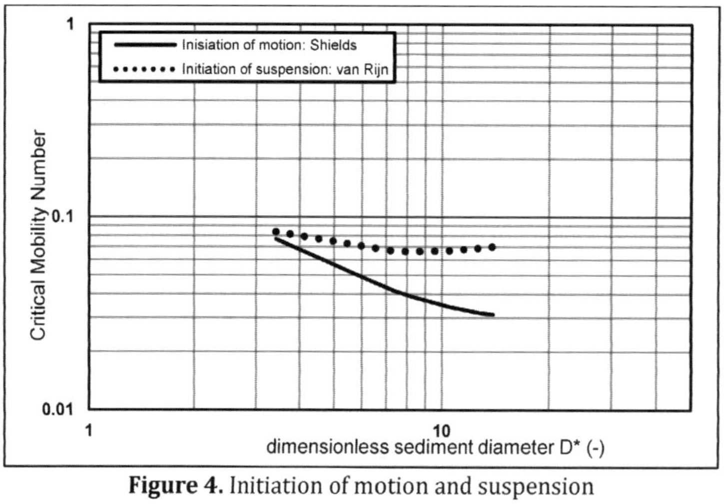

Figure 4. Initiation of motion and suspension

The sediment movement initiation values in the waters of Galesong Beach range from 0.031 to 0.077, indicating that sediment begins to move or undergo mobilization. Meanwhile, for sediment suspension in the waters of Galesong Beach, the values range from 0.070 to 0.084, which are relatively lower than 0.1, but still indicate potential for suspension (Table 3).

Table 3. The values for sediment movement initiation and suspension in the waters of Galesong Beach

Mean diameter of sediment particles (mm)	Dimensionless sediment particles (-)	Shield number theta critical; Shields	Shield number theta critical; van Rijn
0.088	3.421	0.077	0.084
0.190	7.387	0.041	0.067
0.215	8.359	0.038	0.066
0.270	1.497	0.034	0.067
0.330	1.830	0.032	0.069
0.360	13.996	0.031	0.070

4. Conclussion

Based on the research conducted, the following conclusions can be drawn:

- The particle size distribution analysis shows that the dominant sediment type in the waters of Galesong Beach is fine to very fine sand, with a fall velocity ranging from 0.01 to 0.19 m/s.
- The sediment movement initiation analysis indicates a critical mobility number ranging from 0.031 to 0.077, suggesting that the sediment is undergoing movement or mobilization. Meanwhile, sediment suspension values range from 0.070 to 0.084, indicating a high potential for sediment suspension.

References

[1] Abuodha J O Z 2003 Grain size distribution and composition of modern dune and beach sediments, Malindi Bay coast, Kenya *J. African Earth Sci.* **36** 41–54
[2] de Andrade T S, Sousa P H G de O and Siegle E 2019 Vulnerability to beach erosion based on a coastal processes approach *Appl. Geogr.* **102** 12–9
[3] Karamma R, Hatta M P, Mustari A S, Manda S P, Ilda A and Ihsan M 2022 Numerical Modelling of Shoreline changes of Galesong, Takalar's coast *IOP Conf. Ser. Earth Environ. Sci.* **1117**
[4] Najafi S, Dragovich D, Heckmann T and Sadeghi S H 2021 Sediment connectivity concepts and approaches *Catena* **196** 104880
[5] Van Rijn L C 2011 Coastal erosion and control *Ocean Coast. Manag.* **54** 867–87
[6] van Rijn L C 1984 Sediment Transport, Part II: Suspended Load Transport *J. Hydraul. Reseacrh* **110** 1613–41
[7] van Rijn L C 1993 *Principles of Sediment Transport in Rivers, Estuaries and Coastal Seas* (Amsterdam, the Netherlands: Aqua Publisher)
[8] Bayhaqi A and Dungga C M A 2015 Distribusi butiran sedimen di pantai Dalegan, Gresik, Jawa Timur *Depik* **4** 153–9

IOP Conf. Series: Earth and Environmental Science **1525** (2025) 012022 doi:10.1088/1755-1315/1525/1/012022

Unraveling tsunami hazard potential along the central Makassar Strait thrust

Muh. Faqih Raihan[1*], Muhammad Alimuddin Hamzah Assegaf[1]

[1] Department of Geophysics, Faculty of Mathematics and Natural Sciences, Hasanuddin University, Makassar, 90245, Indonesia

[*]E-mail: faqihraihan.akarbadi@gmail.com

Abstract. This study investigates the tsunami hazard potential along the Central Makassar Strait Thrust through numerical modeling of an earthquake scenario. A finite fault model representing a moment magnitude (Mw) 7.8 earthquake was developed using realistic fault geometry and a uniform slip distribution, based on empirical scaling relationships. The simulation employed the Cornell Multi-grid Coupled Tsunami Model (COMCOT) to analyze tsunami wave propagation and run-up across three different scenarios. Bathymetric data from BATNAS was utilized to ensure accurate representation of the seafloor topography. Results indicate that a uniform coseismic slip of 2.5 meters can generate vertical crustal deformation exceeding 6 meters. Maximum run-up values reach approximately 5 meters, particularly in Central Mamuju, where coastal topography and bay morphology significantly amplify wave energy. These findings underscore the region's high tsunami risk due to short wave arrival times and localized amplification effects. This research highlights the importance of site-specific hazard assessments and the urgent need to improve early warning systems and community-based preparedness strategies in tsunami-prone coastal areas.

Keyword. Tsunami hazard, Central Makassar Strait Thrus, Early warning systems

1. Introduction

Sulawesi, located at the center of the Indonesian archipelago, lies at the convergence of three major tectonic plates: the Indo-Australian, Eurasian, and Philippine plates. This tectonic setting has profoundly influenced the island's geological evolution, resulting in its distinctive multi-armed morphology shaped by complex deformation processes. These processes reflect the collision and interaction of crustal elements originating from the Sundaland continental promontory of Eurasia to the north and the Australian plate to the south [1].

The intricate tectonic framework of Sulawesi has given rise to significant crustal deformation, creating a network of active faults and subduction zones that mark the region as one of the most geodynamically active areas in the Indonesian archipelago. This dynamic environment is characterized by heightened seismic activity, with frequent earthquakes and tectonic uplifts serving as evidence of ongoing plate interactions (Figure 1) [1][2][3][4][5].

As a result, Sulawesi is particularly vulnerable to natural hazards such as earthquakes and tsunamis, where several active fault systems converge. The region experienced a series of significant seismic events in the late 1960s. The tectonic stress generated along its fault systems

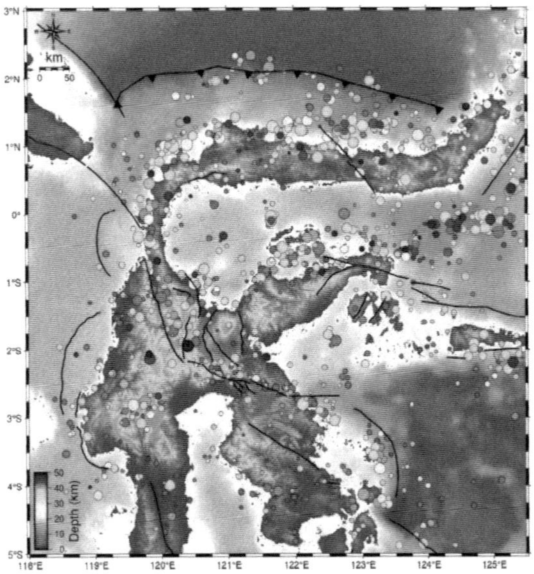

Figure 1. The tectonic setting of Sulawesi, along with recorded earthquake events from 1970 to 2024, has been analyzed to understand seismic activity in the region. The study focuses on earthquakes with magnitudes ranging from 3.0 to 7.5 and depths between 10 and 50 km, providing critical insights into the tectonic dynamics and seismic hazard potential of the area.

and subduction zones not only amplifies seismic events but also increases the potential for tsunamigenic activity, especially from undersea earthquakes or submarine landslides. A powerful earthquake struck near Central Sulawesi on August 23, 1967, triggering a tsunami that caused widespread damage. On August 14,1968, the devastating earthquake which had a magnitude 7.4 also generated a destructive tsunami near Donggala, Central Sulawesi. Then, on February 23, 1969, a magnitude 7.0 earthquake struck the coastal region of Majene in West Sulawesi, producing a tsunami that caused a significant coastal damage [6][7]. These factors highlight the urgent need for continuous research and improved disaster preparedness to mitigate the risks posed by Sulawesi's volatile tectonic setting.

In this context, modeling tsunamis in the Central Makassar Strait Thrust particularly important. This area is known for its active fault zones and potential for large seismic events, making it a critical region to understand tsunami dynamics. By focusing on this region (Figure 2), the study aims to assess the tsunami potential, providing valuable data that can support disaster risk reduction efforts and contribute to a better understanding of tsunami hazards specific to the Central Makassar Strait.

2. Methodology

In this study, we aim to simulate the potential tsunami that could be generated by seismic activity along the Central Makassar Strait Thrust with a length of 197.62 km. According to PusGeN (2017) [8], this Thrust zone (Figure 2) has the potential to generate an earthquake with a magnitude greater than 7. Building upon this assessment, our simulation focuses on the Central Makassar Strait Thrusting for analysis.

IOP Conf. Series: Earth and Environmental Science **1525** (2025) 012022 doi:10.1088/1755-1315/1525/1/012022

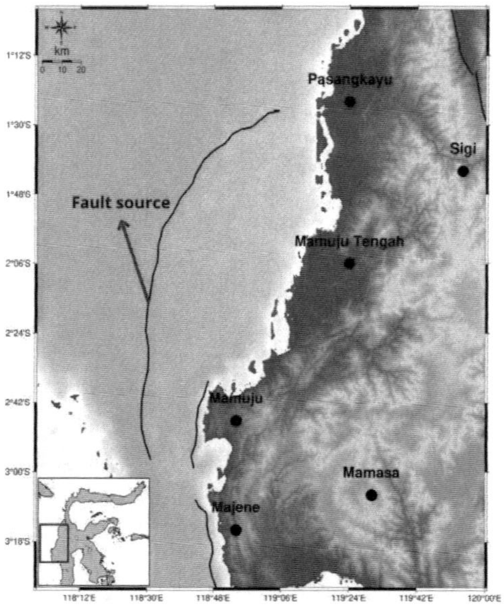

Figure 2. The study area is located within the Central Makassar Strait Thrust region, selected due to its tsunami potential associated with active fault activity. The map highlights the names of regions within the study radius, along with key topographic features and the delineation of fault segments.

To determine the fault geometry parameters such as width, magnitudo and slip, we applied the empirical relationships established by Wells and Coppersmith (1994) (Table 1) [9]. These calculations provide essential inputs for defining the fault geometry in tsunami modeling. Tsunami simulations were conducted using one scenario based on slip magnitudes 2.5 m, utilizing the Cornell Multi-grid Coupled Tsunami Model (COMCOT), a robust numerical model initially developed by Liu et al. (1995) [10][11].

Table 1. Deformation Calculations

Equation	Coefficients and Standard Errors	
	a	b
$M_w = a + b.\log(SRL^*)$	5.00	1.22
$\log(RW^*) = a + b.M_w$	-1.61	0.41
$M_0 = 10^{1.5(M_w+10.7)}.10^{-7}$		
$D = \dfrac{M_0}{\mu.A}$		

*SRL: Surface Rupture Length (km).
*RW: Downdip Rupture Width (km).

For accurate analysis of tsunami wave propagation, we utilized bathymetric data sourced from the official Indonesian Geospatial Portal, BATNAS (https://tanahair.indonesia.go.id/portal-web/). This bathymetric dataset allows for detailed tracking of wave propagation patterns across the Central Makassar Strait Thrust, providing critical insights into the potential impact of a tsunami in this region. Through this approach, the study aims to enhance understanding of tsunami risks and inform disaster preparedness efforts.

3. Result and Discussion

A comprehensive understanding of tsunami potential in the study area is crucial for evaluating the maximum tsunami wave height in affected regions. To develop tsunami models for each scenario, we assume that the earthquake-induced deformation (displacement) is homogeneously distributed across the fault zone. The fault geometry parameters, including dip angle, rake angle, and focal depth, are based on data published by the Indonesian National Centre for Earthquake Studies (PusGeN, 2017) [8]. Tsunami wave height analysis was conducted by placing tide gauges along the coastline of West Sulawesi, covering Pasangkayu, Central Mamuju, Mamuju, and Majene. The place of the tide gauges can be seen in Figure 3.

Based on deformation calculations used (Table 1), an Mw 7.8 earthquake generates a maximum displacement of 2.5 meters. This value serves as the primary input for tsunami modeling, where the displacement is uniformly distributed across predefined locations for each analyzed scenario. The homogeneous displacement approach remains widely used in tsunami modeling, particularly for estimating the lower-bound tsunami wave heights. In this study, the Mw 7.8 scenario represents the maximum credible earthquake expected in the region, derived from empirical relationships and historical seismic activity, particularly referencing events that occurred in the late 1960s in Sulawesi. This scenario is considered the worst-case event that could occur in the future, providing a conservative estimate of potential tsunami hazards in the study area.

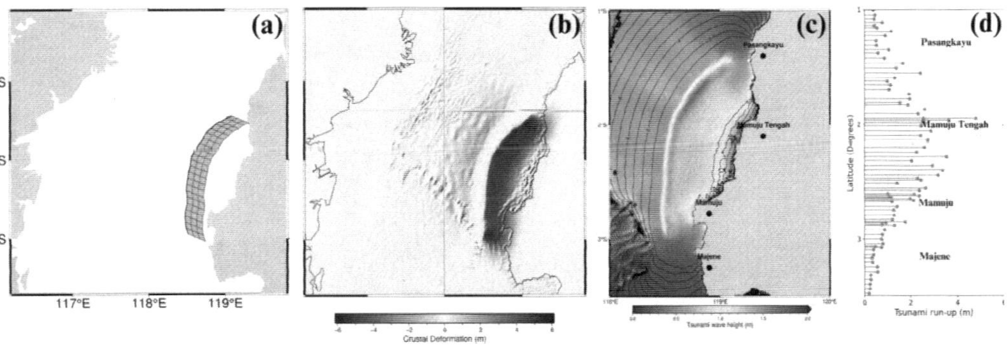

Figure 3. Simulation of tsunami parameters resulting from an earthquake source off the western coast of Sulawesi. (a) Fault geometry used in the simulation, represented as segmented grids along the fault zone; (b) Crustal deformation distribution due to slip on the fault plane; (c) Tsunami wave height distribution; (d) Tsunami run-up distribution along the coastline.

To evaluate the tsunami hazard potential along the coast of West Sulawesi, we constructed a finite fault model representing a Mw 7.8 megathrust earthquake, although there is no information regarding when such an event might occur. The fault geometry was discretized into a series of rectangular sub-faults with variable depths ranging from 18 km in the updip section to 45 km in

IOP Conf. Series: Earth and Environmental Science **1525** (2025) 012022 doi:10.1088/1755-1315/1525/1/012022

the downdip, reflecting the realistic slab interface morphology. A uniform slip of 2.5 meters was applied across the rupture plane to simulate a homogeneous coseismic displacement scenario (Figure 3a).

The simulated crustal deformation field (Figure 3. (b)) displays a broad offshore zone of positive vertical displacement, with uplift reaching up to approximately 6 meters, consistent with rupture along a shallow crustal thrust fault. The deformation is indicative of compressional tectonics, with the overriding block experiencing significant uplift due to fault slip concentrated at shallow depths. The overall pattern of deformation is typical of large thrust fault earthquakes, with the highest vertical displacements occurring near the shallow up dip segment of the rupture.

Tsunami modeling results (Figure 3. (c)) reveal that coseismically displaced seafloor generated tsunami waves exceeding 2 meters in some coastal areas. In particular, Mamuju, Pasangkayu and its surrounding areas were affected by wave heights between 1.5 and 2.0 meters, while Central Mamuju also experienced significant tsunami wave levels. Bathymetric focusing and coastal geometry strongly influenced nearshore wave height amplification. Run-up analysis (Figure 3. (d)) further highlights tsunami impacts. Maximum run-up values reached up to ~5 meters, especially in Central Mamuju where the coastal slope was steep and wave energy was effectively channeled. The spatial distribution of run-up along the coast suggests that bathimetry played important roles in modulating tsunami wave height, with local amplification occurring in narrow bays.

These results underscore the considerable tsunami hazard posed by a Mw 7.8 event along the Central Makassar Strait Thrust. The modeled tsunami impact driven by realistic fault geometry, variable rupture depth, and uniform slip demonstrates the potential for significant Tsunami wave height, warranting further site-specific risk assessment and preparedness planning.

Tsunami simulations reveal considerable variations in tsunami wave heights. These differences are likely governed by a combination of factors, including submarine topography, shoreline configuration, and the proximity of the fault to the study area. This suggests that, beyond the influence of submarine topography, the bay's geometry may have played a crucial role in wave amplification. When the tsunami wavelength resonates with the bay's natural oscillation period, constructive interference and resonance can amplify wave heights as they propagate into the bay, a phenomenon also observed in other tsunami-prone embayments [12][13].

This result aligns with the findings of Qiang Qiu (2019) [14], which highlight that in near-field regions, tsunami wave heights are primarily controlled by source geometry and the fault's rupture characteristics. Conversely, in far-field locations, tsunami wave transformations are predominantly influenced by bathymetric variations and coastal morphology [14].

4. Conclusion

We conducted tsunami modeling for the Central Makassar Strait Thrust using a deterministic scenario based on a magnitude 7.8 earthquake with a finite fault model featuring a uniform slip of 2.5 meters. The fault geometry, extending from ~18 km to 45 km in depth, produced significant crustal deformation, with maximum vertical displacement exceeding 6 meters, sufficient to trigger a large-scale tsunami. The simulation results indicate that this fault system is capable of generating tsunami waves with the first wave arrival occurring under 1 minutes after the rupture, depending on the distance to the source. The run-up distribution along the coastline reveals maximum values exceeding 4 meters, particularly impacting coastal areas of Central Mamuju. These findings highlight the critical vulnerability of coastal communities in West Sulawesi due to the extremely limited evacuation time, which poses substantial challenges for effective emergency response. The results underscore the urgent necessity for enhanced early warning systems, and community-based preparedness initiatives to mitigate the devastating impact of future tsunami events in the region.

References

[1] R Hall, *Australia–SE Asia collision: Plate tectonics and crustal flow in Hall*, in *The SE Asian Gateway: History and Tectonics of the Australia–Asia Collision*, R M A Cottam and M E J Wilson (eds), London: Geological Society, 2011, pp 75–109 https://doi.org/10.1144/SP355.5

[2] W B Hamilton, *Tectonics of the Indonesian Region*, vol **1078**, Washington, DC: US Government Printing Office, 1979

[3] A H Satyana, T Faulin and S N Mulyati, Tectonic evolution of Sulawesi area: Implications for proven and prospective petroleum plays, in *Proceedings of the 36th HAGI and 40th IAGI Annual Convention and Exhibition*, Makassar, 2011

[4] E A Silver, R McCaffrey, Y Joyodiwiryo and S Stevens, Ophiolite emplacement by collision between the Sula Platform and the Sulawesi Island Arc, Indonesia, *J. Geophys. Res. Solid Earth* **88**(B11), pp 9419–9435, 1983 https://doi.org/10.1029/JB088iB11p09419

[5] I R Pranantyo, A Cipta and H A Shiddiqi, Source reconstruction of the 1969 Western Sulawesi, Indonesia, earthquake and tsunami, *Pure Appl. Geophys.* **180**, pp 1765–1783, 2023 https://doi.org/10.1007/s00024-022-03064-2

[6] I R Pranantyo, A Cipta, H A Shiddiqi and A R Gusman, Source reconstruction of the 1969 Majene, Sulawesi earthquake and tsunami, *IOP Conf. Ser.: Earth Environ. Sci.* **873**(1), 012043, 2021 https://doi.org/10.1088/1755-1315/873/1/012043

[7] I M Watkinson and R Hall, Fault systems of the eastern Indonesian triple junction: Evaluation of Quaternary activity and implications for seismic hazards, *Geol. Soc. Lond. Spec. Publ.* **441**(1), pp 71–120, 2017 https://doi.org/10.1144/SP441.4

[8] PuSGeN, *Peta Sumber Bahaya Gempa Indonesia Tahun 2017*, 1st ed, Bandung: PuSGeN, 2017

[9] D L Wells and K J Coppersmith, New empirical relationships among magnitude, rupture length, rupture width, rupture area, and surface displacement, *Bull. Seismol. Soc. Am.* **84**(4), pp 974–1002, 1994 https://doi.org/10.1785/BSSA0840040974

[10] P L F Liu, Y S Cho, M J Briggs, U Kanoglu and C Synolakis, Runup of solitary waves on a circular island, *J. Fluid Mech.* **302**, pp 259–285, 1995 https://doi.org/10.1017/S0022112095004095

[11] R P Felix, J A Hubbard, K E Bradley, K H Lythgoe, L Li and A Switzer, Tsunami hazard in Lombok and Bali, Indonesia, due to the Flores back-arc thrust, *Nat. Hazards Earth Syst. Sci.* **22**, pp 1665–1682, 2022 https://doi.org/10.5194/nhess22-1665-2022

[12] A B Rabinovich, Spectral analysis of tsunami waves: Separation of source and topography effects, *J. Geophys. Res. Oceans* **102**(C6), pp 12663–12676, 1997 https://doi.org/10.1029/97JC00479

[13] E A Okal and C E Synolakis, Source discriminants for near-field tsunamis, *Geophys. J. Int.* **158**(3), pp 899–912, 2004 https://doi.org/10.1111/j.1365-246X.2004.02347.x

[14] Q Qiu, L Li, Y J Hsu, Y Wang, C H Chan and A D Switzer, Revised earthquake sources along Manila Trench for tsunami hazard assessment in the South China Sea, *Nat. Hazards Earth Syst. Sci.* **19**(7), pp 1565–1583, 2019 https://doi.org/10.5194/nhess-19-1565-2019

IOP Conf. Series: Earth and Environmental Science **1525** (2025) 012023 doi:10.1088/1755-1315/1525/1/012023

Characterization Geochemical Gas While Drilling Impacted Determination Porosity- Applied in Deltaic Sedimentation Multilayer Hydrocarbon

Desianto Payung Battu[1*], Adi Tonggiroh[1], Meutia Farida[1]

[1] Geological Department, Universitas Hasanuddin, Makassar, Indonesia

[*]E-mail: battudp23d@student.unhas.ac.id

Abstract. Uncertainty in interpretation of the existence and characteristics of each reservoir layer will greatly influence the level of success in the drilling process. Formation evaluation through wireline logging operation can obtain information about reservoir characteristics, one of which is calculating rock porosity, is sometimes not carried out due to limitations in drilling operations which are closely related to drilling costs which are increasingly expensive. The Semberah Field is located in a delta depositional environment which has reservoir layers that have been producing since the 1970s. Drilling activities that produce data in the form of gas drilling ratio and wireline logging have been widely carried out but still have problems in analyzing rock porosity values. To improve level confidence and limitation data affected drilling operation during attempt porosity analysis thorough wireline logging, gas drilling ratio data from 10 wells was taken and then several gas chromatography ratios were calculated, namely wetness, balance, and fluid mobility or estimated potential porosity (FMPP). Then, all these ratios were compared with the effective porosity (phie) from the petrophysical calculation. FMPP apparently has a sensitivity value that matches the phie value from Petrophysical calculations, namely around 70.25%. This gas drilling ratio method is expected to reduce the uncertainty of porosity calculations due to geological environment and in the future can be used as data to determine reservoir characteristics without having to use expensive wireline logging data.

Keywords: Chromatography, Wetness, Balance, Fluid Mobility or Estimated Potential Porosity

1. Introduction

Expressing ration between the void pore spaces and whole rock volume is typically expressed as Rock porosity. It explained for capacity of reservoir filled by certain volume of hydrocarbon. Reservoir and production management will be attracted to determine by actual porosity to quantified volume of hydrocarbon.

During reach reservoir, drilling operation is penetrate subsurface data and bring up all the rock crushed include with mud together gas molecule trapped inside it. Gas data components

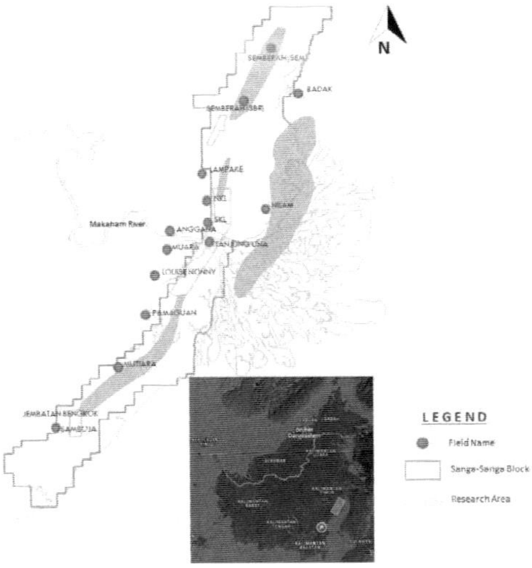

Figure 1. Research Area in Semberah Field represented on yellow shaded part of Northern Sanga- Sanga Block

extracted can be correlated with the volume in fraction or percentage of pore spaces inside the lithology corresponded as porosity indication [1].

Gas detection technique was continuously improved form past decade especially on gas trap efficiency, thermo isolated, chromatography composition reading detector, constant condition [2]. Quantity of gas released is depend on volume of rock crushing, rate of penetration, differential pressure accumulate knows as Total Gas [3].

Traditional method was measured the porosity using log analysis from Logging While Drilling or Wireline Logging while other method was direct measurement with core analysis. Accuracy both methods really affected by calculation of capacity storage of the petroleum resource volume. It continues impacted for reserve estimations that caused for field development especially for mature block like on research area. Determined porosity based on logging data can result on large errors due to tool-calibration issues, borehole conditions and drilling fluid compositions. Improve on utility of mudlogging data can support and gain porosity's information based on both methods above.

2. Research Area

The Semberah Field is located in the northern part of the Sanga-Sanga block together with other fields where Pertamina's work area as the operator of this block is operationally divided by the Mahakam River, namely the northern and southern parts.

2.1 Geological Background

Sanga-Sanga block is part of Mahakam Delta lies on north of Kutai Basin where Semberah field include Sanga-Sanga block on North West discovery on 1974, produce oil and gas with peak production time around 1994-2000 as shown on Figure 1.

This block characterized by Samarinda anticlinorium with a series of highly prolific North-East and South-West trending anticline. Hydrocarbon accumulations are eventually found developed as delta plain-delta front reservoir, which related to distributary channel and mouth bar deposit Mid-Late Miocene [4]. Its source rocks from coal, carbonaceous shale, and marine mudstone completed as structural and stratigraphic trap [5].

2.2 Facies Sedimentation

The sediment types in the Sanga-Sanga block are primarily associated with marine, deltaic, and

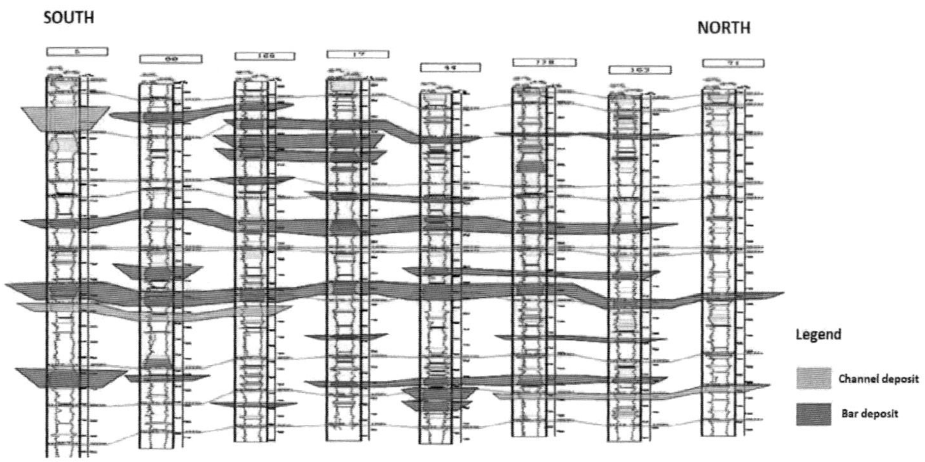

Figure 2. Cross section showing from South to North Semberah Field change from channel to bar deposit reflected change from blue to red colour

fluvial environments, which are typical of regions that can accumulate significant hydrocarbon deposits.

Depositional process was influenced by combination of fluvial, deltaic and shallow marine environments [6] thus will affected by reservoir quality together with petroleum system.

A comprehensive from number log drilled on Semberah field was developed to identified characteristic whole reservoir. A cross section was modified to shown lateral connectivity reservoir layer from South to North field with changes in sediment type from channels dominant to bar reflected by changing from blue to red colour shown on figure 2 above. Cross section above generated by wireline that systematically attempt at Semberah field to analysis formation evaluation, thus log data will utilize as well to gain information for further porosity analysis.

3. Material and Method

Analysed was started with gathering wireline and gas ratio drilling data from 4 wells in the Semberah field. Logging data such as gamma ray (GR), resistivity (Res), density (Dens), neutrons (Neu) together with gas chromatography such as methane (C1), ethane (C2), propane (C3), normal and iso butane (C4), normal and iso pentane (C5) were taken as main data on this research illustrated on Figure 3.

Next step was conditioning data consist of synchronization of depth shifts that arise due to differences in cable and pipe stretch same as well with efficiency equipment that used to separate

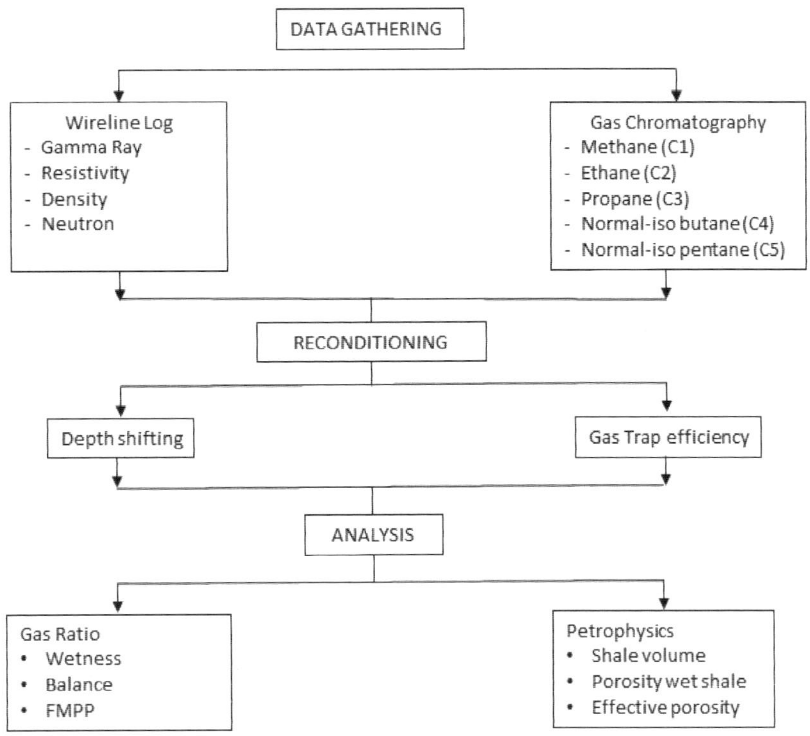

Figure 3. Flow Chart illustrated three main step combination for research

between drilling fluid and gas drilling on the surface called as Gas Trap. The equipment actually design with constant agitation that ideally create constant rotation during separation. The primary objective of the extraction device is to obtain a gas air mixture with the same relative composition as the gas entrained in the mud. One of parameter and terms applying to the process of extracting gas from mud is trap efficiency, it consists of fraction of gas actually extracted from mud.

Analysis was continue by compare result between gas ratio and petrophysis determination for porosity calculation, where the gas ratio equation that used are wetness, balance, FMPP whereas several log data such as gamma ray, density , neutron are used to identified shale volume, porosity wet shale, effective porosity for petrophysis analysis purposed as mentioned on Figure 3

3.1 Gas Ratio

Gas ratio is a comparison between 2 or more types of alkane gas formed from Gas While Drilling (GWD) extraction [7]. Observation with specific analysis gas trend allowing to gain interpretation hydrocarbon bearing zones to non-potential hydrocarbon [8].

Estimated Potential Porosity (FMPP) is a comparison between alkanes with different molecular sizes that release from pore space of rock at different speeds. This is a qualitative

IOP Conf. Series: Earth and Environmental Science **1525** (2025) 012023 doi:10.1088/1755-1315/1525/1/012023

indication of fluid mobility, in sand and shale intercalations, a fluid mobility of more than 10 indicates sand and can be calibrated locally.

$$FMPP = C1+C2/C4+C5 \quad (1)$$

Wetness (WH), basically ratio that provides direct information on changes in reservoir type and contact between different types of fluids. This data is plotted into a depth-based log so that it can be used as a reference for qualitative log interpretation.

$$WH = 100 \times (C2+C3+C4+C5/C1+C2+C3+C4+C5) \quad (2)$$

Balance (BH), is a direct comparison between light gas and heavy hydrocarbon gas. This BH analysis value is used to support the Wetness Ratio plot to increase the accuracy of the reservoir analysis.

$$BH = C1+C2/C3+C4+C5 \quad (3)$$

where:

C1 = methane (ppm)
C2 = ethane (ppm)
C3 = propane (ppm)
C4 = butane (ppm)
C5 = pentane (ppm)

3.2 *Petrophysics calculation*

Porosity calculations for the research area were carried out using calculations for siliciclastic rocks. Wireline log data such as gamma ray, density and neutron are used to calculate each parameter used in the formula below to calculate the total porosity of the rock:

$$Vsh = GRlog - GRmin / GRmax - GRmin \quad (4)$$
$$\phi Ncorr = \Phi N - (\Phi NSH/0.45) \times 0.13 \times Vsh \quad (5)$$
$$\phi D = p_{ma} - p_b / p_{ma} - p_{fb} \quad (7)$$
$$\phi Dcorr = \Phi D - (\Phi NSH/0.45) \times Vsh \quad (8)$$
$$\Phi e = ((\phi Ncorr/ \phi Dcorr)/2) - (Vsh \times \Phi sh) \quad (9)$$
$$\phi t = \phi e + Vsh \times \phi \text{ Wet Shale} \quad (10)$$

where:

GRlog = Gamma ray reading on shally-sand (API)
GRmax = Gamma ray maximum reading at clear Shale (API)
GRmin = Gamma ray minimum reading at clear Sandstone (API)
ϕN = Neutron porosity (v/v)
ΦNSH = Neutron porosity from high GR at shale (0.3-0.45) (v/v)
$\phi Ncorr$ = Neutron porosity correction (v/v)
ϕD = Density, porosity (g/cm3)
p_{ma} = Density from shale (g/cm3)
p_b = Density from shally formation (g/cm3)

ρ_f	=	Density fluid (g/cm3)
ϕDcorr	=	Density porosity correction (g/cm3)
ϕt	=	Total porosity (v/v)
ϕe	=	Effective porosity (v/v)
Vsh	=	Volume shale (v/v)
ϕ Wet Shale	=	Total porosity wet clay (v/v)

4. Analysis and Discussion

Figure 4 and 5 are shown comparison between wireline log and gas ratio chromatography where each of them was used to identified porosity with several equation that explained above. Where the reservoir zone has been stratigraphically correlated so that it can show the vertical relationship and its availability from each existing well.

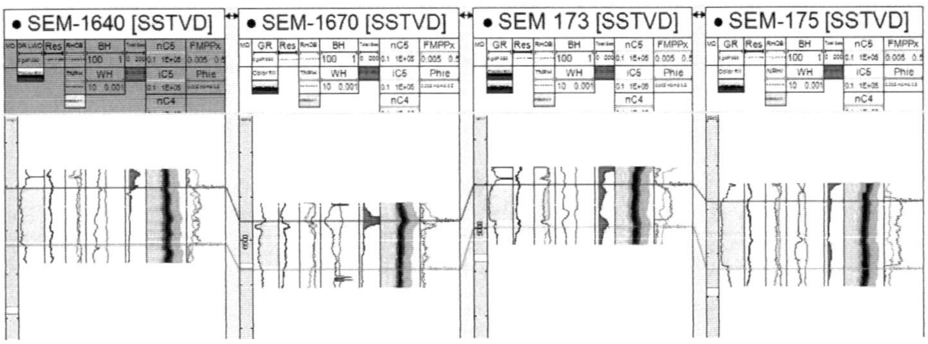

Figure 4. Porosity analysis comparison between Gas Ratio versus Petrophysics at Zone 1

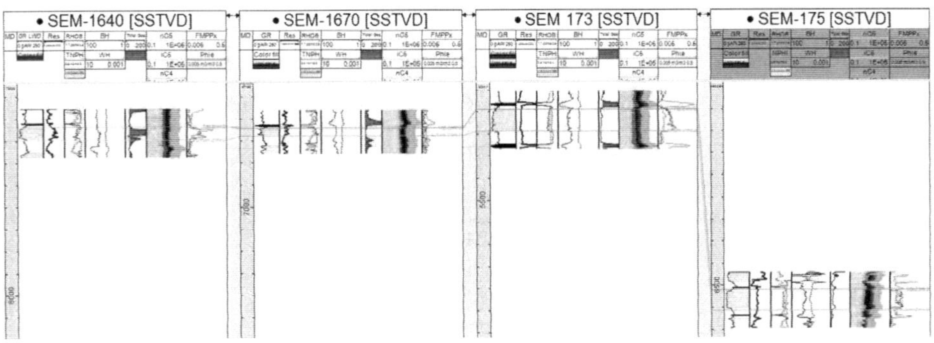

Figure 5. Porosity analysis comparison between Gas Ratio versus Petrophysics at Zone 2.

There are 8 columns in each of figures 4 and 5 that try to provide an overview of the data and porosity analysis from both the gas ratio method and the petrophysis method. From the right of each figure there is an MD (Measured Depth) column explaining the depth, the GR-Res -TNPH (Neu)-RHOB column is the column where the wireline log data is. The next columns in turn are the gas chromatography ratio data, namely the BH-WH- Total Gas (TG) column and chromatography, namely C1-C5. While the last column is a column that explains the results of the porosity analysis using the petrophysis method (Phie) and the gas ratio method (FMPPx).

Quality Control for depth adjustment and gas equipment is also displayed from the log. In order to avoid confusion in selecting interest zones, especially to differentiate between porous and non-porous rocks where porous rocks are shown in the GR column with yellow shading which are then interpreted as sandstone, some gas ratio formula data is also attached to interpret lithological characteristics.

Interpretation of the existence of reservoir in zone 1 (Fig.4) in each well using both gas ratio and petrophysis methods shows similar analogy as in GR Res value with TG value that increases shown from yellow shading that appears in GR column and red shading in TG column although red shading in TG column will be corrected against the trend of convex concave value in WH BH column and changes in FMPPx and Phie value in the last column as a result of porosity analysis. The existence of reservoir zone 2 (Fig.5) is very well controlled by the emergence of coal lithology before penetrating the reservoir layer which is marked by a significant increase in the value change in TG column.

While the trend of WH-BH values when compared with the trend of values in the RHOB-TNPH column, namely when yellow shading appears in the RHOB-TNPH column it will be followed by a trend that tends to experience a concave shape in zone 1 (Fig.4) although this value trend is inconsistent in several wells due to the control and function of the reservoir characteristics shown from the porosity analysis in the FMPPx-Phie column. Likewise in figure 5 which shows the inconsistency between the trend of WH-BH and RHOB-TNPH values due to being controlled by the reservoir characteristic value itself, one of which is the porosity value depicted in the FMPPX- Phie column as a result of comparing porosity values between the gas ratio and petrophysis methods. Changes in the thickness of the yellow shading in the RHOB-TNPH column and the concave-convex trend in the WH-BH column will be directly proportional to the magnitude of the porosity value analyzed in the FMPPx-Phie column.

Comparison of porosity values displayed in the FMPPx-Phie column from both figures 4 and 5 shows the varying porosity values of each reservoir. For comparison of porosity values from both the gas ratio and petrophysis methods from each well and each zone can be seen in the following table:

Table 1. Comparison porosity value

Well Code	Porosity Value (v/v)			
	Zone 1		Zone 2	
	Gas Ratio	Petrophysics	Gas Ratio	Petrophysics
SEM–164	0.04 – 0.18	0.03 – 0.19	0.05 – 0.32	0.03 – 0.21
SEM–167	0.03 – 0.09	0.01 – 0.24	0.04 – 0.22	0.01 – 0.22
SEM–173	0.06 – 0.17	0.01 – 0.23	0.04 – 0.10	0.02 – 0.23
SEM–175	0.06 – 0.14	0.006 – 0.16	0.03 – 0.10	0.05 – 0.14

Table 1 above shows a comparison of porosity value variations in each reservoir zone with the 2 methods used. The distribution of porosity values from each zone and each method used experienced a maximum difference in well SEM-167 zone 1, namely in the upper range of values around 0.15% while in zone 2 the minimum distribution value was around 0.03% which had the smallest difference between the petrophysical method and the gas ratio.

Table 2. Diffrentiation percentage ratio between Porosity Calculation based on Petrophysics and Gas Ratio Method

Well Code		Population	Cumulative differentiate	% Ratio
SEM–164				
	Zone 1	69	3762	55
	Zone 2	41	2006	49
SEM–167				
	Zone 1	56	5294	95
	zone 2	29	1715	59
SEM–173				
	Zone 1	51	3528	69
	Zone 2	56	4954	88
SEM–175				
	Zone 1	41	2323	54
	Zone 2	21	4498	93

[a] Average ratio calculation is 70.25%

Each chromatography and wireline log data that has been processed using gas ratio analysis or petrophysics is then compared statistically to obtain the level of compatibility between the two. Each reservoir zone that will be analyzed is first correlated based on its structural marker, then the gas chromatography and wireline log data per 1 ft interval are analyzed which are then known in the form of the actual number of data populations that provide a picture of the thickness of each reservoir. Where it is known that the highest population is in SEM-164, namely 69 populations, which illustrates that zone 1 in the well has a thickness of 69 ft.

Then each difference per 1 ft of data depth is summed in the form of a percentage which is recorded in the cumulative differentiation column of each reservoir zone per well analyzed.

Before taking the average of each percentage ratio, the highest porosity calculation suitability was in well SEM-167 in zone 1, which was around 95%, while the smallest value was in well SEM-175 in zone 1, which was around 54% suitability between the petrophysical and gas ratio methods. From data population collection of around 169 data obtained ratio of porosity calculations with gas ratios to petrophysics calculations of average around 70.25%.

5. Conclusion

The impact of drilling activities on the calculation of rock porosity through wireline logging data can be confirmed through the calculation of gas ratio porosity. From the results of data processing carried out in the research area by comparing the petrophysical method with the gas ratio, it turns out to provide results that can increase our level of confidence in conducting porosity analysis. Time and cost efficiency should be possible considering that sometimes due to well conditions that do not allow for wireline logging work, so the solution to calculate porosity with gas ratio analysis can be done.

6. Acknowledgement

The authors would like to thank Pertamina Hulu Indonesia Regional 3 Zona 9 for granting permission to publish this paper, especially Well Operation Petrophysics and Data Management Hendra Halomoan Pasaribu, Dwi Kurniawan Said, Danny Nursasono, Rangga Mahardika Khairully. Fruitful discussion and material preparation are with support from Petrophysics team Yunita Meilany, Nur Amalia Murtrijito, Jasmine Mustika Sukmawati.

References

[1] Flores Vitoria, et.al (2024), Maximizing Mud Logging Data Value: Sw And Porosity Prediction for The Cabeças Formation - Parnaı́ba Basin, SPWLA 65th Annual Logging Symposium, Brazil

[2] S. A Baylis , K.Hall (1994), The analysis of the C1--C5 components of natural gas samples using gas chromatography-combustion-isotope ratio mass spectrometry, Org. Geochem. Elsevier, Great Britain

[3] Beda. G (1999), Gas While Drilling (GWD); A Real Time Geologic and Reservoir Interpretation Tool, SPWLA

[4] B.C Duval, G,Choppin de Janvry, Bernard Loiret (1992),TOTAL; The Mahakam Delta Province: An Ever-Changing Picture and a Bright Future, 24th Annual OTC, Houston Texas

[5] Gultom, Luhut (2010), VICO Indonesia; Integrated 3D Reservoir Model and Petrophysical Study to Optimize Field Development in Low Permeability Deltaic Reservoir, Badak Field-Indonesia, SPE Asia Pacific, Australia 40th Annual Logging Symposium, Italy-France

[6] Salahuddin and Lambiase,2013, Sediment Dynamics and Depositional Systems of The Mahakam Delta, Indonesia, Journal of Sedimentary Research, v. 83, 503-521

[7] Buckle, Philip et all, 2022, Utilization of Digitalized Numerical Model Derived from Advanced Mud Gas Data For Low Cost Fluid Phase Identification, Derisking Drilling And Effective Completion Plan In Depleted Reservoir, SPWLA 63rd Annual Logging Symposium, Norway

[8] Ahsan, Mohammad et all, 2021, Advanced Gas While Drilling GWD Comparison with Pressure Volume Temperature PVT Analysis to Obtain Information About the Reservoir Fluid Composition, a Case Study from East Kuwait Jurassic Reservoir, SPE, Dubai

[9] Mode Wilfred, Ayonna et.al (2014), The Application of Chromatographic Gas Ratio Analysis in Reservoir Fluid Evaluation of "Beta" Field in the Congo Basin, Journal Geological Society of India, V.84

[10] Crampin, Tom, 2013, Application of Advanced Mud Gas Logging for Improved Hydrocarbon Phase Determination in a Highly Depleted Reservoir, IPTC 17088, Beijing

[11] Archie, G.E., 1942, The electrical resistivity log as an aid in determining some reservoir characteristics, *Transactions of the American Institute of Mining, Metallurgical and Petroleum Engineers*, **146**(01), 54-62.

IOP Conf. Series: Earth and Environmental Science **1525** (2025) 012024 doi:10.1088/1755-1315/1525/1/012024

Real-Time Environmental Monitoring of A Greenhouse and A Poultry Farm

Warni Warni*, Ratnalia Ratnalia, Siti Fathana Azh Zahra and Halmar Halide

Department of Geophysics, Hasanuddin University, Makassar, Indonesia

*warni21h@student.unhas.ac.id

Abstract. Environmental monitoring based on the Internet of Things (IoT) is the key technology to monitoring the environment in real-time using advanced sensors to acquire data efficiently. The research aims to develop microclimate monitoring using Arduino with a BME280 sensor and a TEMT6000 sensor located in a greenhouse and a poultry farm (closed house and open house). The BME280 sensor measures temperature, humidity, and air pressure. The TEMT6000 measures light intensity. The tool is designed to acquire data in a large database directory using Influx DB Cloud. The data is sent to Grafana for visualization in a dashboard. The research's findings show successful real-time environmental monitoring of acquiring data can be seen both numerically and visually. The tool has inherent simplicity, affordability, and adaptability. The tool is easily implemented in a variety of contexts.

1. Introduction

Recently, the developing sectors of agriculture and farming have drawn central attention, encouraging productivity in these sectors and enhancing quality improvement and productivity products in the sectors. Surely, real-time environmental monitoring plays an essential role. In total, environmental monitoring is used to observe the quality of the environment. Weather conditions have effects on important sector activities including farming, agriculture, transportation, etc. Weather information accuracy in a sector is important [1].

The need for real-time, in-situ, and high-accuracy data drives the development of research in the field of instrumentation. In particular, several applications such as greenhouses, livestock pens, storage warehouses, and others, require 24-hour environmental monitoring. This measurement needs is tough to fulfill using conventional methods [2]. Monitoring systems provide opportunities to improve production management decisions, so current technology such as the Internet of Things (IoT) is considered a solution that can solve this problem [3]. On the other hand, the abundant availability of electronic components in the form of sensors and controller boards on the market at fairly affordable prices is also a driving factor for the development of research in the field of instrumentation. Thus, it allows the development of systems that are more efficient and responsive to changing environmental conditions [2].

Arduino is an open-source electronics prototyping platform that is flexible and easy to use. Arduino can sense the surrounding environment by receiving input signals from various sensors and can operate automatically after being run by a computer. By implementing an IoT (Internet of Things)-based monitoring system, the data obtained will be processed and entered into a

IOP Conf. Series: Earth and Environmental Science **1525** (2025) 012024 doi:10.1088/1755-1315/1525/1/012024

database, where the data can then be monitored directly (in real-time) [4]. A BME280 and A TEMT6000 are sensors used in the research with accurate data and real-time. Significantly, the system contributes to managing a greenhouse and a poultry farm. It can increase the efficiency of productivity sectors.

2. Method

The block diagram in Figure 1 illustrates the system design process for real-time environmental monitoring. BME280 and TEMT6000 are sensors used in designing a tool. A BME280 sensor measures temperature, humidity, and air pressure. The sensor parameters have been a temperature resolution of 0.01°C, a humidity accuracy of 3%, and an air pressure of 1 hPa, supporting the Serial Peripheral Interface (SPI) and Inter-Integrated Circuit (I2C) interfaces [5]. Additionally, TEMT6000 is a sensor that measures light intensity and functions as a transistor. It detects the light density and reflects the analog voltage signal with a supplying voltage range from 3.3V to 5.5V [6]. Arduino integrates a combination of sensors. It is creating a framework for real-time environmental monitoring.

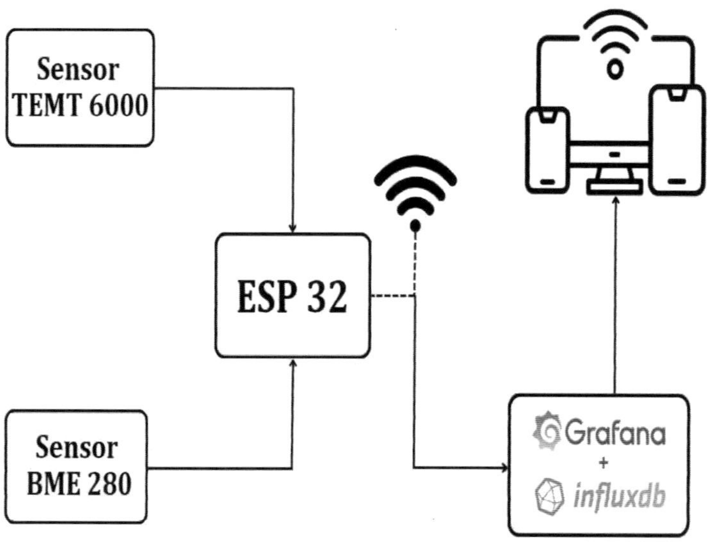

Figure 1. Block Diagram

The tool was tested in a Hasanuddin University library. The data sensors connect to a programable microcontroller (ESP32) using the Arduino IDE [7]. The interval for sending data is 15 seconds. The system needs a power supply to run the tool and an internet network to send data to Influx DB Cloud, a large directory database connecting to Grafana to show a real-time monitoring dashboard numerically and visually. The dashboard shows parameters in a time series, a standard gauge visualization, a table of contents, and a visualization of the Indonesia map. The data can be accessed on a computer, a PC, and a phone. Also, a dashboard on Grafana shows data parameters that can be downloaded in a CSV file and require time in a historical monitoring environment. After testing the tool, real-time monitoring will be continued in the faculty of agriculture's greenhouse and animal science's poultry farm (closed house and open house) at Hasanuddin University.

IOP Conf. Series: Earth and Environmental Science **1525** (2025) 012024 doi:10.1088/1755-1315/1525/1/012024

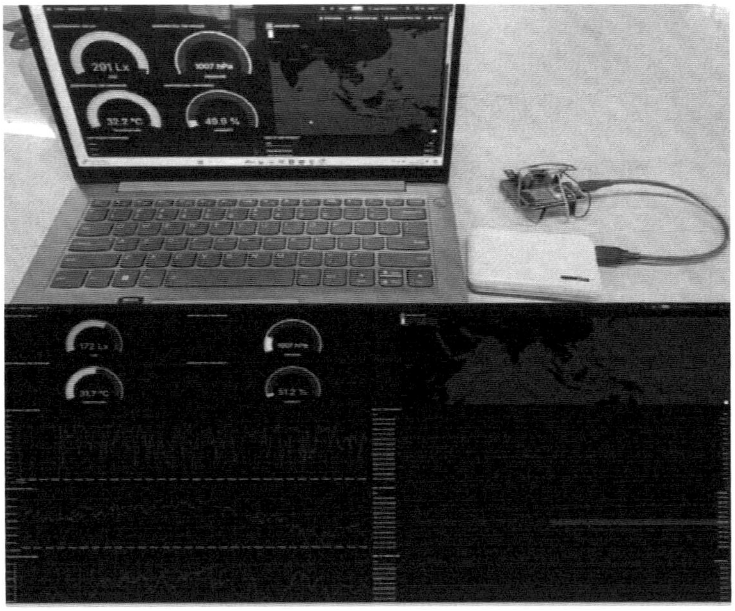

Figure 2. Display on Hardware

3. Result and Discussion

Real-time environmental monitoring in a greenhouse and a poultry farm results in real-time data changes at main parameters temperature, humidity, air pressure and light intensity. Each location has a different characteristic environment. Parameter changes are affected by factors such as weather, time and activity happening in a monitoring area. The record data below is shown on the Grafana dashboard. The record data of real-time monitoring was done on 3rd October 2024 in a greenhouse and 4th October in a poultry farm.

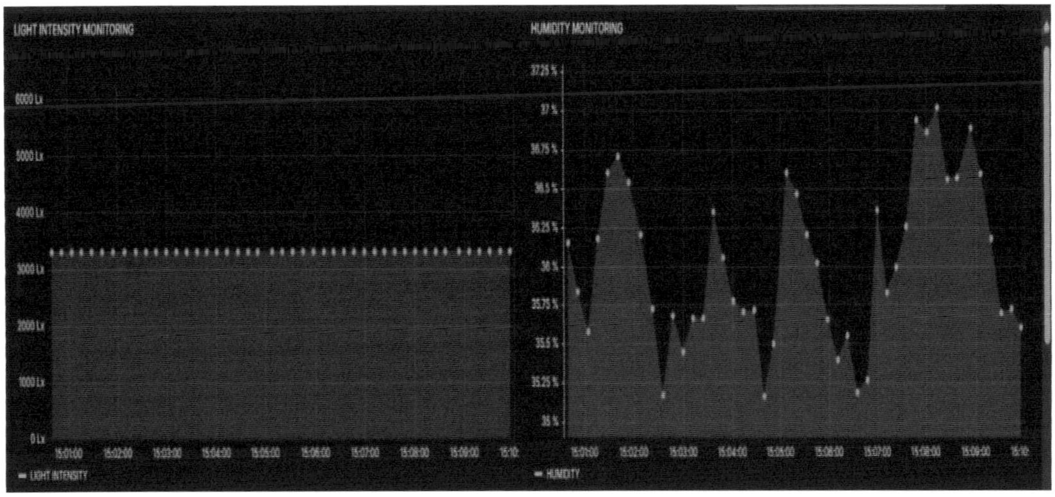

Figure 3. Real-time monitoring in a greenhouse. Left and right indicate monitoring of light Intensity and humidity, respectively.

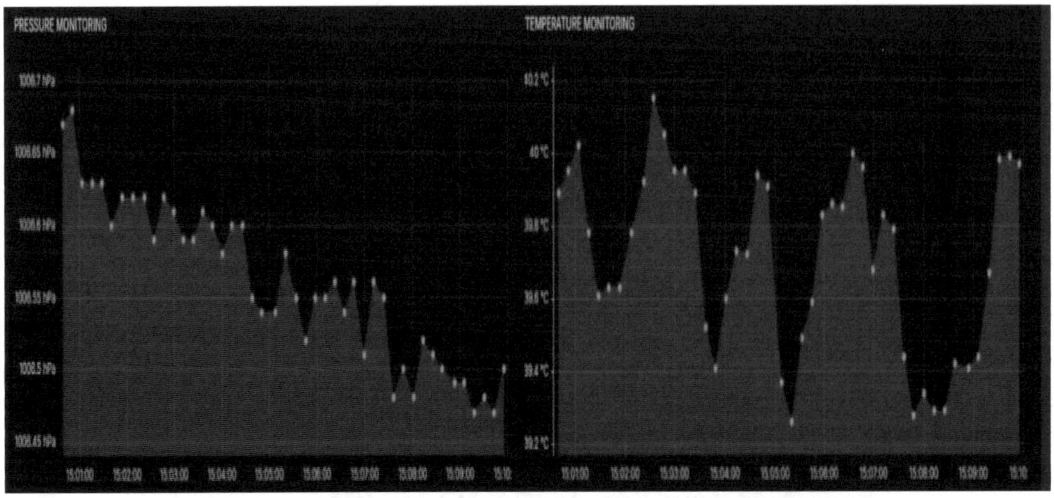

Figure 4. Real-time monitoring in a greenhouse. Left and right indicate monitoring of pressure and temperature, respectively.

Based on monitoring data collected in a greenhouse between 15:01:00 and 15:10:00 in Figure 3 and 4, A light intensity of 3000 lux was observed within a reasonable and acceptable range for typical daytime lighting conditions of a greenhouse designed for cultivating plants that require indirect or partial sunlight. Humidity that ranges from 35.5% to 37.5% is relatively low, so it may be worth paying attention to, especially if the cultivated plant requires higher humidity to reduce the risk of plant stress. Although bounds are still within an acceptable range, a sign of a shift in the outside weather is a drop in air pressure from 1006.8 hPa to 1005.4 hPa. The majority of plants thrive at temperatures between 25°C and 30°C, so temperatures that range between 39.4°C and 40.2°C are extremely high and may harm some plants [8]. Therefore, greater attention must be devoted to temperature and humidity to maintain the ideal circumstances for plant growth.

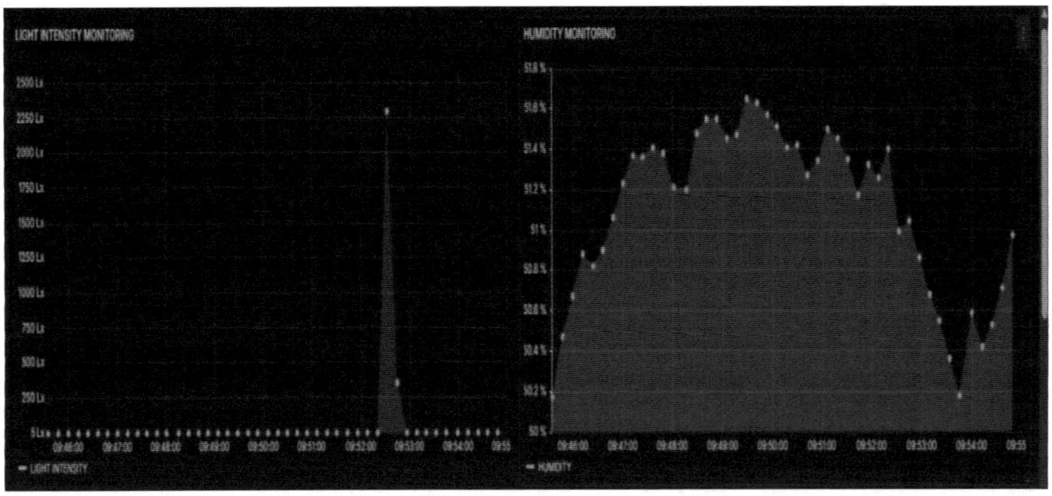

Figure 5. Real-time monitoring in a closed house poultry farm. Left and right indicate monitoring of light Intensity and humidity, respectively.

IOP Conf. Series: Earth and Environmental Science **1525** (2025) 012024 doi:10.1088/1755-1315/1525/1/012024

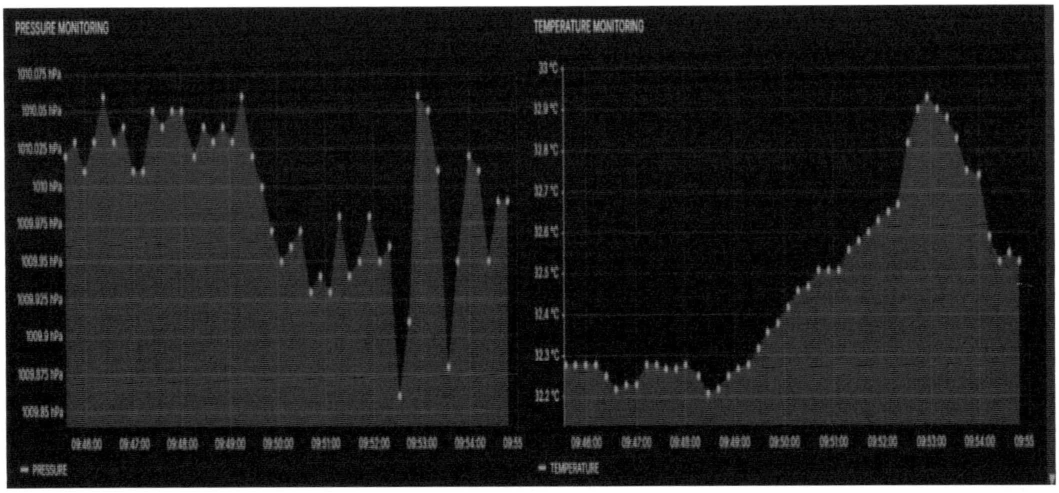

Figure 6. Real-time monitoring in a closed house poultry farm. Left and right indicate monitoring of pressure and temperature, respectively.

The same environmental parameters were monitored in the closed-house poultry farm with data collected between 09:46:00 and 09:55:00 in Figure 5 and 6. This monitoring was conducted to evaluate environmental conditions in a controlled poultry facility, where climatic parameters are regulated to support optimal livestock development. The resulting graph shows that temperature, humidity, and air pressure remained relatively stable, indicating favorable conditions for poultry health. Although the lighting intensity was set to a low level, an anomalous spike was observed due to a deliberate test in which the sensor was brought closer to the light source to assess its sensitivity. Closed-house poultry systems are equipped with ventilation and climate control mechanisms designed to maintain thermal comfort and prevent heat stress. Failure of these systems may lead to adverse effects on poultry health and egg quality [9][10].

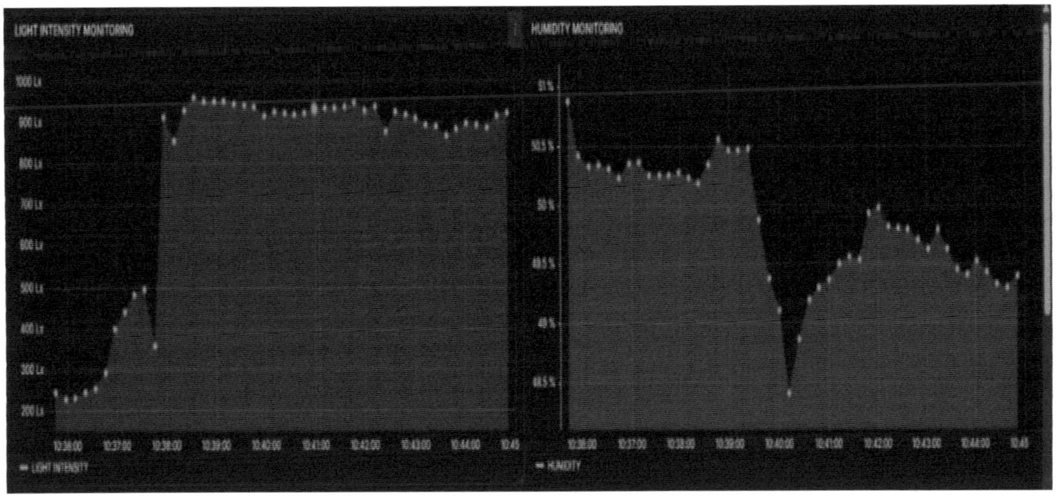

Figure 7. Real-time monitoring in an open house poultry farm. Left and right indicate monitoring of light Intensity and humidity, respectively.

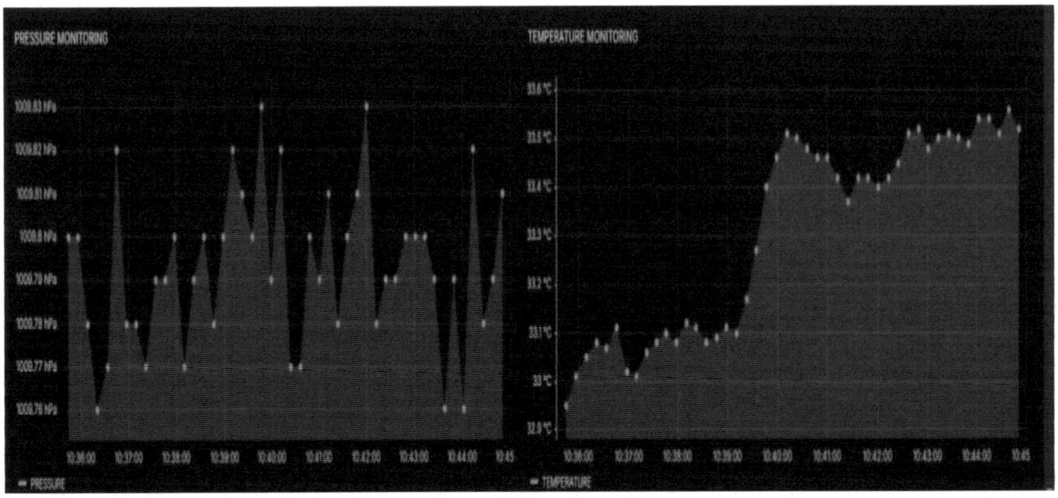

Figure 8. Real-time monitoring in an open house poultry farm. Left and right indicate monitoring of pressure and temperature, respectively.

The results of real-time monitoring of temperature, humidity, pressure, and light intensity in the open-house poultry farm revealed noticeable fluctuations across parameters. Unlike closed-house systems, open-house poultry farms are highly affected by external environmental factors such as weather, as no climate control mechanisms are implemented. Data collection was conducted between 10:36:00 and 10:45:00 local time in Figure 7 and 8. According to the monitoring graph, temperature remained relatively stable between 32.9°C and 33.5°C, which falls within the ideal range for poultry growth. Humidity levels ranged from 48% to 51%, which is also considered optimal for chicken development. Air pressure fluctuated slightly between 1009.78 hPa and 1009.83 hPa due to external influences, but remained within a normal range that poses no risk to livestock. Light intensity varied between 200 and 1000 lux, with a notable spike from 350 lux to 900 lux, attributed to changing sunlight conditions. Although this intensity is still within acceptable limits, prolonged exposure may increase the risk of stress in poultry [11]. Additionally, open-house poultry farms lack environmentally friendly features due to unstructured ventilation systems, resulting in limited control over air quality and increased potential for environmental pollution in surrounding areas [12].

Based on real-time monitoring analysis in greenhouse and poultry farms (closed house and open house). Environmental conditions in each location are different, thus having a major impact on plant development and chicken health. These factors, which are all important in creating a conducive habitat for plants and animals, include changes in temperature, humidity, light intensity, and air quality. Data from the sensor shows good performance in recording these environmental parameters, provides accuracy and timely information, and underscores the importance of adaptive control so that growing conditions remain optimal at all three locations. Thus, the application of this sensor technology can help in making better decisions to increase agricultural and livestock productivity sustainably.

4. Conclusion

An IoT-based environmental monitoring system that combines BME280 and TEMT6000 sensors with Arduino and ESP32 works well for real-time monitoring of greenhouse and poultry farm conditions. The system can provide reliable data on temperature, humidity, pressure, and light intensity, critical for maintaining optimal conditions for plant growth and poultry health. Monitoring results show that dynamic environmental factors, such as light intensity and humidity, greatly influence productivity in the agricultural and livestock sectors. Therefore, applying this technology supports more informed decision-making to increase sustainable productivity in both sectors.

5. Acknowledgments

We want to sincerely thank everyone who helped and supported us during our research, especially our supervisor. In addition, also gratitude to the organizing committee that made this research possible.

References

[1] Narayana TL, Venkatesh C, Kiran A, J CB, Kumar A, Khan SB, et al. Advances in real time smart monitoring of environmental parameters using IoT and sensors. Heliyon [Internet]. 2024;10(7):e28195. Available from: https://doi.org/10.1016/j.heliyon.2024.e28195
[2] Rustami E, Fitria Adiati R, Zuhri M, Arif Setiawan A. Uji Karakteristik Sensor Suhu Dan Kelembaban Multi-Channel Menggunakan Platform Internet Of Things (IOT). Berk Fis [Internet]. 2022;25(2):45–52. Available from: https://ejournal.undip.ac.id/index.php/berkala_fisika/article/view/47903
[3] Mörth O, Eder M, Holzegger L, Ramsauer C. IoT-based monitoring of environmental conditions to improve the production performance. Procedia Manuf [Internet]. 2020;45(2019):283–8. Available from: https://doi.org/10.1016/j.promfg.2020.04.018
[4] Shirsath PDO, Kamble P, Mane R, Kolap A, More PRS. IOT Based Smart Greenhouse Automation Using Arduino. Int J Innov Res Comput Sci Technol. 2017;5(2):234–8.
[5] Sari IP, Gunawan AAN, Wibawa IMS, Putra IK, Yusuf M. Design of Radiosonde Based on Arduino Pro Mini Using BME280 Sensor. Tech Innov Eng Res Vol 5. 2023;(July):145–57.
[6] Siddagangaiah S. A Novel Approach to IoT Based Plant Health Monitoring System. Int Res J Eng Technol [Internet]. 2016;3(11):880–6. Available from: https://www.irjet.net/archives/V3/i11/IRJET-V3I11154.pdf
[7] Akbar A, Zaenudin Z, Mutaqin Z, Samsumar LD. IoT-Based Smart Room Using Web Server-Based Esp32 Microcontroller. Formosa J Comput Inf Sci. 2022;1(2):79–86.
[8] Hatfield JL, Prueger JH. Temperature extremes: Effect on plant growth and development. Weather Clim Extrem. 2015;10:4–10
[9] Fattah AH, Faridah R, Amalia AHN, Khaeruddin K. Pengaruh Pengaturan Suhu dan Kelembaban di Kandang Closed House Terhadap Performa Broiler. Musamus J Livest Sci. 2023;6(1):12–20.
[10] Igbokwe NA. Effects of environmental heat stress on reproduction and its management in chickens. Niger Vet J. 2018;39(2):101.
[11] Kustiawan E, Rukmi DL, Imam S, Permadi SO. The Study of Lighting Intensity on The Production Peak of Layer Chicken in UD. Mahakarya Farm Banyuwangi. J Ilmu Peternak Terap. 2019;3(1):14–8.
[12] Mulia SB, Erdani Y, Febrian MR, Alfian RF. Rancang Bangun Miniatur Sistem Kontrol Dan Monitoring Suhu Kandang Close House Berbasis Arduino Uno. TEDC. 2022;16(2):116–25.

IOP Conf. Series: Earth and Environmental Science **1525** (2025) 012025 doi:10.1088/1755-1315/1525/1/012025

Mineralogical and Chemical Characteristics of Lateritic Nickel Profiles at Morowali Regency, Central Sulawesi, Indonesia

Lasti[1]*, Irzal Nur[1], Sri Widodo[1],

[1] Mining Engineering Department, Faculty of Engineering, Hasanuddin University, 91711, Indonesia

*E-mail: lastiminers@gmail.com

Abstract. The aim of this study is to analyze the mineralogical and chemical characteristics of lateritic nickel profiles at the Mining Permit area of Morowali Regency, Central Sulawesi. Two profiles of nickel laterite were arranged from eight core samples of two drill holes. For mineralogical study, petrographic analysis were performed for two bedrock samples, while XRD analysis were conducted for eight bedrock, saprolite, and limonite samples. For chemical analysis, the whole eight samples were analyzed using XRF methode. The results showed the nickel laterite in the study area consists of four layers, namely bedrock, rocky saprolite (bluezone saprolite), saprolite, and limonite. The bedrock is serpentinized harzburgite which is composed of olivine, enstatite and lizardite. For the two profiles, the bluezone saprolite are composed of antigorite, enstatite, lizardite, quartz, and magnetite. The saprolite zone consist of enstatite, goethite, antigorite, magnetite, and quartz. While the limonite zone are composed of goethite, magnetite, and hematite. Chemically, the bedrock is characterized by significant concentration of SiO_2 and MgO which reflects the presence olivine. pyroxene, and serpentine in the bedrock. The Ni grade in the saprolite zone is much higher than in the bluezone saprolite, around 1.45 to 1.39% and 0.23 to 0.20% respectively. Meanwhile, the limonite zone is very typically characterized by high Fe content, which ranges from 73.23 to 72.12%.

1. Introduction

Sulawesi, the worlds producer of nickel resources, is leading the rapid growth of the nickel industry in Indonesia. The laterite nickel deposits in Sulawesi formed from the world-scale lateritisation of the East Sulawesi Ofioite under active tectonic conditions and a tropical rainforest climate and are widely exposed to the surface (1,2,3). The study area is located on the island of Sulawesi in Bahomotefe Village, Morowali Regency with commodities of nickel laterite deposits currently being exploited by PT Haverst Consturction Mining Service in the form of an exploration IUP. The regional geology of this area refers to the Geological Map of the Bungku Sheet system at a scale of 1:250,000 (4) (Figure 1) with the East Sulawesi Ofiolite Belt Ultramafic Complex Formation (Ku) of limestone age composed of harzburgite, lherzolite, wehrlit, websterlit, serpentinite, peridotite, dunite, and gabbro (4,5,6).

One significant area, the Bahodopi Morowali area, has mineral characteristics in the limonite zone dominated by oxide hydroxide mineral groups, namely goethite and hematite, and there is chlorite and quartz in small amounts. The saprolite zone is dominated by serpentine group minerals, namely antigorite minerals, in addition there are enstatite minerals with a large composition, chlorite minerals and small amounts of quartz and magnetite. The bedrock zone is dominated by lizardite minerals, the main minerals are spinel and enstatite and there are magnetite minerals. The abundance of major and minor elements and compounds in the profile of nickel laterite deposits has varying grades (7).

In the Morowali area of the Northeast arm of Southeast Sulawesi, there are high-grade saprolite-type ore bodies 10 to 20 m thick. Such cases show that topographic relief tends to govern the Ni grade distribution and thickness of deposits, high grade intervals tend to occur in places where garnierite joints and veins are dense (1). Differences in the characteristics of a laterite nickel deposit area are related to differences in the mineralogical composition and chemical content of the laterite nickel deposit profile.

The diversity of mineralogical and chemical characteristics can be known based on physical properties (actual visual), chemical properties (X-Ray Flourscence Analysis) and observations of optical properties (X-Ray Diffraction and petrographic analysis) to determine the characteristics of lateritic nickel deposits in the study area. Studies on the mineralogy and geochemistry of laterite deposits on the islands of Sulawesi and Kalimantan have been carried out by several researchers (8,9,10,11,12,13,14) so that this research is focused on analysing the mineralogy and chemistry of the laterite nickel deposit profile of PT Harvest Construction Mining Service in Morowali Regency, Central Sulawesi.

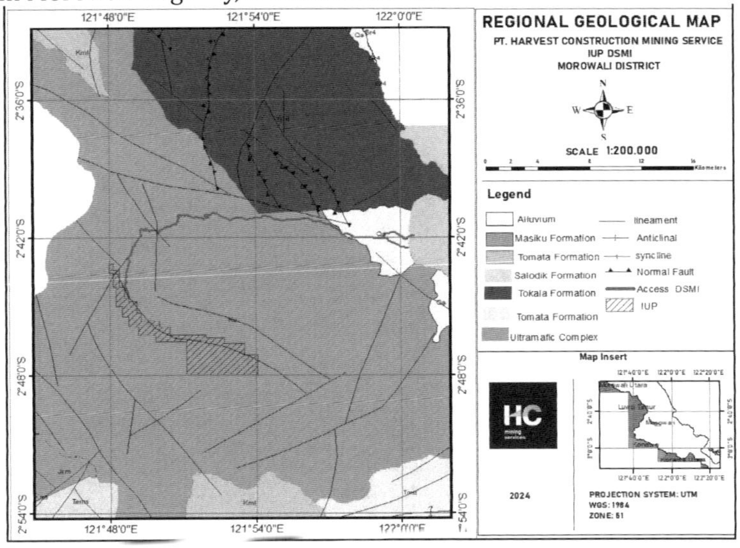

Figure 1. Regional Geological Map of IUP PT. PT. Harvest Construction Mining Service in Morowali Regency (4).

2. Methods

This research was conducted using two methods: field data collection and laboratory analysis. Field data collected consisted of logging descriptions (actual visuals) of two laterite nickel deposit profiles, namely hole-id DL 1361 0 to 24 m thickness and hole id DC 1763 0 to 22 m thickness (Figure 2). These types of laterite profiles were then visualised using Strater5 software by drawing the laterite nickel deposit profile at each borehole point. Laboratory activities consisted of two types of preparation: thin section samples (bedrock samples) for petrographic analysis and 200 mesh size fraction samples for XRD (X-Ray Diffraction) and XRF (X-Ray Flourscence) analysis.

Petrographic analysis was carried out using microscopic polarisation in the form of thin section observations (this section) to identify microscopic minerals as well as optical properties and structures in rocks that can be used as a basis for determining the type of rock name. The analysis was carried out in the preparation room, Department of Geological Engineering, Faculty of Engineering, Hasanuddin University using a Nikon Eclipse LV 100N Pol light refraction type microscope.

X-Ray Diffraction analysis was carried out to detect the set of minerals in the sample using the Bragg equation method. The results of XRD (X-Ray Diffraction) analysis, diffractograms are processed using match!4 software which is then interpreted to obtain mineral compounds and determine the set of minerals in the research sample. The output of match!4 software is an analysis report in the form of mineral phases, list speak, diffraction pattern graphs and others. This analysis was carried out at the Mining Engineering Department's Mining Materials Processing Laboratory, Hasanuddin University using a Shimidzu Maxima-X XRD 7000 type XRD instrument.

XRF (X-Ray Flourscence) analysis was conducted to determine the major, minor and trace elements and compounds in the nickel laterite sediment profile samples. Major chemical compounds such as MgO, Fe_2O_3, SiO_2, Al_2O_3, minor elements and compounds such as Ni, Co, Cr_2O_3, MnO, CaO and trace compounds such as TiO_2, K_2O, P_2O_5, SO_3, Na_2O in units of percent content. The content of elements and compounds contained in the sample is detected through X-ray absorption. This analysis was carried out at the Laboratory building of PT Sucofindo Makassar Branch.

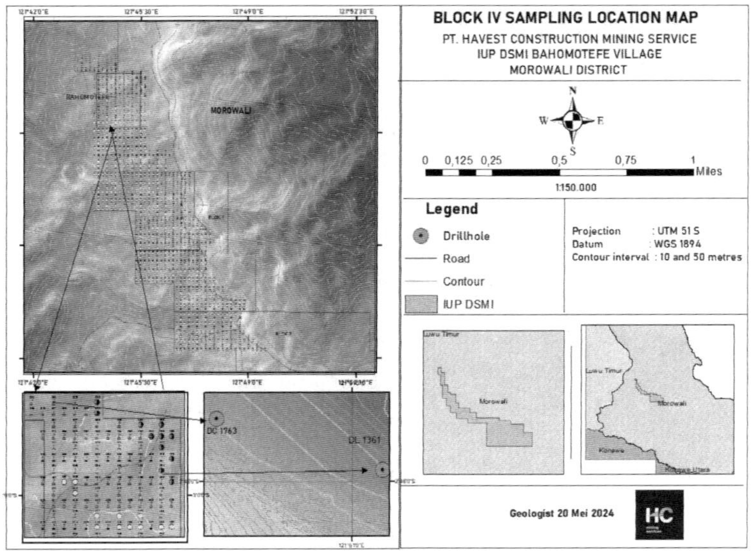

Figure 2. Sampling location map

3. Results and Discussion

3.1 Mineralogical Analysis

3.1.1 Petrographic Analysis

Petrographic analysis of sample DC 1763 shows serpentinised harzburgite rock with massive structure and secondary alteration grade of about <50.5%. The main minerals consist of olivine, orthopyroxene, clinopyroxene, and opaque minerals (magnetite). The rock has several textures, including protogranular and exsolution lamellae in the pyroxene minerals. Olivine is green to purplish in colour with high birefringence, some small grains (0.02 to 0.08 mm) separated by a mesh texture containing serpentine-like alteration minerals. Other olivine crystal sizes reach 1.5 mm. Orthopyroxene is white to grey-brown with low birefringence and subhedral-anhedral shape (grain size up to 1 mm). Some crystals are replaced by serpentine at the edges and surfaces. Clinopyroxene exsolution on olivine is yellowish green to reddish orange with

IOP Conf. Series: Earth and Environmental Science **1525** (2025) 012025 doi:10.1088/1755-1315/1525/1/012025

moderate relief and oblique extinction. Opaque minerals are scattered among the olivine and partly truncated by veins. Magnetite is mostly seen in pyroxene minerals and between crystals, some of which are cubic as major inclusions. Microscopic observation of sample DC 1763 as shown in Figure 3.

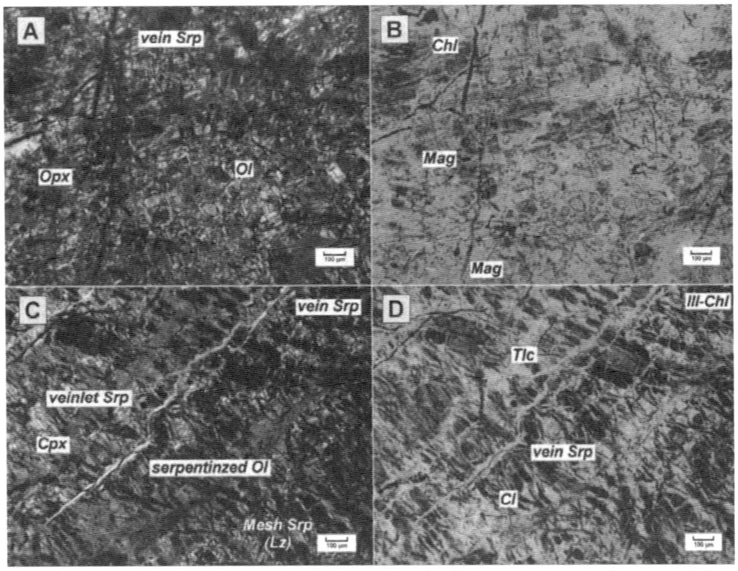

Figure 3. Photomicrographs of bedrock sample DC 1763 showing mineral assemblages of olivine (Ol), Orthopyroxene (Opx), Clinopyroxene (Cpx), veinlet serpentine, lizardite (lz), illite (ill), chlorite (Chl), talc (Tlc), clay (Cl), and magnetite (Mag). Figure A and C are cross nicol; Figure B and D are parallel nicol.

Thin section microscopic observations of Sample DL 1361 show that it is an ultramafic igneous rock called serpentinised harzburgite. This rock has a massive structure, but has undergone secondary alteration of about 56%. This alteration is characterised by changes in its mafic minerals to serpentine (greenish yellow in colour), illite, chlorite, iddingsite and opaque minerals. Mineralogically, serpentinised harzburgite is composed of orthopyroxene, clinopyroxene, olivine and opaque minerals. The texture is protogranular with characteristic exsolution lamellae, especially in pyroxene minerals. Olivine in these rocks is green to purplish in colour with high birefringence. Grain size is variable, with some small grains (0.05 to 0.5 mm) separated by a mesh texture containing serpentine-like alteration minerals. Orthopyroxene is white to grey-brown with low birefringence. It is subhedral-anhedral in shape with mineral grain sizes up to 2.15 mm. Clinopyroxenes exsolution on olivine are yellowish green to reddish orange in colour. Opaque minerals are scattered among the olivine and partly truncated by veins. Microscopic observation of sample DL 1361 as shown in Figure 4.

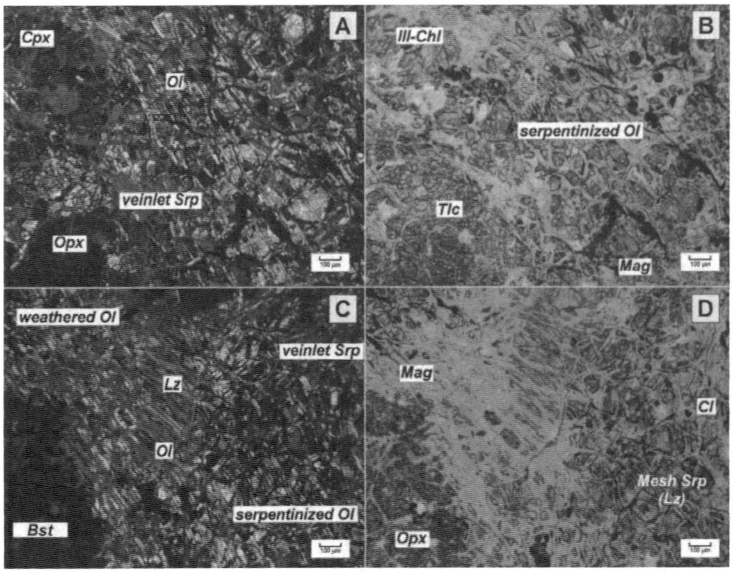

Figure 4. Photomicrographs of bedrock sample DL 1361 showing mineral assemblages of olivine (Ol), Orthopyroxene (Opx), Clinopyroxene (Cpx), veinlet serpentine, lizardite (lz), bastite (Bst) illite (ill), chlorite (Chl), talc (Tlc), clay (Cl), and magnetite (Mag). Figure A and C are cross nicol; Figure B and D are parallel nicol.

3.1.2 XRD Analysis (X-Ray Diffraction)

The XRD (x-ray diffraction) test results consist of 8 samples of DC 1763 and DL 1361. The following is a description of the XRD analysis results from both drill points in the study area:

3.1.2.1 DC 1763

This analysis was carried out on samples from each profile of nickel laterite deposits consisting of bedrock zones, bluezone saprolite zones, saprolite zones and limonite zones. Mineral occurrence diffractograms from XRD analysis of X-ray diffraction (Figure 5).

1. The bedrock zone shows a mineral composition dominated by the mineral antigorite ($H_{79}Mg_{48}O_{147}Si_{34}$) with a mineral composition presentation of 48.9%. It is followed by lizardite ($H_4Mg_3O_9Si_2$) at 41.9%, spinel (Al_2MgO_4) at 8.1%, and enstatite ($MgSiO_3$) at 1.1%.
2. The bluezone saprolite zone shows a diverse mineral composition, with antigorite ($H_{62}Mg_{48}O_{147}Si_{34}$) as the dominant mineral with a mineral composition presentation of 40.8%. This is followed by enstatite ($MgSiO_3$) at 25.3%, lizardite ($H_4Mg_3O_9Si_2$) at 21.4%, quartz (SiO_2) at 11.1%, and magnetite (Fe_3O_4) at 1.5%.
3. The saprolite zone shows a diverse mineral composition, with enstatite ($MgSiO_3$) as the dominant mineral with a mineral composition presentation of 70.8%. Followed by goethite ($FeO(OH)$) with a percentage of 16.6%, antigorite ($H_{62}Mg_{48}O_{147}Si_{34}$) at 9.1%, magnetite ($Fe_3O_4$) reaching 2.1%, and quartz (SiO_2) at 1.3%.
4. The limonite zone shows a mineral composition dominated by goethite ($FeO(OH)$), with a percentage of 89.9%. Followed by magnetite (Fe_3O_4) with a mineral composition of 6.1%, and hematite (Fe_2O_3) at 3.9%.

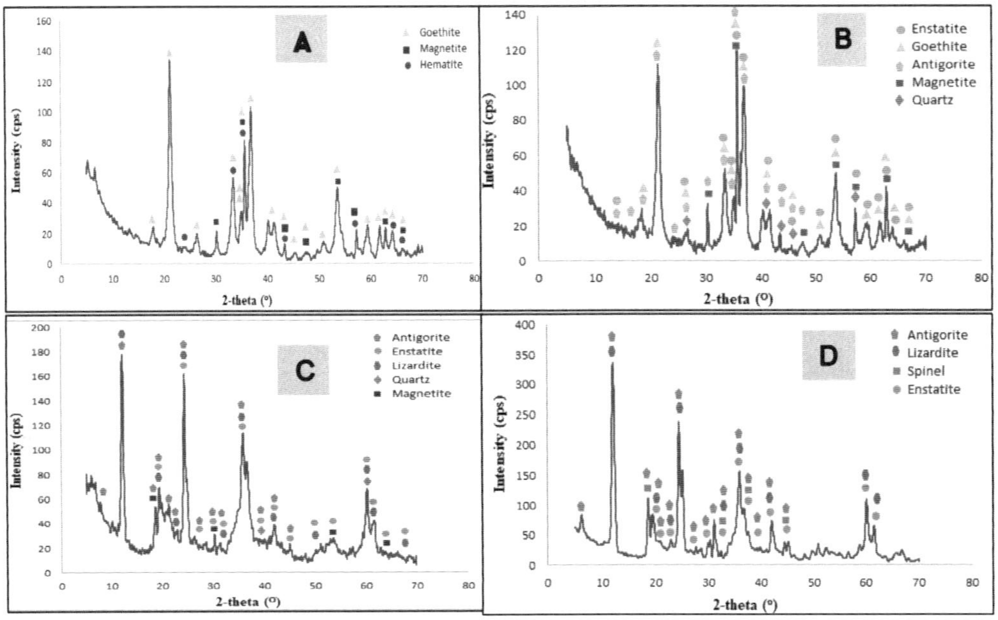

Figure 5. XRD Diffractograms of sample DC 1763. (A) limonite, (B)saprolite, (C) bluezone saprolite, (D)bedrock

3.1.2.2 DL 1361

This analysis was carried out on samples from each profile of nickel laterite deposits consisting of the bedrock zone, bluezone saprolite zone, saprolite zone and limonite zone, the mineral occurrence diffractogram from XRD analysis of X-ray diffraction (Figure 6).

1. The bedrock zone shows that the minerals that have the most dominant composition presentation are the antigorite mineral $H_{62}Mg_{48}O_{147}Si_{34}$ 30.7%, the spinel mineral Al_2MgO_4 26% which is a group of oxide group minerals, the lizardite mineral $H_4Mg_3O_9Si_2$ 22.9% and the enstatite mineral MgO_3Si 20.4%.
2. The bluezone saprolite zone has the most dominant composition presentation, the antigorite mineral $H_{79}Mg_{48}O_{147}Si_{34}$ 76.1%. Lizardite mineral $H_4Mg_3O_9Si_2$ 8.9%. Mineral enstatite MgO_3Si 7.7%, Mineral magnetite Fe_3O_4 3.7%. Quartz mineral SiO_2 3.6%.
3. The saprolite zone with mineral presence shows that, the mineral that has the largest composition presentation is the mineral goethite Fe_3O_4 48.3%. Antigorite mineral $H_{62}Mg_{48}O_{147}Si_3$ 47.2%. Quartz mineral SiO_2 38.2%. enstatite mineral MgO_3Si 3.4%. Magnetite mineral Fe_3O_4, 2.8%.
4. The limonite zone the mineral composition of which is dominated by the mineral goethite (FeOOH) with a percentage reaching 77.5% indicates strong oxidising conditions in the limonite zone, the mineral hematite (Fe_2O_3) which is a stable iron oxide mineral under high oxidising conditions, has a percentage of 12.6%, and the mineral magnetite (Fe_3O_4) is a magnetic iron oxide mineral found in small quantities (9.9%).

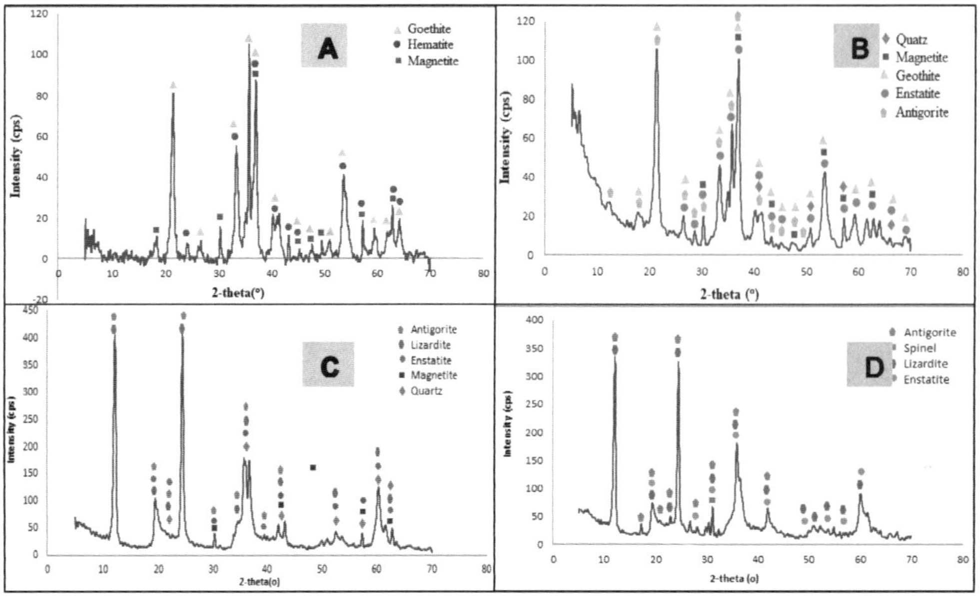

Figure 6. XRD Diffractograms of sample DL 1361. (A) limonite, (B)saprolite, (C) bluezone saprolite, (D)bedrock

3.2 Chemical Analysis

The results of chemical analysis of the samples from the three layers from the two drill holes are shown in shown in Table 1.

Table 1. Chemical composition of samples from the two drill holes

	HOLE ID							
	DC 1763				DL 1361			
CHEMICAL COMPOSITION	LIM 0 - 3	SAP 3 -14	BZ SAP 14 -16	BRk 16 -22	LIM 0 - 3	SAP 3 -7	BZ SAP 7 - 20	BRK 20 - 24
Major Elements								
Fe_2O_3	75.21	64.12	9.41	8.51	73.23	68.09	9.43	8.61
SiO_2	2.61	6.42	40.20	40.90	0.70	6.34	40.15	40.15
MgO	0.72	2.05	38.50	37.76	0.70	2.04	38.42	37.70
Al_2O_3	10.30	7.13	-0.09	-0.04	10.20	6.97	-0.08	-0.04
Minor Elements								
Ni	0.56	1.49	0.24	0.29	0.53	1.45	0.23	0.26
MnO	0.24	0.37	0.09	0.14	0.20	0.31	0.09	0.11
CaO	0.14	0.32	-0.09	0.54	0.10	0.29	-0.05	0.52
Cr_2O_3	2.56	2.41	0.41	0.41	2.47	2.38	0.39	0.39
Co	0.18	0.15	0.08	0.07	0.14	0.10	0.04	0.03
Trace Elements								
TiO_2	0.07	0.09	0.01	0.01	0.05	0.07	0.01	0.01
K_2O	0.01	0.01	0.01	0.01	0.01	0.01	0.01	0.01
P_2O_5	0.06	0.04	0.00	-0.01	0.05	0.02	0.00	-0.01
SO_3	0.28	0.30	0.00	0.00	0.26	0.28	0.00	0.00
Na_2O	0.05	0.01	0.06	0.06	0.02	0.01	0.04	0.04

Abbreviations: LIM= Limonite, SAP= Saprolite, BZ SAP= Bluezone Saprolite, BRK= Bedrock

3.2.1 DC 1763

The DC 1763 nickel laterite deposit profile of 0 to 22 metres thickness obtained a bedrock zone of 17 to 22 metres, a bluezone saprolite zone of 13 to 17 metres, a saprolite zone of 3 to 13 metres and a limonite zone of 0 to 3 metres, as well as major, minor and trace element and compound grade presentations (Figure 7).

1. The bedrock zone is the lowest part of the nickel laterite deposit profile with a thickness of 19 to 24 metres. Based on visual observation in the field, this zone is characterised by black and greenish-greyish colour with a low degree of weathering, and is rich in pyroxine and olivine minerals. The serpentine mineral content is slightly abundant, indicating a low degree of serpentinisation. The chemical composition of the deposit consists of major chemical compounds of MgO 37.76%, Fe_2O_3 8.51%, SiO_2 40.90% and Al_2O_3 0.04% and minor elements and compounds of Ni 0.29%, Co 0.07%, Cr_2O_3 0.41%, MnO 0.14% and CaO 0.54% and traces of TiO_2 0.01%, K_2O 0.01%, P_2O_5 -0.01%, SO_3 0.00% and Na_2O 0.06%.
2. The bluezone saprolite zone located at a thickness of 13 to19 metres produces a greyish green to blackish colour due to the presence of iron-rich olivine and pyroxine minerals. However, there are other mineral contaminations such as white asbestos, maghemite and goethite. The chemical composition of the deposit consists of major chemical compounds of MgO 38.5%, Fe_2O_3 9.41%, SiO_2 40.20% and Al_2O_3 0.09% and minor elements and compounds of Ni 0.24%, Co 0.08%, Cr_2O_3 0.41%, MnO 0.09% and CaO -0.09% as well as traces of TiO_2 -0.01%, K_2O 0.01%, P_2O_5 0.00%, SO_3 0.00% and Na_2O 0.06%.
3. The saprolite zone with a thickness of 3 to 13 metres, has some characteristic visual features; medium grain size between 1 mm to 5 mm in diameter indicating medium weathering. The yellowish-brown colour of this zone is due to the abundance of iron minerals, such as goethite, manganese minerals. The chemical composition of the deposit consists of major chemical compounds of MgO 0.72%, Fe_2O_3 64.12%, SiO_2 6.41% and Al_2O_3 7.13% and minor elements and compounds of Ni 1.49%, Co 0.15%, Cr_2O_3 2.41%, MnO 0.37% and CaO 0.32% and traces of TiO_2 -0.09%, K_2O 0.01%, P_2O_5 0.04%, SO_3 0.3% and Na_2O 0.01%.
4. The limonite zone is located at the very top of the laterite nickel deposit profile, with a thickness of 0 to 3 metres. Visual characteristics that show Manganese are known to produce black colour, pink to dark red colour caused by high iron cap/iron oxide (Fe) content, high weathering rate with very fine rock grain size (<1 mm diameter), no serpentinisation process occurs because it does not contain serpentine minerals. The abundant mineral in this zone is the mineral hematite. In addition to hematite, this zone also contains manganese minerals in fairly abundant quantities and a small amount of maghemite minerals. The chemical composition of the deposit consists of major chemical compounds of MgO 0.72%, Fe_2O_3 75.21%, SiO_2 2.61% and Al_2O_3 10.31% and minor elements and compounds of Ni 0.56%, Co 0.18%, Cr_2O_3 2.56%, MnO 0.24% and CaO 0.14% and traces of TiO_2 -0.07%, K_2O 0.01%, P_2O_5 0.06%, SO_3 0.28% and Na_2O 0.05%.

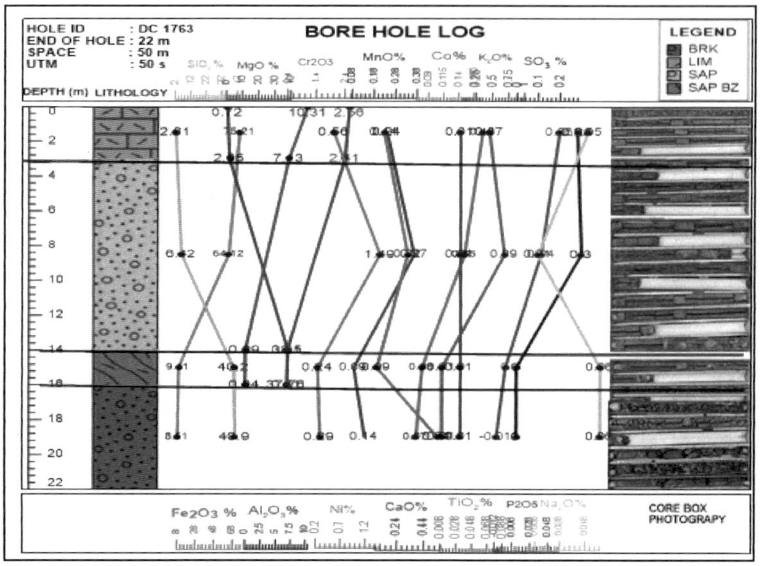

Figure 7. Laterite nickel profile hole id DC 1763 showing vertical geochemical characteristics

3.2.2 DL 1361

DL 1361 nickel laterite deposit profile 0 to 24 metres thickness obtained bedrock 19 to 24 metres, saprolite bluezone 7 to 19 metres, saprolite 3 to 7 metres and limonite zone 0 to 3 metres, as well as major, minor and trace element and compound grade presentation (Figure 8).

1. The bedrock zone is the lowest part of the nickel laterite deposit profile with a thickness of 19 to 24 metres. Based on visual observations in the field, this zone is characterised by black and greenish-greyish colour with a low degree of weathering, and is rich in pyroxine and olivine minerals. The serpentine mineral content is slightly abundant, indicating a low degree of serpentinisation. The chemical composition of the deposit consists of major chemical compounds of MgO 37.70%, Fe_2O_3 8.61%, SiO_2 40.89% and Al_2O_3 -0.04% and minor elements and compounds of Ni 0.26%, Co 0.03%, Cr_2O_3 0.39%, MnO 0.11% and CaO 0.52% and traces of TiO_2 0.01%, K_2O 0.01%, P_2O_5 -0.01%, SO_3 0.00% and Na_2O 0.04%.
2. The bluezone saprolite zone found at a thickness of 7 to 19 metres is characterised by a high degree of serpentinisation, with the weathering rate classified as intermediate, the colour of this zone is greyish green to blackish, suggesting a high nickel content rich in serpentine minerals. In addition to serpentine, the zone also contains abundant olivine minerals and small amounts of pyroxine minerals. The chemical composition of the deposit consists of major chemical compounds of MgO 38.42%, Fe_2O_3 9.43%, SiO_2 40.15% and Al_2O_3 -0.08% and minor elements and compounds of Ni 0.23%, Co 0.04%, Cr_2O_3 0.39%, MnO 0.09% and CaO -0.05% and traces of TiO_2 -0.01%, K_2O 0.01%, P_2O_5 0.00%, SO_3 0.00% and Na_2O 0.04%.
3. Saprolite zone thickness 3 to 7 meters actual visual yellowish brown field that is (goethite and maghemite). The low level of serpentinization indicates that the bedrock originally did not contain many serpentine minerals. This zone is contaminated with manganese rocks originally containing manganese, talc is a soft mineral and white there. The chemical composition of the deposit consists of major chemical compounds such as MgO 2.04%, Fe_2O_3 68.09%, SiO_2 6.34% and Al_2O_3 6.97% and minor elements and compounds such as Ni 1.45%, Co 0.10%, Cr_2O_3 2.38%, MnO 0.31% and CaO 0.29% and traces of TiO_2 -0.07%, K_2O 0.01%, P_2O_5 0.02%, SO_3 0.28% and Na_2O 0.01%.
4. The limonite zone with a thickness of 0 to 3 meters is a zone at the very top of the laterite nickel deposit profile. This zone is rich in iron and has several visual characteristics, namely the size of very fine rock grains with a diameter of about < 1 mm, high weathering rates, no

serpentinization, and swelling 0.25 meters to 2 to 3 meters thick due to high water content. The pink to dark red color comes from hematite, while the brownish yellow and reddish brown colors come from goethite is an iron mineral mineral commonly found in Limonite zones. Maghemite and manganese can also be found in this zone, but in smaller quantities. The chemical composition of the deposit consists of major chemical compounds such as MgO 0.7%, Fe_2O_3 73.23%, SiO_2 2.79% and Al_2O_3 10.2% and minor elements and compounds such as Ni 0.53%, Co 0.14%, Cr_2O_3 2.47%, MnO 0.2% and CaO 0.14% and traces of TiO_2 -0.05%, K_2O 0.01%, P_2O_5 0.05%, SO_3 0.26% and Na_2O 0.02%.

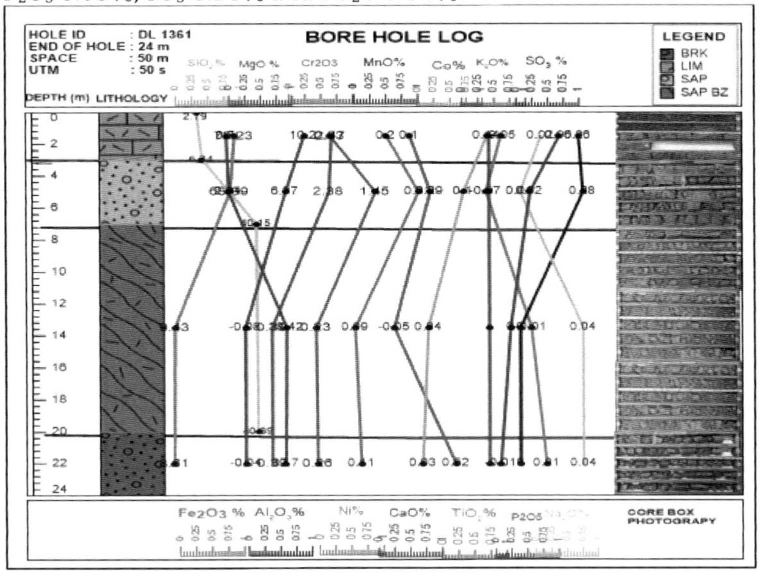

Figure 8. Laterite nickel profile hole id DL 1361 showing vertical geochemical characteristics

4. Conclusion

Petrographic analysis of bedrock the two samples, DL 1361 and DC 1763, the bedrock is a serpentinised harzburgite, composed of olivine, orthopyroxene (enstatite), clinopyroxene, serpentine, opaque minerals, chlorite, talc and clay minerals. XRD analysis on the two profiles showed that the Bedrock zone is composed of antigorite, spinel, lizardite and enstatite. The saprolite bluezone is composed of antigorite, enstatite, lizardite, quartz and magnetite. The saprolite zone consists of goethite, antigorite, enstatite, magnetite and quartz. Meanwhile, the limonite zone is composed of goethite, magnetite, and hematite.

X-Ray Fluorescence (XRF) analysis of nickel laterite deposits at DL 1361 and DC 1763 shows that Fe_2O_3 is the main component with the highest concentration in the limonite layer (73.23% and 72.12%). Nickel (Ni) accumulated most in saprolite (1.45% and 1.39%) Co was more concentrated in limonite (0.14% and 0.10%), while MgO was highest in the saprolite bluezone (38.42% and 36.40%). Al_2O_3 is highest in limonite (10.20% and 10.12%), Cr_2O_3 is concentrated in limonite (2.47% and 2.21%), and SiO_2 is highest in bedrock (40.89% and 40.73) and CaO is highest in bedrock (0.52% and 0.41).

Acknowledgement

The authors wish to express a sincere gratitude to Dean of Engineering Faculty, Hasanuddin University for the financial support through "The 2024 Faculty of Engineering Laboratory Based Education (LBE) Research Grant".

References

(1) Choi Y., and Kim. B., 2023. Characteristics and Controlling Factors on Nickel Laterite Deposits in Sulawesi, Indonesia. The Korean Society Of Economic And Environmental Geology, 56(3): 343-363.

(2) Gifaricandrabayu, M., Obrin Trianda, O., Rizqi, F. H. A., 2023. Karakteristik Endapan Nikel Laterit Pada Daerah X, Kabupaten Kolaka Utara, Provinsi Sulawesi Tenggara. Prosiding Nasional Rekayasa Teknologi Industri dan Informasi XVIII. Tahun 2023 (ReTII), November 2023, pp. 601~609, ISSN: 1907-5995.

(3) Sufriadin, Fauth S., Tindoilo P.D., Nur I., Purwanto, Amalia R., Irfan U.R., Basir D.N., Otake T., 2024. Thermal Beneficiation of a Nickel Laterite Ore from the Obi Island of North Maluku, Indonesia Using Corncob Char as Reductant. Advances in Science and Technology. Vol. 141, pp 95-102.

(4) Simanjuntak T. O., Rusmana E., Supanjono B. J. dan Koswara A., 1993. Peta Geologi Lembar Bungku, Sulawesi. Pusat Pengembangan dan Penelitian Geologi.

(5) Hasria., Ramadhan, M. A., Okto,A., Bahdad, M., La Ode Ngkoimani, O. L., Azzaman, A. M., 2022. Analisis Petrografi dan Geokimia Batuan Ultramafik Kompleks Ofiolit Kecamatan Andowia Kabupaten Konawe Utara, Provinsi Sulawesi Tenggara Jurnal GEOSAPTA Vol. 08 No.02. p-ISSN: 2460-3457 e-ISSN: 2527-5844.

(6) Lediyantje Lintjewas, L., Setiawan, I., Andrie Al Kausar, K. A. A., 2019. Profil Endapan Nikel Laterit Di Daerah Palangga, Provinsi Sulawesi Tenggara. Jurnal RISET Geologi Dan Pertambangan, Vol.29, No.1, Juni 2019, 91-104

(7) Pasolon A.R., Ilyas A., dan Widodo S., 2022. Analisis Karakteristik Mineralogi dan Geokimia Berdasarkan Zona Profil Endapan Nikel Laterit (Studi Kasus: Blok X PT Ang and Fang Brother, Site Lalampu, Kecamatan Bahodopi, Kabupaten Morowali, Provinsi Sulawesi Tengah. Jurnal Geomine. Vol. 10, hal. 01-12.

(8) Nurhawaisyah, S.R., Jafar, N., Bakri, S., Artiningsih, A., Widodo, S., 2021. A petrographic study of bauxite of Kenco area, Landak District, West Kalimantan Province. IOP Conference Series: Earth and Environmental Science. 921, 1-8

(9) Sufriadin, Sri Widodo, Meinarni Thamrin, Akane Ito, Tsubasa Otake. 2021. The Latowu Ultramafic Rock-Hosted Iron Mineralization in the Southeastern Arm Sulawesi, Indonesia: Characteristics, Origin, and Implication for Beneficiation. International Journal on Advanced Science Engineering Information Technology. 11, 987-993.

(10) Sufriadin, Widodo, S., Nur, I., Ilyas, A., Ashari, M.Y., 2020. Extraction of Nickel and Cobalt from Sulawesi Limonite Ore in Nitric Acid Solution at Atmospheric Pressure. IOP Conf. Series: Materials Science and Engineering, 875, 1-6.

(11) Sufriadin, Widodo, S., Thamrin, M., Maulana, A., Ito, A., Otake, T., 2020. The nature of ultramafic rocks from Sulawesi, Indonesia and their suitability for CO2 sequestration. IOP Conf. Series: Earth and Environmental Science, 589, 1-8.

(12) Sufriadin, Widodo, S., Trianto, M., 2019. Beneficiation of Lateritic Iron Ore from Malili Area, South Sulawesi, Indonesia Using Magnetic Separator. IOP Conf. Series: Materials Science and Engineering, 619, 1-7.

(13) Sufriadin, Widodo, S., Mantung, M., 2019. Beneficiation Study of Iron Ore from Pakke area of Bontocani Subdistrict, Bone Regency of South Sulawesi, Indonesia Using Roasting and Magnetic Separation Methods. International Journal of Engineering and Science Applications, 2, 67-71

(14) Sufriadin1, Widodo, S., Ito, A., Otake, T., and Sanematsu, K., 2018. Geochemical study of ultramafic rocks from Latowu area of North Kolaka, Southeast Sulawesi and its implication for CO2 sequestration. Journal of degraded and mining lands management, 5, 1403-1408.

Analysis of Carbon Absorption Potential of Above-Ground Forest in the Karst Area of Maros Bantimurung Bulusaraung National Park

Syafiqatul Fuady[1], and Muhammad Arsyad[1]

[1] Department of Physics, State University of Makassar, 90224, Indonesia

*E-mail: syafiqatulf24@gmail.com

Abstract. This research was conducted in January 2024 with the aim of analyzing the carbon absorption potential of forests vegetation on the ground surface in the Karst Area of Maros Bantimurung Bulusaraung National Park. The research data used consists of 31 years of rainfall and air temperature data (1993-2023) obtained from ERA-5 satellite rainfall imagery, as well as biomass and necromass data obtained from direct field measurements. Data collection was conducted in the flatland forest area with a total of 30 plots, each measuring 20m x 20m (400 m^2). The location of the carbon storage measurement plots was determined using purposive sampling method. Carbon data analysis was performed using allometric method. Based on the results of data analysis, the Karst Area of Maros Bantimurung Bulusaraung National Park has an average air temperature of 26.4°C and an average annual rainfall of 2,864.13 mm/year. Thus, this area is classified as a Humid/Wet climate zone. Based on its climate zone, the biomass of vegetation on the surface of the Karst Area of Maros Bantimurung Bulusaraung National Park is 407.443 tons/ha, with a total carbon reserve of 191.498 tons C/ha, and carbon absorption of 702.160 tons/ha. The Karst Area of Maros Bantimurung Bulusaraung National Park covers an area of 9,610.71 ha, resulting in a total carbon stock is 1,840,432.412 tonsC and total carbon absorption is 6,748,252.177 tons. The calculation of carbon absorption can be used to measure the contribution of forests in stabilizing the global climate and to formulate further actions to combat climate change.

Keywords: Babul NP, Biomass, Carbon Absorption, Carbon dioxide, and Climate Change.

1. Introduction

The Fifth Assessment Report 2014 from the Intergovernmental Panel on Climate Change (IPCC) notes that human activities have increased the Earth's temperature by about 1.1°C from the late 1800s to 2020, with projections of an increase of up to 5.4°C without mitigation. Since 1990, global warming has increased by 1.1 W/m^2, with 80% caused by CO_2. Indonesia is the tenth-largest carbon emitter in the world, contributing 1.69% of the global total emissions, which is approximately 589.5 million tons in 2020 [1].

The measurement of forest carbon absorption can illustrate the forest's ability to absorb CO_2 from the atmosphere, while carbon reserves depict the CO_2 that is stored and not released through

IOP Conf. Series: Earth and Environmental Science **1525** (2025) 012026 doi:10.1088/1755-1315/1525/1/012026

the respiration process [2]. Land clearing causes the release of CO_2 emissions into the atmosphere. Therefore, reducing deforestation and forest degradation can significantly curb CO_2 emissions into the atmosphere [3-4]. Carbon accounting is the calculation of carbon reserves stored in forests. The amount of carbon calculated follows the SNI 7724: 2011 standard where the carbon content is estimated to be 47% of the biomass [5-6]. Above ground biomass (AGB) and tree C stock were calculated based on an allometric model developed by [7], which is based on the amount of rainfall in a given area [8].

Karst ecosystems cover about 10-20 % of the total land area on Earth [9-10], having a significant role in ecology but susceptible to the impacts of mining activities. In South Sulawesi, there is the Maros-Pangkep Karst Area, most of which is located within the Bantimurung Bulusaraung National Park, covering an area of approximately 43,750 ha [11]. In 1997 IUCN designated karst areas as regions that must be preserved, thus requiring sufficient information to protect them.

Mitigating climate change is a global commitment that affects all countries. Indonesia, as a tropical country with a large portion of its territory being ocean, has a significant influence on global weather and climate conditions [12], plays an important role in reducing emissions from deforestation, degradation, and forest carbon absorption. The concept of Reducing Emission from Deforestation and Forest Degradation (REDD+) offers financial incentives for countries that successfully maintain forest conservation as carbon absorbers and storers [13]. This research was conducted to support those efforts.

2. Methodology

2.1 Time and location of the research

This research was carried out in January 2024 in the Karst Area of Maros Bantimurung Bulusaraung National Park. Data collection was performed at three locations: first, in the Tourist Area of Bantimurung Bulusaraung National Park, which consists of 10 data collection plots; second, in Leang-leang, Bantimurung District, which consists of 11 data collection plots; and third, in Labuaja Village, Cenrana District, which consists of 9 data collection plots. The research location map can be seen in Figure 1.

2.2 Data and data sources

Research data consists of primary and secondary data. Primary data is collected directly in the field, including tree biomass and necromass. Laboratory measurements include the wet and dry weight of leaves and stems of plants with a diameter of <5 cm and litter, conducted in the Synthesis Room on the 3rd floor of the Physics Department, FMIPA UNM. Secondary data comes from ERA-5 satellite imagery, covering monthly rainfall and air temperature in the Karst Area of Maros Bantimurung Bulusaraung National Park from 1993 to 2023.

2.3 Procedure

The research procedure begins with preparation, including field observations to obtain a clear picture of the research area. Data collection was carried out by creating 30 sample plots of 20m x 20m (400 m^2) using the purposive sampling method. Next, measure the tree trunk circumference to obtain data on the diameter at breast height at a height of 1.3 m to determine the tree biomass, and the trunk circumference and height of dead trees within the plot [8]. Sampling of understory plants and litter was conducted in the dominant area of the plot, in a subplot measuring 0.5m × 0.5m. Then, the total weight of the understory plant and litter samples

IOP Conf. Series: Earth and Environmental Science **1525** (2025) 012026 doi:10.1088/1755-1315/1525/1/012026

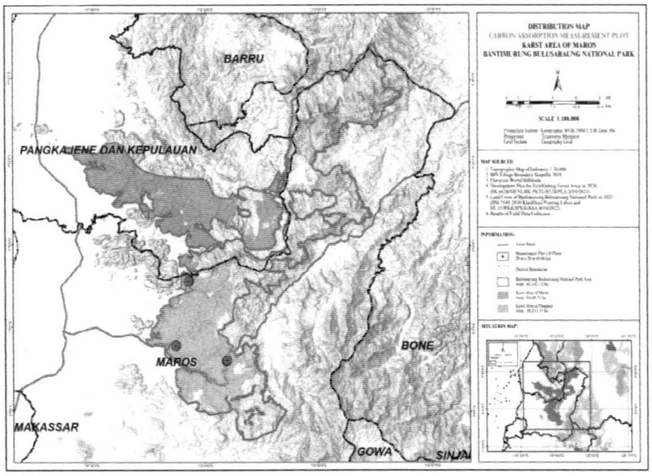

Figure 1. Research location

was measured, and approximately 100 grams were taken for drying in an oven at a temperature of 85°C until a constant weight was achieved.

2.4 Data analysis

Analysis of rainfall and air temperature data in the Karst Area of Maros Bantimurung Bulusaraung National Park is conducted by averaging monthly-scale. Above-ground biomass (AGB) and tree C stock were calculated based on the allometric model developed by [5] in Table 1.

Table 1. Allometric Equations for Estimating Tree Biomass in Forests Based on Climate Zones

Rainfall (mm/year)	Allometric Formula
Dry (<1500)	1. $(AGB)_{est} = 0.112\ (\pi D^2 H)^{0.916}$
	2. $(AGB)_{est} = \pi \times \exp(-0.667 + 1.784 \ln(D) + 0.207(\ln(D))^2 - 0.0281\ (\ln(D))^3)$
Humid (1500-4000)	1. $(AGB)_{est} = 0.0509\ (\pi D^2 H)$
	2. $(AGB)_{est} = \pi \times \exp(-1.499 + 2.148 \ln(D) + 0.207(\ln(D))^2 - 0.0281\ (\ln(D))^3)$
Wet (>4000)	1. $(AGB)_{est} = 0.0776\ (\pi D^2 H)^{0.94}$
	2. $(AGB)_{est} = \pi \times \exp(-1.239 + 1.980 \ln(D) + 0.207(\ln(D))^2 - 0.0281\ (\ln(D))^3)$

Equation number 1 in table 1 is used when the data consists of tree height and diameter. While equation number 2 is used when only tree diameter data is available. Where $(AGB)_{est}$ is above-ground tree biomass (kg/tree), D is DBH (diameter at breast height) (cm), H is height of the tree (m), and π is wood density(g/cm^3).

To calculate the total biomass of understorey plants and litter, the following equation is used

$$Total\ BK = \frac{BK\ sub\ example\ (g)}{BB\ sub\ example\ (g)} \times Total\ BB\ (g)$$

where BK is biomass (dry weight) (g) and BB is wet weight (g). The calculation of total woody necromass biomass is performed using the equation:

$$BK = \frac{\pi H D^2}{40} \times f$$

where BK is biomass (g), π is specific gravity of wood necromass (g/cm^3), H is length/height of necromass (cm), D^2 is diameter necromass (cm) and f is correction factor.

IOP Conf. Series: Earth and Environmental Science **1525** (2025) 012026 doi:10.1088/1755-1315/1525/1/012026

The calculation of total carbon in plants can be determined by adding the total tree biomass to the total necromass biomass using the equation:

$$C = (Total\ tree\ biomass + total\ necromass\ biomass) \times 0.47$$

where C is total carbon reserves in each plot (ton C/ha) and 0,47 is the percentage value of carbon content in plants. By measuring the amount of stored carbon, it can illustrate the amount of CO_2 absorbed by plants using the equation:

$$Carbon\ Absorption = \frac{Mr\ CO_2}{Ar\ C} \times C$$

where $Carbon\ Absorption$ is absorption of carbon dioxide (CO_2) (ton/ha), $Mr\ CO_2$ is relative molecular mass (44), and $Ar\ C$ is relative atomic mass (12).

3. Results and Discussion

3.1 Analysis of the climate characteristics of the Karst Area of Maros Bantimurung Bulusaraung National Park

Climate analysis of the Karst Area of Maros in Bantimurung Bulusaraung National Park was conducted to calculate tree biomass using allometric equations in table 1. This analysis includes rainfall and air temperature from 1993 to 2023. Data were obtained from ERA-5 satellite imagery at the following three coordinates 05°01'08,7" S 119°41'17,4" E, 04°57'19,0" S 119°42'01,6" E, and 5°02'00,0" S 119°44'32,0" E, at an altitude of 85-350 m, as shown in Figure 2

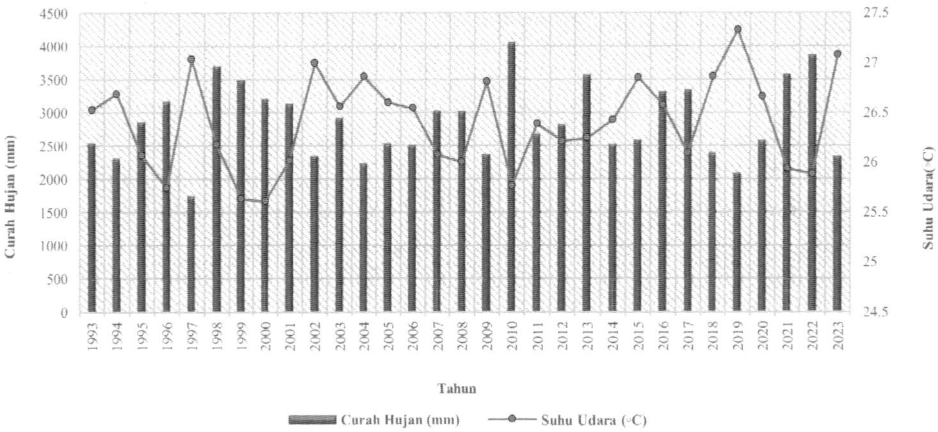

Figure 2. Annual Rainfall and Air Temperature Graph of the Karst Area of Maros in Bantimurung Bulusaraung National Park from 1993-2023 using ERA-5 satellite imagery data

Based on Figure 2, the air temperature and annual rainfall over 31 years (1993-2023) in the Karst Area of Maros Bantimurung Bulusaraung National Park have an average annual air temperature of 26.4°C, while the average annual rainfall is 2,864.13 mm/year. Thus, the climate zone is Humid with an annual rainfall between 1,500-4,000 mm/year. Therefore, to calculate tree biomass, the Humid equation number 2 in Table 1 is used.

3.2 Forest carbon absorption of the Karst Area of Maros Bantimurung Bulusaraung National Park

The calculation of carbon reserves and absorption in this study consists of 4 carbon pools, namely trees, understory vegetation, woody necromass, and litter. Data collection from 30 plots covering an area of 1.2 ha is considered to represent the entire homogeneous Karst Area of Maros Bantimurung Bulusaraung National Park. The calculation results can be seen in Table 2.

Tabel 2. Biomass, Carbon Reserves, and Carbon Sequestration in Land Cover in the Form of Forests in the Karst Area of Maros Bantimurung Bulusaraung National Park (ton/ha)

Plot	Tree	Undergrowth	Woody Necromass	Litter	Vegetation Biomass	Carbon Reserves	Carbon Absorption
1	498.114	27.504	0.016	47.360	572.995	269.308	987.461
2	508.218	3.520	0.019	38.259	550.016	258.507	947.861
3	407.054	12.438	0.052	43.200	462.744	217.490	797.462
4	418.244	19.402	0.004	61.962	499.611	234.817	860.996
5	433.755	14.933	0.005	55.411	504.104	236.929	868.739
6	295.703	15.266	0.227	46.594	357.790	168.161	616.591
7	385.410	21.672	0.180	36.941	444.202	208.775	765.509
8	600.962	11.424	0.065	33.267	645.719	303.488	1112.789
9	432.542	6.720	0.013	85.118	524.394	246.465	903.705
10	731.729	9.229	0.037	18.355	759.350	356.894	1308.613
11	948.492	11.194	0.035	31.334	991.055	465.796	1707.918
12	131.935	9.498	0.005	21.638	163.076	76.646	281.035
13	227.356	3.840	0.014	6.080	237.290	111.526	408.930
14	360.403	10.024	0.011	37.869	408.307	191.904	703.649
15	117.757	3.840	0.002	2.400	123.999	58.279	213.691
16	95.120	1.760	0.044	23.856	120.780	56.766	208.144
17	342.598	3.360	0.020	24.891	370.870	174.309	639.132
18	235.004	6.275	0.017	24.461	265.758	124.906	457.989
19	1055.905	8.736	0.021	39.014	1103.677	518.728	1902.003
20	116.256	1.280	0.018	8.784	126.339	59.379	217.723
21	123.543	4.640	0.014	11.677	139.874	65.741	241.049
22	315.089	1.120	5.682	22.824	344.716	162.016	594.060
23	236.660	3.040	0.045	38.259	278.004	130.662	479.094
24	173.351	5.440	0.104	57.568	236.462	111.137	407.503
25	185.365	6.560	0.034	23.962	215.920	101.482	372.102
26	285.981	6.080	1.248	29.131	322.440	151.547	555.672
27	203.829	3.840	0.026	17.013	224.708	105.613	387.247
28	476.321	5.920	0.109	10.893	493.243	231.824	850.022
29	233.398	1.920	0.053	21.878	257.249	120.907	443.327
30	428.754	6.400	2.973	40.464	478.592	224.938	824.773
Average	366.828	8.229	0.370	32.015	407.443	191.498	702.160

Based on Table 2 it is known that the average aboveground vegetation biomass in the Karst Area of Maros Bantimurung Bulusaraung National Park is 407.443 tons/ha, with a total carbon reserve of 191.498 tons C/ha and carbon absorption of 702.160 tons/ha. Covering 9,610.71 ha, the total carbon reserve is 1,840,432.412 tons C and total carbon absorption is 6,748,252.177 tons. The amount of carbon reserves obtained from this research is quite significant. Trees are the largest carbon sinks, therefore if trees are cut down, their ability to absorb carbon will significantly decrease, and the stored carbon will be released and become emissions in the air [4].

According to the Ministry of Environment and Forestry (KLHK) in 2020, the economic potential of carbon trading over the next 5 years is estimated to reach Rp 350 trillion, with a scenario of selling carbon credits at US$ 5 per ton in the carbon market [1]. The economic value of carbon in the Karst Area of Maros Bantimurung Bulusaraung National Park is US$ 9.2 million. At the current exchange rate of Rp. 15,745.11, this is about Rp. 144,889,053,862.

Carbon reserves and carbon absorption are interconnected, more carbon reserves mean more carbon absorption. This data supports environmental policies to reduce greenhouse gas emissions and promote ecosystem preservation. Carbon absorption of the forest vegetation above ground in the Karst Area of Maros Bantimurung Bulusaraung National Park contributes to climate change mitigation. Recommended mitigation measures include forest conservation, ecosystem restoration, renewable energy use, and community education.

4. Conclusion

The biomass of the Karst Area of Maros Bantimurung Bulusaraung National Park is 407.443 tons/ha, with a total carbon reserve of 191.498 tons C/ha, and carbon absorption of 702.160 tons/ha. Covering 9,610.71 ha, the total carbon reserve is 1,840,432.412 tons C and carbon absorption is 6,748,252.177 tons. The economic value of carbon in this area is estimated at US$ 9.2 million, or Rp. 144,889,053,862. The potential carbon absorption of the forest vegetation above ground contributes to climate change mitigation. Recommended climate change mitigation measures include forest conservation, ecosystem restoration, renewable energy use, and community education.

5. References

[1] *Indonesia Carbon Trading Handbook.* 2022. Katadata Insight Center.

[2] Manese, M. A., Hamidun, M. S., & Baderan, D. W. 2023. Serapan karbon atas permukaan tanah untuk konservasi Hutan Lindung Gunung Damar Gorontalo. *Jurnal Penelitian Saintek*, 99-109

[3] Sugiarto, A., Utaya, S., Sumarmi, Bachri, S., & Shrestha, R. P. 2024. Estimation of carbon stocks and CO2 emissions resulting from the forest destruction in West Kalimantan, Indonesia. *Environmental Challenges*, 1-14

[4] Azham, Z. 2015. Estimasi Cadangan Karbon pada Tutupan Lahan Hutan Sekunder, Semak dan Belukar di Kota Samarinda. *Jurnal AGRIFOR*, 325-338

[5] *Pengukuran dan Perhitungan Cadangan Karbon-Pengukuran Lapangan untuk Penaksiran Cadangan Karbon Hutan (ground based forest carbon accounting). SNI 7724.* 2011. Jakarta: Badan Standarisasi Nasional

[6] Manuri, S., C.A.S. Putra dan A.D. Saputra. 2011. Tehnik Pendugaan Cadangan Karbon Hutan. Merang REDD Pilot Project, German International Cooperation – GIZ. Palembang

[7] Chave, J., Andalo , C., Brown, S., Cairns, M. A., Chambers, J. Q., Eamus, D., . . . Yamakura, T. 2005. Tree Allometry and improved estimation of carbon stocks and balance in tropical forests. *Oecologia*

[8] Muluneh, M. G., & Worku, B. B. 2022. Carbon storages and sequestration potentials in remnant forests of different patch sizes in northern Ethiopia: an implication for climate change mitigation. *Agriculture & Food Security*, 1-38

[9] Arsyad, M., Mukarramah, D., Susanto, A. 2023. Studi Tentang Sifat Fisis dan Kuat Uji Tekan Batuan Karst Kawasan Gua Leang Londrong Taman Nasional Bantimurung Bulusaraung. *Jurnal Ilmu Lingkungan*, 789-795

[10] Wang, K., Zhang, C., Chen, H., Yue, Y., Zhang, W., Zhang, M., . . . Fu, Z. (2019). Karst landscapes of China: patterns, ecosystem processes and services. *Landscape Ecol*, 2743-2763.

[11] Ahmad, A., & Hamzah, A. S. 2016. *Database Karst Sulawesi Selatan.* Makassar: Badan Lingkungan Hidup Daerah Provinsi Sulawesi Selatan

[12] Arsyad, M., Fitriani, F., Tiwow, V. A., Palloan, P., Sulistiawaty, S., & Susanto , A. 2022. The Effects of El Nino Southern Oscillation on Rainfall in the Karst Area of Maros, National Park Bantimurung Bulusaraung South Sulawesi and its Impact on Flood Disasters. *Jurnal Ilmu Lingkungan*, 772-778

[13] Wibowo, A., Samsoedin, I., Nurtjahjawilasa, Subarudi, S., & Muttaqin, M. Z. 2013. Petunjuk Praktis Menghitung Cadangan Karbon Hutan. *Pusat Penelitian dan Pengembangan Perubahan Iklim dan Kebijakan.*

IOP Conf. Series: Earth and Environmental Science **1525** (2025) 012027 doi:10.1088/1755-1315/1525/1/012027

Refunctionalization of Post-Mining Land According to Land Capability Classes in the Tapunggaya area, North Konawe

Muhardi Mustafa[1]*, Deniyatno[1], Wahab[1], Moh. Suriyaidulman Rianse[1], Fitra Saleh[2]

[1] Mining Engineering Department, Halu Oleo University, 93232, Indonesia
[2] Geography Department, Halu Oleo University, 93232, Indonesia

*E-mail: muhardy65@uho.ac.id

Abstract. This study was conducted in the Tapunggaya area, Molawe district, North Konawe, Southeast Sulawesi. The study aims to classify post-nickel mining land capabilities and provide recommendations for land use after mining activities. The method used in this research is the matching technique, which involves aligning the classification system with the actual field conditions. Several parameters for land capability classification were considered, including soil texture, effective soil depth, surface slope, soil drainage, erosion, permeability, rocks and gravel, and susceptibility to flooding or inundation. The results of the matching process revealed varying land capability levels. Three land capability classes were identified in the study area: class IV, covering 3.11 hectares; class VI, covering 2.54 hectares; and class VII, covering 9.14 hectares. Based on land capability classifications and the corresponding land use recommendations, the recommended use for class IV land is forest, which can be designated as a nature reserve or for limited production. Grazing land is divided into limited, moderate, and intensive grazing, while agricultural use is restricted and difficult to implement. The recommendation for class VI land is limited to forestry and grazing. Likewise, class VII land is only suitable for forest and limited grazing.

1. Introduction

Mining is a complex, high-risk, long-term industry requiring advanced technology, substantial capital, and regulatory compliance across multiple sectors. Due to its significant environmental impact, mining demands comprehensive planning from the initial stages through post-mining activities, including reclamation. As outlined in Act No. 4 of 2009 on mineral and coal mining, reclamation is essential for restoring and improving environmental quality and ecosystem functionality. Neglected post-mining land can reduce regional revenue and impact local environmental services.

Land resource development planning is crucial for sustainable land use, both now and in the future. Land evaluation informs land-use policies, supporting rational planning for optimal utilization. Given the limitations of post-mining land, not all land uses are feasible. Therefore, this study proposes a plan for post-mining land functionalization based on land capability classes in North Konawe. This research aims to guide post-mining land use based on land capability classification, aiming to support reclamation efforts that effectively restore and improve environmental quality, as mandated by Law No. 4 of 2009.

2. Literature Review

Land capability classification provides a framework for assessing land use suitability by evaluating soil properties, slope, drainage, and other factors. According to Arsyad, land capability classes range from Class I (highly suitable for various uses) to Class VIII (limited to conservation uses), where soil quality, erosion risk, and other constraints intensify with each class. Reclamation success relies on aligning these classes with appropriate land uses, thereby maximizing land productivity while mitigating environmental impacts [1]. Studies by Chaney et al. highlight the role of revegetation and erosion control in stabilizing degraded lands, especially through the use of leguminous plants and phytoremediation strategies [2].

Land evaluation is a planning tool for sustainable land use management. Food and Agriculture Organization (FAO) states that comprehensive land evaluation considers both current land conditions and prospective environmental impacts of reclamation interventions, thereby ensuring that land restoration contributes positively to local ecosystems [3].

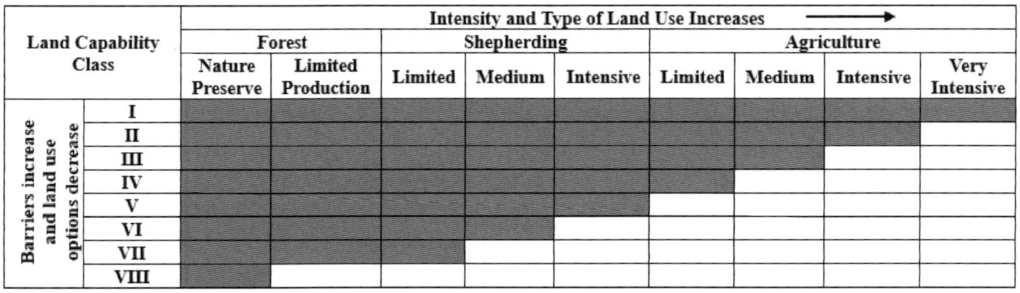

Figure 1. Land Capability Class Scheme Based on Land Use.

3. Data and Methods

3.1 Data Collection

Data were collected through primary surveys and secondary sources. Primary data included:

- Soil and Terrain Analysis: Soil samples were collected to assess texture, permeability, and effective soil depth. Terrain data, such as slope and drainage, were obtained using drone-based photogrammetry and topographic analysis through DEM (Digital Elevation Model) processing.
- Climatic and Hydrological Data: Meteorological records provided insights into local climate patterns, while hydrological data indicated flood-prone areas.

Secondary data included geological maps, thematic land system maps, and land-use plans from relevant government agencies.

3.2 Classification of Land Capability

Land was classified into three primary capability classes within the study area:

- Class IV: Land with moderate constraints, suitable for limited agriculture, forestry, and grazing. Requires soil conservation practices to mitigate erosion.
- Class VI: Severe constraints limiting use to grazing and certain types of forestry.

IOP Conf. Series: Earth and Environmental Science **1525** (2025) 012027 doi:10.1088/1755-1315/1525/1/012027

- Class VII: Very severe limitations, restricted to conservation efforts due to high erosion and drainage issues.

3.3. Analysis Process

The analysis process begins with the processing of the obtained data, with the following steps:

- Satellite Image Interpretation: Satellite image interpretation is carried out based on the results of field surveys to match the collected data with field conditions. This process is important to minimize interpretation errors and strengthen data accuracy.
- Laboratory Analysis: Laboratory analysis examines soil samples to determine soil texture and other relevant parameters.
- Data Processing and Analysis: The data analysis process uses spatial interpolation techniques, buffer creation, and overlay analysis of the obtained land data and information. All collected data is extracted using a geographic information system (GIS) to produce maps and visualizations that depict land conditions according to the predetermined land capability classification.

The results of the match between land use requirements and land quality are combined with an environmental impact assessment and a socioeconomic analysis to produce a land capability class appropriate for each relevant land use.

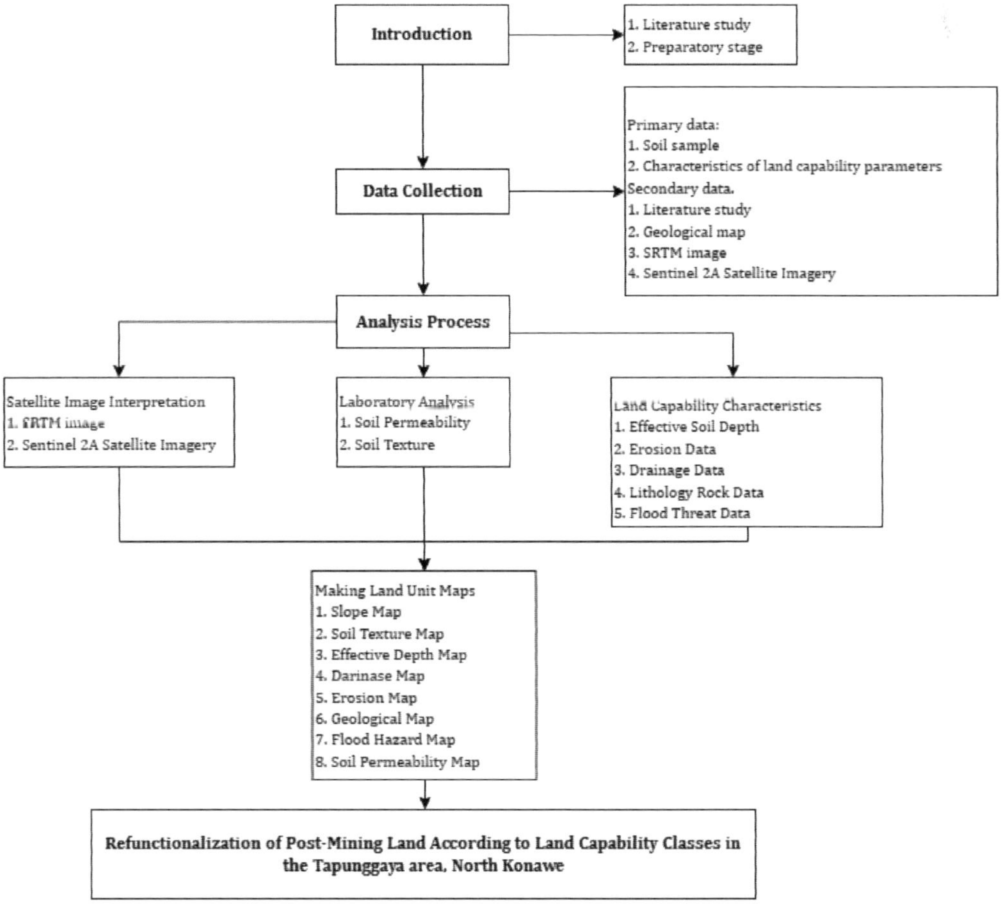

Figure 2. Research Flow Chart

IOP Conf. Series: Earth and Environmental Science **1525** (2025) 012027 doi:10.1088/1755-1315/1525/1/012027

4. Result

4.1 Soil and Topographical Characteristics

The predominant soil type in the study area is clay loam with subtypes of loam and silty clay. Laboratory results show a significant percentage of clay loam soil (68.58%) with a high susceptibility to erosion, especially on sloped areas. Topographically, the study area features varied slope classes, ranging from flat (0-3%) to steep (>15%). These characteristics significantly influence the soil's capability to support plant growth and agricultural activities [4].

4.2 Erosion and Drainage Constraints

Field data indicated that erosion is a significant constraint across all classes, with Classes IV and VI showing moderate to severe erosion levels. Class VII land, characterized by high erosion and limited drainage capacity, is deemed unsuitable for agriculture. Detailed erosion risk maps were produced to identify zones requiring targeted conservation practices [5].

4.3 Land Use Recommendations by Class

The following recommendations were developed for each land capability class:

- Class IV: Suitable for controlled agricultural activities, particularly legumes and root crops that enhance soil structure. Erosion control measures, such as terracing and mulching, are necessary.
- Class VI: Limited to grazing and selective forestry, with emphasis on leguminous plants like Leucaena and Acacia to enrich soil nitrogen and control erosion [2].
- Class VII: Reserved for natural conservation and afforestation. Phytoremediation using species like Sunflower and Jatropha is recommended to stabilize soil and mitigate heavy metal contamination [6].

5. Discussion

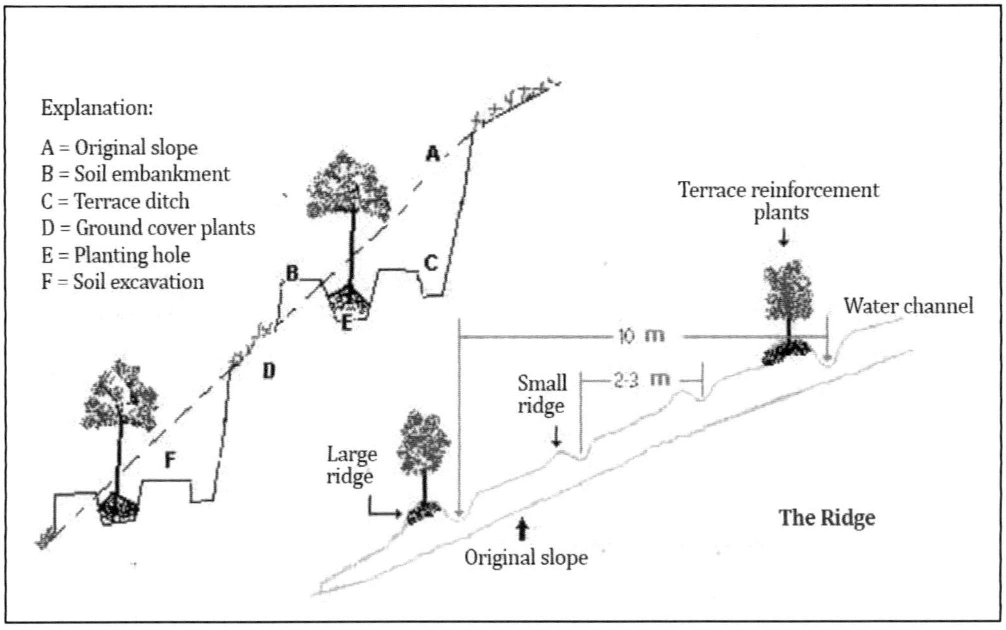

Figure 3. Terrace Bench and Ridge Form Scheme.

The results emphasize the importance of tailored reclamation strategies based on land capability classifications. For Class IV areas, the primary concern is maintaining soil integrity through erosion control. Implementing soil conservation practices, such as strip cropping and contour plowing, will enhance land stability. Class VI areas, with greater limitations, require specialized plant species adapted to low nutrient levels and acidic conditions. Studies by Suprapto support the use of acid-tolerant species to address soil acidity in post-mining landscapes [7].

The inclusion of leguminous plants is crucial for nutrient-poor soils. For Class VII land, leguminous species like *Pueraria javanica* can provide cover, reduce erosion, and contribute organic matter to restore soil quality [8]. Phytoremediation plants, including *Helianthus annuus*, have proven effective in absorbing heavy metals, making them suitable for areas with metal-contaminated soils [2].

6. Conclusion

This study highlights the need for structured land capability assessments in guiding post-mining land reclamation efforts. The findings underscore the critical role of erosion control, soil quality improvement, and appropriate vegetation selection in stabilizing degraded lands and enhancing land productivity.

6.1 Implications for Sustainable Land Use

Reclamation success depends on understanding and mitigating the unique limitations of each land class. Properly managed, reclaimed land can support sustainable agriculture, forestry, or conservation while meeting environmental regulations. Implementing these findings in North Konawe could serve as a model for other mining-affected regions in Indonesia, promoting sustainable land use practices across the country.

6.2 Future Research Directions

Further research should investigate long-term soil health improvements and ecological resilience following reclamation. The role of microbial communities and organic amendments in post-mining land restoration warrants exploration to improve soil stability and fertility in mined lands.

References

[1] Arsyad S. Konservasi Tanah dan Air. Bogor: IPB Press; 2006.
[2] Chaney RL, Malik M, Li YM, Brown SL, Brewer EP, Angle JS, et al. Phytoremediation of soil metals. Curr Opin Biotechnol 1997;8:279–84. https://doi.org/10.1016/S0958-1669(97)80004-3.
[3] Food and Agriculture Organization (FAO). A Framework for Land Evaluation. vol. 32. Rome: Soil Resources Development and Conservation Services, Land and Water Development Division; 1976.
[4] Arsyad S. Konservasi Tanah dan Air. Bogor: IPB Press; 1989.
[5] Aziz S. Evaluasi Kemampuan Lahan dan Pendugaan Erosi untuk Arahan Pemanfaatan Lahan di Sub DAS Juwet dan Dondong, Gunung Kidul Yogyakarta. Universitas Gadjah Mada, 2008.
[6] Rande S. Analisis Kesesuaian Lahan Bekas Tambang Batubara pada PT Asia Multi Invesama di Kabupaten Tebo Provinsi Jambi. PROMINE 2018;4. https://doi.org/10.33019/promine.v4i1.101.
[7] Suprapto SJ. Pemanfaatan dan permasalahan endapan mineral sulfida pada kegiatan pertambangan. Buletin Sumber Daya Geologi 2006;1:34–9.
[8] As-Syakur A. Evaluasi Zona Agro-klimat dari Klasifikasi Schimidt Ferguson Menggunakan Aplikasi Sistem Informasi Geografi (SIG). Jurnal Pijar MIPA 2008;3:17–22.

IOP Conf. Series: Earth and Environmental Science **1525** (2025) 012028 doi:10.1088/1755-1315/1525/1/012028

Water Turbidity Qualitative Analysis Based on Sentinel Imagery in Jeneberang River Downstream, Makassar, South Sulawesi

Putu Abel Nugraha Ardyan[1] and Redondo[2]

[1,2] Departement of Geological Engineering, Hasanuddin University, Makassar, Indonesia

E-mail: abelardyan@gmail.com

Abstract. Jeneberang River watershed, especially in the downstream part, there is a decrease in clarity and an increase in water turbidity which is influenced by the presence of particles (sediment, algae and pollutants) that are suspended by the river natural process or because of the busy activities along the river, such as rock and sand mining activities, ports, settlements and industrial activities around the river. The analysis method using satellite imagery is used because it is considered very efficient and effective in saving time, energy and cost. The research begins by inputing sentinel – 2A imagery data and Google Earth Engine as data processor. After that, followed by analyzing the Normalized Difference Water Index (NDWI) and the Normalized Difference Turbidity Index (NDTI) to determine the presence of surface water and water turbidity distribution from accumulation of silt and waste which is NDWI and NDTI are represented through the color index. Based on the results of the research in the NDWI and NDTI index and map, coastal area and the area along the river directly adjacent to the land were dominantly red, then on the transitional part between river and sea were dominantly green. This means that activity at the coastal and the river area directly adjacent to the land greatly affects the level of water turbidity.

1. Introduction

Nowadays, the problem of water turbidity has become a very urgent problem that needs to be controlled and overcome. The result of changes in water turbidity itself can occur due to human activity or due to the natural process of the river itself which results in water mixed with particles such as sediment, algae or pollutants.Simply put, water turbidity is water that is contaminated with particles such as sediment, algae or pollutants. The level or amount of particles that are present or mixed and even dominate the water can be used as a paramater to determine the level of water clarity.

The usual way to analyze the level of water turbidity is by mapping the river body and taking samples in the field and analyzing or testing field samples in the laboratory. However, due to time and cost limit, this research analyzes using remote sensing methods by conducting preliminary mapping to indicate water turbidity using sentinel imagery. Sentinel imagery data can be an alternative option as an initial analysis which relies on the spectral response of the Earth's surface. Sentinel imagery datacan then be processed according to the needs, which this time the focus is to determine the condition of water turbidity. Therefore, in this study, water turbidity analysis was carried out efficiently using the results of sentinel image processing.

IOP Conf. Series: Earth and Environmental Science **1525** (2025) 012028 doi:10.1088/1755-1315/1525/1/012028

2. Research Area

The study on water turbidity using Sentinel imagery was carried out in the Jeneberang River, specifically focusing on its downstream section. The Jeneberang River, located in South Sulawesi Province, Indonesia, stretches approximately 75-80 km from east to west, originating from Mount Bawakaraeng and Mount Lompobattang before emptying into the Makassar Strait. The river flows through eight districts and one city within the province.

As the primary river of the Jeneberang Watershed, it runs along the southern arm of Sulawesi Island, descending the western slopes of Mount Lompobattang, a stratovolcano with a loosely compacted rock structure. The summit of this volcano features a distinct crater. The geological characteristics of the Jeneberang River watershed are predominantly composed of alluvial deposits from rivers, lakes, and coastal areas along its course. The rock formations present in the area include the Camba Formation, consisting of marine sedimentary rocks and volcanic materials such as breccias, lahars, tuffs, and conglomerates, as well as the Lombobattang Formation found in the upper reaches of the river.

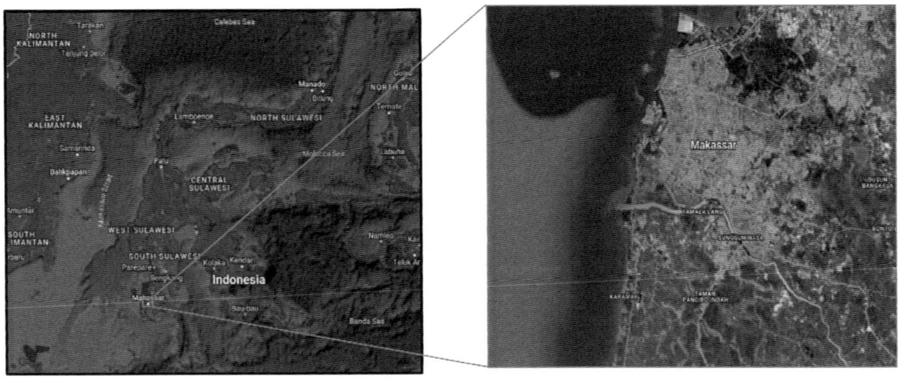

Figure 1 Location of Jeneberang River Downstream as Research Area

As shown in **Figure 1**, the downstream section of the Jeneberang River is situated in the city center, surrounded by densely populated residential areas. The water conditions in this part of the river significantly impact the quality of life of the local community. Therefore, assessing the water quality in the downstream section of the Jeneberang River is essential to understand its effects and ensure better environmental management.

3. Methodology

In this research, data processing was carried out with the help of the Google Earth Engine using the javascript programming language. The data used is sentinel-2A satellite imagery published from June 2024 to September 2024. Sentinel-2A setellite imagery data is high-resolution satellite imagery that can be used for a variety of applications, including agriculture, forestry and disaster mapping [1,2]. Sentinel-2A is a satellite carrying the Multi-Spectral Instrument (MSI) which takes images of the Earth by sampling 13 spectral bands across the visible and infrared spectrum. Sentinel-2A has a spatial resolution of 10 meters per pixel and has a global review frequency of five days.

3.1 Calculation of Normalize Difference Wetness Index (NDWI)

To monitor turbidity variations in the river, Sentinel-2 optical remote sensing data was analyzed by examining changes in reflectance, which are directly linked to turbidity levels. The Normalized

Difference Water Index (NDWI) leverages reflected near-infrared (NIR) radiation and visible green light to enhance water features while suppressing soil and terrestrial vegetation. In this context, RgR_g and RnirR_{nir} represent reflectance in the green and NIR bands, respectively. These wavelengths were selected to maximize water reflectance in the green spectrum while minimizing NIR reflectance from water, benefiting from the high NIR reflectance of vegetation and soil [5].

NDWI is commonly used to distinguish water bodies from dry land and is particularly effective for water surface mapping. Since water bodies exhibit low radiation and strong absorption in the visible infrared spectrum, NDWI enhances water-related features efficiently. While beneficial for developed areas, it can sometimes overestimate water coverage. The Normalized Water Difference Index (NDWI) is calculated using the formula NDWI = (Green - NIR) / (Green + NIR). This satellite-derived index is useful for highlighting water bodies, with Sentinel-2A imagery utilizing Band 3 for the green spectrum and Band 8 for the NIR spectrum.

3.2 Calculation of Normalize Difference Water Turbidity Inded (NDTI)
Where Rr represents reflectance in the red band, pure water generally reflects more in the green wavelength than in the red. Higher turbidity levels result in increased values of the Normalized Difference Turbidity Index (NDTI), while lower turbidity levels produce lower NDTI values. The reflectance of the green and red bands is used to estimate suspended sediment concentration within stratified water columns. To verify the results, NDTI mapping was conducted using a normalized ratio of red and green band reflectance. Observations indicate that the red and near-infrared (NIR) bands are particularly sensitive to turbidity estimation [4,6].

These spectral bands are highly useful for quantitatively assessing turbidity, especially when field data for optically deep water is unavailable. The study suggests that remote sensing techniques can serve as a preliminary approach for estimating water quality in the absence of ground-based measurements. NDTI represents turbidity levels using a color index ranging from red to green. In Sentinel-2A imagery, Band 3 corresponds to the green spectrum, while Band 4 represents the red spectrum.

4. Result and Discussion

The image processing results from Google Earth Engine (GEE) generate an NDTI (Normalized Difference Turbidity Index) image with pixel values ranging from -1 to 1. However, when displayed using standard image viewing software, the output appears only in black and white, which makes it challenging to distinguish subtle variations in the NDTI index values. This limitation hampers the ability to interpret turbidity distribution effectively, as the differences between high and low turbidity regions are not visually prominent. As a result, the image can only be analyzed in a basic manner, without clear differentiation between various turbidity levels [3].

When the image is displayed against a white background, it becomes slightly clearer, revealing black areas on the right, bottom, and as scattered black dots, which correspond to water bodies with specific NDTI values. This suggests that standard image viewers interpret the NDTI output as binary pixel values, rather than as a continuous range, restricting their ability to depict turbidity variations in the downstream section of the Jeneberang River. This misrepresentation of data prevents a comprehensive assessment of turbidity conditions, which is crucial for environmental monitoring and water quality management.

To address this issue and enhance visual interpretation, the use of specialized geographic information system (GIS) software is necessary. Such software is designed to handle spatial data more effectively, allowing for the application of color gradients that accurately reflect the full range of NDTI values. In this study, open-source ArcGIS software was employed to visualize the

NDTI index values more precisely. By leveraging the advanced visualization capabilities of GIS software, the turbidity levels could be represented in a more detailed and analyzable manner, facilitating a better understanding of sediment distribution and water quality in the study area.

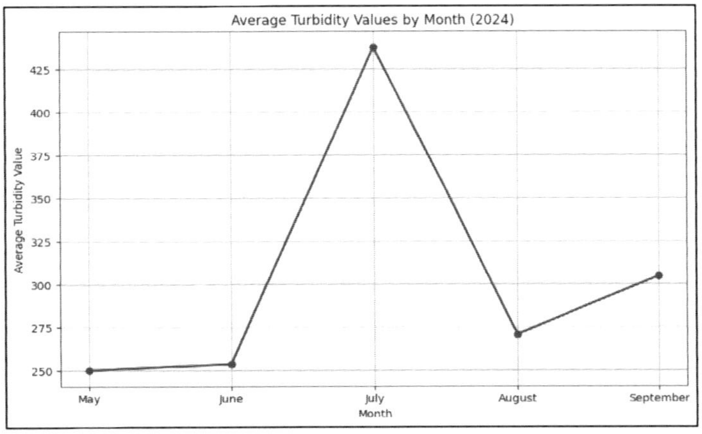

Figure 2 Average Turbidity Value May 2024 – September 2024

Based on data processing using Google Earth Engine which is summarized in **Figure 2**, the average estimated value of water turbidity in the lower reaches of the Jeneberang River for five months (May 2024 - September 2024) by NDTI count is found at around 303.36. In May 2024 the average value is 250.17, in July 2024 the average value is 253.57, in June 2024 shows an average value of 437.80, in August the average value is 270.71 and in September the average value is 304.56. Based on the analysis of water turbidity in the downstream of the Jeneberang river in the last 5 months, it shows an increase turbidity levels from May (lowest value) until June (highest value), then decrease on August and increase on September.

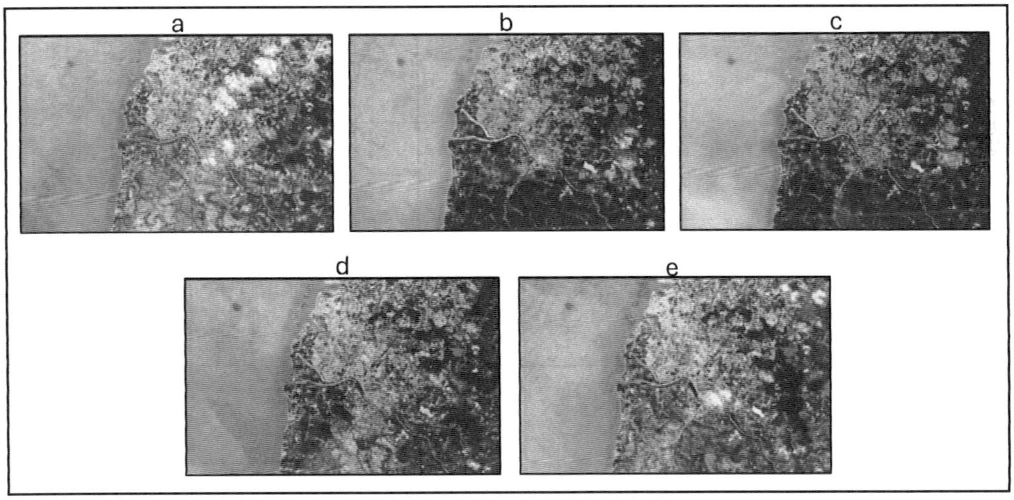

Figure 3 (a). NDTI of May, (b). NDTI of June, (c). NDTI of July, (d). NDTI of August, (e). NDTI of September.

The NDTI map of the lower Jeneberang River shows that turbidity is relatively consistent across different regions and months. However, higher turbidity levels are detected in downstream water bodies near the main dam, as illustrated in **Figure 3**. The highest turbidity concentrations

IOP Conf. Series: Earth and Environmental Science **1525** (2025) 012028 doi:10.1088/1755-1315/1525/1/012028

are observed in transition zones between land and the river or coast. While the index maps show minimal differences, particularly in July, the average turbidity value curve indicates that July has the highest turbidity levels, significantly differing from other months. This inconsistency may be attributed to the Sentinel-2A satellite's data collection interval of every five days, which may not fully capture short-term variations in turbidity. To enhance accuracy and validate these remote sensing findings, field surveys, water sampling, and laboratory analysis are recommended. These steps would provide a more reliable assessment of turbidity levels in the Jeneberang River.

5. Conclusion

The NDTI values obtained from Google Earth Engine (GEE) are read as binary pixel values, making it difficult to visually interpret turbidity conditions. Therefore, spatial GIS software is required for effective visualization. The analysis indicates that turbidity levels are highest in July, followed by a decline in August, and then an increase again in September. Turbidity is present across almost the entire area, but the highest concentrations are observed in the transition zone between land and water.

References

[1] Caballero, I., Steinmetz, F., G., N., 2018. Evaluation of the first year of operational Sentinel-2A data for retrieval of suspended solids in medium- to high-turbidity waters. Remote Sensing

[2] Johnson RW 1975 Quantitative sediment mapping from remotely sensed multispectral data Remote sensing of earth resources 4 565–576

[3] Ritchie JC, Zimba PV, Everitt JH 2003 Remote sensing techniques to assess water quality Photogrammetric Engineering Remote Sens 69 695–704

[4] S Bid and G Siddique 2019 Identification of seasonal variation of water turbidity using NDTI method in Panchet Hill Dam, India Model. Earth Syst. Environ 5 1179–1200,

[5] X Xu, H Fan, X Chen, and C Mi 2020 Estimating low eroded sediment concentrations by turbidity and spectral characteristics based on a laboratory experiment Environ Monit Assess 19

[6] R Foteh, V Garg, B R Nikam, M Y Khadatare, S P Aggarwal, and A S Kumar 2018 Reservoir Sedimentation Assessment Through Remote Sensing and Hydrological Modelling Journal of the Indian Society of Remote Sensing 46 1893–1905

IOP Conf. Series: Earth and Environmental Science **1525** (2025) 012029 doi:10.1088/1755-1315/1525/1/012029

Risk Management Analysis in the Construction of Solar Power Plants (PLTS) at Ibu Kota Nusantara, Indonesia

Isradi Zainal[1], Abdul Zain[2], and Andi Surayya Mappangile[1]

[1] Department of D4K3, Faculty of Vocation, Balikpapan University, Balikpapan, Indonesia
[2] Energy Systems Engineering Study Program, BJ Habibie Institute of Technology, Indonesia

E-mail: abd_zain@ith.ac.id

Abstract. The establishment of a Solar Power Plant (PLTS) in IKN (Ibu Kota Nusantara) Indonesia is grounded in the vision of IKN as a sustainable World City, which emphasizes the implementation of new and renewable energy sources within the IKN region. Based on this, the Government is building a Solar Photovoltaic (PV) type Solar Power Plant (PLTS) with a capacity of 50 MW. To ensure that PLTS construction runs smoothly, according to expectations and with minimal risks, risk management analysis is needed. The aim of this research is to determine the source of risk and propose risk mitigation. The method used is interviews and direct observation. The research results show that there are 15 priority risk sources and 3 mitigation actions that can be proposed to overcome priority risk sources.

1. Introduction

IKN (Ibu Kota Nusantara) Indonesia is a newly proposed capital city that aims to utilize renewable energy sources. To initiate this plan, the Ministry of Energy and Mineral Resources has begun by setting up a Solar Power Plant (PLTS).

Solar Power Plants (PLTS) are power plants that deliver solar thermal energy into electrical energy [1]. PLTS directly converts solar energy in the form of light into electrical energy without using any fuel through solar cell or Photovoltaic (PV) technology [2]. The PLTS system requires an open space that is exposed to solar radiation and free from objects or shadows that can block solar radiation from hitting the solar panels. PLTS can be utilized economically by using the roof of the building as an area for installing solar panels [1].

For IKN Nusantara, President Jokowi carried out the construction of the PLTS on November 23 2023. The construction of the Solar PV installation was carried out by a subsidiary of PT PLN, namely PT. PLN NR, PLN SC, PLN NPS and PLN NPC. This work was carried out from November to January 2024 and carried out a series of tests and commissioning until February 2024. For the initial stage, the government has provided PLTS of 10 MW in IKN which was inaugurated on 28 February 2024 by President Jokowi and is the first Syncron IKN opened is 89 hectares. Thus, if an additional 40 Megawatts are added, it is possible that around 300 hectares more land will be needed. PLTS IKN 10 MW has received a certificate of synchronous feasibility on February 29 2024 and has received a certificate of operational feasibility for STS 1 and STS 2 on June 3 2023.

Risk management is a structured and systematic process of identifying, measuring, mapping, developing alternative risk treatments and monitoring and controlling risk management [3].

2. Literature Review

2.1 Risk Management

Risk management is a series of activities carried out to minimize the risks faced by various parties, such as organizations, companies, families and society. This activity includes planning, organizing, leadership/coordination and supervision (including evaluation) of risk management programs [3]-[4].

Risk management is an approach taken to risk by understanding, identifying and evaluating a risk. Next, consider what will be done regarding the impacts caused and the possibility of transferring risks to other parties or reducing the risks that occur [4]. The aim of risk management is to maximize the opportunities and consequences of events that have a positive impact and minimize the opportunities and consequences of negative events [4]-[5]. Risk management is a strategy carried out to identify, manage and evaluate all risks in an entity [5].

2.2 Stages of Risk Management

In implementing risk management, there are several stages that must be implemented. Risk identification, risk analysis and risk evaluation are the three main stages in the risk management process [3]. The stages of risk management are as follows:

1. Risk assessment is a structured method that shows whether a company/business has an acceptable level of risk.
2. Risk identification is an effort made to find and identify risks that may arise in the operational activities of a company/business. The aim of risk identification is to determine all risks arising from various factors, such as human, natural and infrastructure factors.
3. Risk analysis is the process of determining risk by analyzing data results from the risk identification process. This stage includes aspects of assessment, characterization, management and procedures related to the risks of a company's business.
4. Risk evaluation is determining risk priorities by comparing risk levels and risk criteria using a risk matrix.
5. Risk treatment is the process of identifying actions that need to be taken to overcome risks and minimize the overall risk impact.

2.3 Project Management Group Progress

In project management, a process is a series of related procedures and activities carried out to achieve the goals of a project [4]. Each process involves the use of various inputs, tools and methods [7]. These processes are interrelated and can be repeated at each project phase. The project management process can be classified into five project management process groups, namely:

1. Initiating, the process of determining the start of a new project or a new phase in an existing project.
2. Planning, the process of determining the project scope, time plan, determining the estimated budget or material costs that need to be monitored as well as the quality requirements and specifications that must be met by the project.
3. Executing, the project implementation process carried out to complete work that has been determined to meet project requirements.

4. Monitoring and Controlling, the process of comparing actual achievements and project progress with plans, predicting deviations, analyzing their causes and overcoming them before they occur.
5. Closing, the project closing process, where all project activities and processes are closed.

3. Research Methods

This research uses a qualitative research approach. Qualitative data were collected through interviews and brainstorming with expert judgment to identify risk events and risk sources in the PLTS project in IKN.

The research began by conducting interviews and brainstorming with expert judgment or experts on PLTS projects and risk management at PT NSSE. The interviews and brainstorming were carried out to find out about the PLTS project activities at PT NSSE. To make project activity mapping easier, project management process groups are used. The results of the identification and mapping of project activities then become input into the risk identification process which includes identifying risk events and risk sources (risk agents).

The research location was carried out at PT NSSE at IKN Nusantara. This research was carried out by means of interviews, brainstorming and distributing questionnaires regarding risks related to PLTS projects. The research was conducted from September 2024 to November 2024.

4. Results and Discussion

4.1. General description of PLTS in IKN

The Solar Power Plant at IKN is planned to have a capacity of 50 MW and is located in Pemaluan Village, Sepaku District, North Penajam Paser Regency, as seen in Figure 1. The developers of this project are PLN NR (51%) and Sembcorp (49%). The PPA scheme is 25 years TOP (BOO). The financing scheme for this project is equity financing and refinancing after COD with an investment value of 64.2 million USD with a PPA rate of 8,499 cUSD/kwh.

The solar power plant (PLTS) at IKN Nusantara is planned to have a capacity of 50 Mega Watt (MW). Groundbreaking for the construction of the PLTS was carried out by President Jokowi on November 23 2023. Construction of the Solar PV installation was carried out by PT PLN's subsidiary, namely PT . PLN NR, PLN SC, PLN NPS and PLN NPC. This work was carried out from November to January 2024 and carried out a series of tests and commissioning until February 2024. For the initial stage, the government has provided a 10 MW PLTS at IKN which was inaugurated on February 29 2024 by President Jokowi and is the first Syncron IKN. The area of land cleared is 89 hectares. The IKN 10 MW PLTS received a certificate of synchronization feasibility on February 29 2024 and received a certificate of operation feasibility for STS 1 and STS 2 on June 3 2024.

Previously electricity needs in Sepaku were supplied from GIS Mobile IKN (Gersik) and G Petung via a 20 KV line. Next, a 150 kV transmission system was built directly to IKN to GIS 4 (Gas Insulated Switchgear), which was then distributed via distribution channels. GIS 4 is connected to GI Kariangau via High Voltage Overhead Channels (SUTT) and underground High Voltage Cable Channels (SKTT). because it has entered the KIPP IKN area where all cables must be planted.

In this PLTS, Smart Grid or a two-way electricity data flow network system with digital communication technology will also be applied. This network system can detect various problems, including faster electricity restoration if an electricity disturbance occurs. Apart from that, there are a number of examples of smart grid-based technology that have the potential to be used in IKN Nusantara, including Advance Metering Infrastructure (AMI) which allows electricity users in IKN to manage and monitor energy consumption in a timely manner.

IOP Conf. Series: Earth and Environmental Science **1525** (2025) 012029 doi:10.1088/1755-1315/1525/1/012029

The condition of the equipment at the time of commissioning can be described in table 1.

Table 1. Equipment condition during commissioning

System	Equipment	QC Checklist Pre-comm	Pre-comm
MV System	MV Swgr Service Breaker	Ok/ Not Ok/ NA O	Ok/ Not Ok/ NA O
	PV MV Swgr	Ok/ Not Ok/ NA O	Ok/ Not Ok/ NA O
	PV Power Transformer	Ok/ Not Ok/ NA O	Ok/ Not Ok/ NA O
	MV Cable	Ok/ Not Ok/ NA O	Ok/ Not Ok/ NA O
STS System	MV Swgr STS	Ok/ Not Ok/ NA O	Ok/ Not Ok/ NA O
	Power Transformer STS	Ok/ Not Ok/ NA O	Ok/ Not Ok/ NA O
	LV Panel STS	Ok/ Not Ok/ NA O	Ok/ Not Ok/ NA O
	MV Cable from/to STS	Ok/ Not Ok/ NA O	Ok/ Not Ok/ NA O
	LV Cable from/to STS	Ok/ Not Ok/ NA O	Ok/ Not Ok/ NA O
PV System	PV Module	Ok/ Not Ok/ NA O	Ok/ Not Ok/ NA O
	Solar DC Cable	Ok/ Not Ok/ NA O	Ok/ Not Ok/ NA O
	PV String Box	Ok/ Not Ok/ NA O	Ok/ Not Ok/ NA O
	PV Inverter	Ok/ Not Ok/ NA O	Ok/ Not Ok/ NA O
	Data logger	Ok/ Not Ok/ NA O	Ok/ Not Ok/ NA O
DG System	Generator control panel	Ok/ Not Ok/ NA O	Ok/ Not Ok/ NA O
	Unit Generator set	Ok/ Not Ok/ NA O	Ok/ Not Ok/ NA O
Control System	Panel RMS	Ok/ Not Ok/ NA O	Ok/ Not Ok/ NA O
	Cable Control/ communication	Ok/ Not Ok/ NA O	Ok/ Not Ok/ NA O
	Sensor/ Instrument	Ok/ Not Ok/ NA O	Ok/ Not Ok/ NA O
Utility/ Auxiliary System	Trafo Auxiliary (UAT)	Ok/ Not Ok/ NA O	Ok/ Not Ok/ NA O
	Utility/ Aux pane	Ok/ Not Ok/ NA O	Ok/ Not Ok/ NA O
	LV Cable	Ok/ Not Ok/ NA O	Ok/ Not Ok/ NA O
Grounding System	Grounding Rods	Ok/ Not Ok/ NA O	Ok/ Not Ok/ NA O
	Lightning Arrester	Ok/ Not Ok/ NA O	Ok/ Not Ok/ NA O
	Bonding & PE	Ok/ Not Ok/ NA O	Ok/ Not Ok/ NA O

4.2. Risk Identification

The data used in this research came from two sources, namely primary data and secondary data. Primary data was obtained through interviews, brainstorming and distributing questionnaires to expert judgment or people who are experts in PLTS projects. The respondents in this study can be seen in table 2 below.

Table 2. Respondent Data

No	Position	Length of working
1	Corporate Secretary	More than 5 years
2	Legal	More than 5 years
3	Procurement	More than 5 years
4	Construction	More than 5 years
5	Technical	More than 5 years
6	Document Controller	More than 5 years
7	Document Controller	More than 5 years

Secondary data comes from scientific publication references that discuss solar power plants (PLTS) [5] and articles sourced from the website of the Indonesian Ministry of Energy and Mineral Resources [6].

The initial stage carried out is risk identification to determine risk events and risk sources. In this research, risk identification was carried out through interviews and brainstorming with expert judgment or people who are experts in PLTS projects at PT NSSE.

4.2.1 Political Risk

Research on solar project-based contractors, their capacities and practices in managing risks is needed to describe the current state of the situation. The main risks associated with solar power projects can be classified into political, financial, social and technical risks which are further classified as follows.

Delays in permits and approvals from government organizations are of particular importance as the risk of obstacles in obtaining regulatory sanctions and achieving financial closure is relatively high. Failure to achieve financial close inflates the funds needed to implement a project if a project does not obtain mandatory approvals. There are several political risks associated with solar power projects in IKN illustrated in Table 3.

Table 3. Political Risk

Risk Classification	**Risk Description**	1	2	3	4	5
Policy Risk	• High costs and unfavorable electricity pricing rules			√		
	• Environmental externalities			√		
	• Institutional barriers and orderliness			√		
	• High rates for consumers			√		
	• Frequent changes in government policy				√	
Licensing, commissioning and risk approval	• Payment guarantee mechanisms are unreliable.		√			
	• The risks result from contracting parties not recognizing their contracts as binding, and the justice system is plagued by lengthy processes, insurmountable delays and "disputed" cases.	√				
	• Problems in obtaining permits and certification			√		
	• Delay in receiving payment			√		
Contract	• The risk arises because the parties to the contract do not recognize its binding nature and the justice system is plagued by large delays that cannot be resolved. "Bank up" cases		√			
	• The payment guarantee mechanism is not reliable			√		

4.2.2 Environmental Risk

There are several environmental risks associated with solar power projects in IKN as shown in Table 4.

Table 4 Environmental Risk Classification

Risk Classification	**Risk Description**	1	2	3	4	5
	• Upcoming environmental legislation that may affect the project		√			

Environment	• Biodiversity protection problems			√		
	• Worker Health and Safety Issues			√		

According to table 4, the solar power project in IKN requires a very large forest area to spread the solar panels up to several hectares to meet the electricity supply target if referring to environmental legislation there are potential risks that are obstacles in its development. There is a problem of reduced biodiversity habitat. Likewise, it will be related to the health and safety of workers.

4.2.3 Financial Risk and Equity

There are several Financial Risks and Equity associated with solar power projects in IKN as shown in table 5. The risks in table 5 will be the government's task in designing financial stability in an effort to complete the relocation of the nation's capital to a new area, especially those related to interest rates, financial incentives and capital.

Table 5. Classification of Financial Risk and Equity

Risk Classification	Risk Description	1	2	3	4	5
Financial and Equity	• The instability of the tax environment in Indonesia often results in significant changes in guidelines in the IKN renewable energy sector.				√	
	• It is difficult to fund working capital because a) banks are hesitant to provide loans to rural entrepreneurs b) banks do not approve raw materials as collateral.			√		
	• Unpredictable borrowing, lending and repayment interest rates.			√		
	• Inappropriate financial incentives			√		
	• High capital losses		√			

4.2.4 Engineering Risk

A number of technical risks in solar power projects in Indonesia can be seen in table 6. The technical risks in table 6 are often obstacles if large projects are implemented in new locations. The majority of heavy and light equipment and labor are brought in from outside the island. Technology, experts, field workers and maintenance result in high costs that require transparency. Likewise, conditions in new locations require certainty of the suitability of the land where the project is implemented.

Table 6. Technical Risk Classification

Risk Classification	Risk Description	1	2	3	4	5
Technology	Technology/joint venture, estimated lead time issues if any.		√			
Operation and Maintenance	Problems such as machine failure and downtime			√		
	The guaranteed return period is high			√		
Procuremen	The indigenization percentage refers to the cost of the project and its impact on the cost of the project				√	
	An increase in the percentage of Natives overall			√		
Human resource risks	Difficulty in finding suitable skilled and unskilled employees to work in this field		√			
Land suitability	Selecting a suitable location Land acquisition issues					√

Transparency Risks	Corruption is common in the administration, bureaucracy and business circles.				√	

5. Conclusion

Risk analysis in the implementation of solar energy projects in IKN using the AHP technique and the conclusions that can be drawn according to the description above are that changes in government policies must be minimized, financial and equity risks in the form of instability in the tax environment in Indonesia often occur significant changes in guidelines in the renewable energy sector of IKN to be improved and transparency risks in the form of corruption that is common in the fields of administration, bureaucracy and business circles to be controlled immediately.

References

[1] Zainal, I., & Zain, A. (2023, June). Study of the Operation and Maintenance System of Solar Power Plant (PLTS) and Diesel Power Plant (PLTD). In *Proceeding ADRI International Conference on Multidisciplinary Research* (Vol. 1, No. 1, pp. 259-270).

[2] I. Jamil, J. Zhao, L. Zhang, S. F. Rafique, and R. Jamil, "Uncertainty Analysis of Energy Production for a 3 × 50 MW AC Photovoltaic Project Based on Solar Resources," International Journal of Photoenergy, p. 1056735, (2019). http://doi.org/10.1155/2019/1056735

[3] Damien Bazin, Nouri Chtourou, Amna Omri (2021), Risk management and policy implications for concentrating solar power technology investments in Tunisia

[4] J. A. B. Wirawan and I. Garniwa, "Risk Analysis Development of Solar Floating Power Plant in The Sea with Monte Carlo Method," in 2018 3rd International Conference on Information Technology, Information System and Electrical Engineering (ICITISEE), 2018, pp. 396-401.

[5] Octa Bimansyah Untoro*, Fakhrina Fahma, Wahyudi Sutopo (2021), Risk Management Based on NPV-at-Risk: A Case Study in Floating Solar Power Plant Project, Jurnal Teknik Industri ISSN : 1978-1431 print | 2527-4112 online Vol. 22, No. 2, August 2021, pp. 245-254

[6] https://www.esdm.go.id/id/publikasi/handbook-of-energy-economic-statistics-of-indonesia Accessed July 2024

[7] Zain, Abdul. "Kendali Sistem Tegangan Baterai Peralatan 20 kV PT PLN Bontang." *Jurnal Teknologi Elekterika* 19, no. 1 (2022): 8-15 . https://doi.org/10.31963/elekterika.v6i1.2807

IOP Conf. Series: Earth and Environmental Science **1525** (2025) 012030 doi:10.1088/1755-1315/1525/1/012030

Assessing the Impact of Land Use Change on Ecosystem Service Using GIS and Remote Sensing Technology: A Review

Rini Anggraini[1,*], Sumbangan Baja[2], Risma Neswati[2]

[1] Post Graduate Student of Regional Planning and Development, Hasanuddin University, Makassar, Indonesia
[2] Department of Soil Science, Hasanuddin University, Makassar, Indonesia

*E-mail: kebumian01@gmail.com

Abstract. This study aims to summarize and evaluate the literature on the use of Ecosystem Service Value (ESV) in analyzing the impact of land use change on the economic value of ecosystems using GIS and remote sensing technology. Using a literature review approach, this study examines various studies that relate ESV to land cover changes induced by human activities such as urbanization and deforestation. The results of the study show that land use change generally reduces the value of ecosystem services, ultimately affecting environmental quality and economic welfare. This finding emphasizes the importance of sustainable land management policies to maintain ecological balance and support sustainable development.

1. Background

Ecosystem service value (ESV) is a monetary estimate of the different services provided by an ecosystem. This concept was introduced by Costanza et al. [1], who proposed a list of ecosystem valuation coefficients by estimating the monetary value of 17 ecosystem services provided by 16 different biomes. Some examples of these services are biodiversity, water filtration, soil retention, and air purification, which are essential for maintaining environmental quality [2]. Ecosystems provide raw materials and other critical functions such as regulating climate, reducing pollution, conserving water resources, and preventing erosion, floods, and fires. All these benefits are collectively known as ecosystem services [3].

Assessing the value of ecosystem services has become an important topic in international research, as these studies combine different disciplines such as ecology, economics, sociology, and geography into a holistic approach. One of the factors that most influences the condition and stability of ecosystems is land use change, often caused by human activities such as urbanization, deforestation, agricultural expansion, and infrastructure development [4].

These changes in land cover have significant impacts on ecosystems, which provide valuable benefits to humans in the form of both natural resources and ecological processes. These benefits are collectively referred to as ecosystem services [5]. Changes in land cover type and land use can affect physical structure, ecological processes, and ecosystem function, so one approach to estimating ESV values is to relate them to land use. This approach highlights the importance of the ecological value of ecosystem services provided by human activities and management [6].

Land use and land cover change are increasingly complex and relevant phenomena in the modern era, especially due to increasing pressure from human activities. Land cover is the physical material appearance of the Earth's surface that shows the relationship between natural and social processes. Land use change is defined as a shift from one type of land use to another. Changes in land cover are often influenced by the dynamics of land use, particularly in relation to the growing needs of the population for space and resources.

The need for food and space for other activities increases with population growth, which in turn drives changes in land use. In particular, urban areas continue to grow at a rapid rate of population growth, leading to an increase in built-up areas. High urbanization is one of the main factors behind the increase in built-up areas in urban areas, driven by the movement of people from rural to urban areas in search of better economic opportunities. The declining number of jobs in rural areas and the shrinking area of agricultural land due to the conversion of land into residential areas have led to many rural communities losing their livelihoods. In addition, the younger generation is increasingly less interested in working in the agricultural sector and prefers to move to the city to find work.

In the face of this rapid change, monitoring and analyzing changes in land use and land cover become very important. Geographic Information System (GIS) is one of the technologies that can be used to monitor these changes. GIS enables the management, storage, and visualization of spatial data, which is important in identifying patterns of land use change [7]. According to Ramadhani, in addition to GIS, remote sensing technology also plays an important role in efficiently monitoring land use change, especially over large areas. Overall, any change in land cover has a direct impact on the value of ecosystem services, which ultimately affects the ecological balance and the economic benefits that humans derive from the environment.

Against this background, the question arises as to how ecosystem service value (ESV) can be applied to land use changes in order to assess the resulting impacts in terms of economic value. The purpose of this paper is to summarize the various literature on land use change using ESV analysis and to provide an overview of how this approach helps to understand and measure the impact of land use change on the value of ecosystem services in a more measurable and structured way.

2. Research methods

A substantial number of studies have employed the method of linking ESV and land use change by utilizing GIS and remote sensing technology. Ecosystem Service Value (ESV) is a technique for assigning a monetary value to the services and goods of an ecosystem. Comprehension of land use and land cover change (LULC) and its repercussions on ESV is imperative to the decision-making process [8]. The capacity of ecosystem services (ES) is significantly influenced by rapid changes in LULC. ES is capable of providing a multitude of services that are indispensable for human survival, livelihoods, and well-being, both in a direct and indirect manner [1]. The benefits of ES to humans manifest in diverse forms, including provisioning (e.g., food or energy yields), regulation (biophysical processes that provide benefits such as climate and air regulation), support (processes that are essential for providing other ecosystem services), and cultural (e.g., recreational, aesthetic, spiritual benefits) (MA in [8]). The provision of ecosystem services occurs when humans derive benefit from the functions and processes of ecosystems. The interaction between humans and nature is evident in the numerous benefits that ecosystem services provide for individual utility and economic development [9].

The land use type has been demonstrated to have a significant impact on the provision of ecosystem services. Landscapes such as forests and savannas are often converted to other uses, including agricultural land, grazing land, and built-up areas. For instance, the estimated global rate of conversion of forest land to agriculture is approximately 13 million hectares per year, with almost 40% of the land surface being used for crop production [10]. As Park et al. [8] demonstrate, the conversion of natural ecosystems to agricultural land, plantations, and urban areas increases the production of food, fiber, wood, and housing but causes a decline in other ecosystem services. The efficient mapping of these changes is facilitated by the integration of remote sensing technology with Geographic Information System (GIS) software.

Land Use Cover (LULC) change is a complex phenomenon influenced by various factors, both natural and anthropogenic. As Schneeberger asserts, this phenomenon is precipitated by various driving factors and actors that accelerate the rate of certain changes. The underlying causes of land change can be classified into two overarching categories: basic causes and direct causes [11]. Direct causes include human activities or actions that directly affect land use, such as urbanization or the expansion of agricultural land. Conversely, underlying causes encompass fundamental processes that exert an indirect influence on land use change, including population dynamics and agricultural policies. The majority of LULC change analysis is conducted using remote sensing at varying scales.

3. Results and Discussion

3.1 Case study of land use change impacts

A considerable body of research has been dedicated to the analysis of land use change, with a particular emphasis on the conversion of forests, agricultural lands, wetlands, and mangrove ecosystems. The principal factors driving these changes include fires, human encroachment, economic expansion, population growth, and urbanization [12][13][14][15].

3.1.1 Forest Land Change

The transformation of land from forests to settlements, plantations, and wetlands has emerged as a pivotal subject in land use change studies, given its substantial environmental ramifications. A substantial body of research has demonstrated that the conversion of forests to urban settlements exerts considerable pressure on natural ecosystems, resulting in a decline in forest cover and the disruption of ecosystem functions. This decline has been shown to increase the risk of environmental damage, especially in the context of urban expansion and the establishment of reservoirs [16][17][18]. Moreover, the conversion of forests into agricultural land and plantations, while yielding positive outcomes such as enhanced agricultural productivity, particularly in the horticultural sector, concomitantly engenders various environmental disasters. These changes can lead to various environmental disasters, including landslides, floods, and the emergence of areas prone to erosion, which in turn exacerbate ecological conditions in the surrounding area [19][20][21][22].

3.1.2 Changes in Plantation Land

The transformation of plantation land into uses such as livestock husbandry, residential areas, educational facilities, Islamic boarding schools, and recreational areas has emerged as a prominent phenomenon. The impetus for this transition is the aspiration of individuals to enhance their income, leading to a shift from plantation activities to these alternative sectors. This phenomenon has been observed since 2000 and has seen a notable escalation in the period 2015–2020 [23]. Consequently, the necessity for public education on the subject of sustainable land management is paramount.

3.1.3 Wetland Change

Peat swamp forest ecosystems are significant because they provide habitat for various types of plants and animals, which support biodiversity. In addition, this ecosystem functions as an effective carbon sink and regulates water management, which plays a significant role in maintaining environmental balance. However, in recent years, many logging activities have been carried out to convert it into plantations and residential areas. Increasing human activity and population are the leading causes of this land conversion. Coupled with the intensification of drought and rising temperatures which accelerate land degradation, decreasing water levels cause the drying of peat substrates [24][25]. The loss of peat swamp forest areas has occurred significantly since the year 2000, with the rate of loss accelerating to 20% in 2011. If this trend continues, the peat swamp forest ecosystem is projected to be completely destroyed before 2060 [26].

3.1.4 Agricultural Land Changes

Population and household growth are the main drivers of the conversion of agricultural land to non-agricultural land such as built-up areas in Indonesia, a phenomenon that is increasingly occurring in various regions [27]. A concrete example is the village of Pasar Baru, where the conversion of agricultural land into residential areas is expected to increase significantly between 2010 and 2019 [28].

Conversion also occurs in the form of agricultural land being converted to tourism land in Latimojong, mining land in Pangkalaero village, and oil palm plantations in Siak Regency. This phenomenon is influenced by several factors, including the deterioration of the quality of agricultural land, which reduces its productivity, and the encouragement of farmers to seek more profitable sources of income. This indicates a significant shift in land-use patterns, affecting the sustainability of the agricultural sector and the welfare of farmers in many regions [29][30][31].

3.1.5 Mangrove Land Change

Changes in mangrove land have become a concern in various studies [32][33]. Environmental changes in coastal areas, coastlines and mangrove ecosystems, are greatly influenced by human activities and natural factors such as sedimentation, abrasion and accretion. Studies in Tamalanrea, Makassar City and At Pantai Bahagia, Bekasi City, reveal the significant impact of changes in coastlines on land use. At the same time, research in Muara Gembong, Bekasi, and East Lampung highlights the decline in mangrove forest cover due to land conversion and economic activities, which impacts the ecological stability and livelihoods of local communities. These three studies emphasize the importance of sustainable management to maintain the balance of coastal ecosystems while meeting socio- economic needs [34][35][36].

Overall, changes in land use are influenced by various factors such as economic needs, population growth, and other human activities. The negative impacts of these changes include environmental degradation, increased risk of natural disasters, and loss of biodiversity. Public education and sustainable land management are therefore essential to reduce these negative impacts.

3.2 Case study of the use of ESV analysis in land use change

Research conducted in China [6][37] examined the impact of land use resulting from human activities on the value of ecosystem services. Both studies found that land use changes were influenced by rapid urbanization, which led to an increase in built-up areas. The land changes that occur are mainly in the form of changes in land cover, where wetlands, water bodies, forests, and grasslands are converted to agricultural and built-up land. Increased economic benefits from agricultural land cannot compensate for ecosystem losses due to wetland degradation [38]. Land reclamation also transforms coastlines into aquaculture ponds, construction sites, and agricultural land, as shown in a case study of Sansha Bay, China [39]. Land change and reclamation reduce ecosystem service value (ESV).

Apart from China, several researchers have also investigated the ecosystem service value (ESV) that is being lost, particularly in agroforestry ecosystems in Ethiopia [40][41]. Anthropogenic activities such as settlement expansion and land degradation have converted forests and grasslands into agricultural land. Apart from settlements, land conversion into agricultural land is also common, especially in watershed areas (WAs), to meet human needs. This land cover change leads to a reduction in the value of ecosystem services. Total ESV losses reached USD 0.551 billion over 30 years, or about USD 18.4 million per year, in the Gojeb River Basin, southwest Ethiopia [42], and a total decline of USD 150,000 over 35 years in the Upper Blue Nile Basin, Ethiopia [43]. Deforestation and reduced ESV values indicate significant ecosystem degradation, reducing environmental functions such as provision of clean air, erosion control, and biological habitat diversity.

In Indonesia, increasing population, housing needs, and urbanization in urban areas are driving land conversion [44][45]. The reduction in the area of vegetation and water bodies converted to built-up land is the main driver of the decline in ecosystem service value (ESV). This decrease is mainly due to the conversion of vegetation and water into built-up land. The total value of ESV decreased from IDR 892.49 billion in 1991 to IDR 684.95 billion in 2022 [22].

Research in Bangladesh shows a significant decrease in ESV of USD 587.51 million due to land use change over three decades [46]. This decline in ESV highlights the urgent need to adopt sustainable conservation, restoration, and land-use policy strategies to ensure the future sustainability of ecosystems. This is supported by research that estimates the total value of ecosystem services in the East African biodiversity hotspot in response to landscape dynamics [8].

Land changes that reduce ecosystem service value (ESV) do not mean that improvements cannot be made. For example, in the Zhifanggou watershed on the Loess Plateau, land-use changes significantly reduced ESV between 1938 and 1958. However, between 1958 and 1999, the ecosystem was successfully restored. The success of this recovery shows the importance of policy-based interventions, such as land restoration and ecosystem conservation, in reducing the negative impacts of land use change [2]. In this case, data were collected using descriptive methods with qualitative, quantitative, or a combination of both approaches. The data obtained were processed using various methods such as geographical information system (GIS) analysis, overlay, and remote sensing to produce data in numerical form. To corroborate these data, direct observations and interviews were conducted at the research site. The results of both types of data are described to clarify the information collected. Identifying land use changes takes a long time, and satellite imagery data is used to monitor these changes at different locations on the Earth's surface. Satellite imagery is used through the process of geometric and radiometric correction. Most of this image data has been obtained from the USGS website, provided by the United States Geological Survey.

The case study uses ecosystem service value (ESV) analysis to assess changes in ecosystem values due to land conversion. Land conversion is mostly caused by human activities such as urbanization, agricultural expansion, and infrastructure development, which have significantly reduced forest cover in various regions. The impacts of these changes include ecosystem degradation, loss of biodiversity, and reduced ecological functions such as water quality and soil fertility. Each significant land use change is calculated for its ESV value according to the research objectives. The ESV value depends on the type of land change occurring and is expressed in terms of the cost of measuring the impact of the change.

The use of global ecosystem service value coefficients may not fully reflect local conditions, which can affect the accuracy of the analysis. The reliance on satellite imagery and aerial photography also has limitations, particularly in detecting subtle changes in land use. In addition, the limited resolution of satellite imagery in certain areas and the lack of detailed field data can limit the ability to conduct more in-depth qualitative analysis. Potential inaccuracies in satellite imagery also arise at small scales of change, particularly in cloudy or restricted access areas.

4. Conclusion

The review found that changes in land use and land cover have a significant impact on the value of ecosystem services (ESV). Increased human activities, such as urbanisation and agricultural expansion, have led to a decline in forests and biodiversity, which ultimately reduces ecosystem value. Technologies such as Geographic Information Systems (GIS) and remote sensing play a critical role in monitoring these changes and providing detailed data to analyse their impacts. The study emphasises the importance of sustainable land management to maintain ecological balance and economic benefits, while also highlighting the role of policies that support environmental conservation.

It is suggested that further research should focus more on a development methodology that takes into account local variability. The use of ESV coefficients that are more specific to the geographical and socio-economic conditions of the research area will increase the accuracy of the analysis results. In addition, increased use of technologies such as high resolution satellite mapping and integration of more detailed field data will help to integrate subtle land use changes and their impacts on ecosystems. This research can also be combined by examining the influence of local environmental policies in conserving ecosystems and reducing negative impacts.

Research on land use change and its impact on ecosystem service values (ESV) has significant benefits and interests in maintaining ecological balance and supporting sustainable development. By understanding the economic value of ecosystem services, this research helps policy makers to develop strategies that take into account the environmental impacts of land use change. This is not only important for the conservation of biodiversity and environmental quality but also provides insight for communities and governments in planning optimal land use to improve human welfare without damaging existing ecosystems..

References

[1] Costanza, R., et al., 1997. The value of the world's ecosystem services and natural capital. *nature* **387** (6630), 253.

[2] Si, J., Nasiri, F., Han, P., & Li, T. (2014). Variation in ecosystem service values in response to land use changes in Zhifanggou watershed of Loess plateau: a comparative study. *Environmental Systems Research*, **3**(1), 1–10. https://doi.org/10.1186/2193-2697-3-2

[3] Zhao, X., Wang, J., Su, J., & Sun, W. (2021). Ecosystem service value evaluation method in a complex ecological environment: A case study of Gansu Province, China. *PLoS ONE*, **16**(2 February), 1–25. https://doi.org/10.1371/journal.pone.0240272

[4] Tolessa T, Senbeta F, Kidane M (2017) The impact of land use/land cover change on ecosystem services in the central highlands of Ethiopia.Ecosyst Service. **23**:47–54. https://doi.org/10.1016/j.ecoser.2016.11.010

[5] Sajid, I. A., Ahmed, M. M., & Sagheer, M. (2011). FPGA-based Householder implementation for efficient eigenvalues calculation. *International Journal of Innovative Computing*, Information and Control, **7**(10), 5939– 5946.

[6] Shao, X., Jing, C., Qi, J., Jiang, J., Liu, Q., & Cai, X. (2017). Impacts of land use and planning on island ecosystem service values: a case study of Dinghai District on Zhoushan Archipelago, China. *Ecological Processes*, **6**(1). https://doi.org/10.1186/s13717-017-0095-3

[7] Perrina, M. G. (2021). Literature Review Geographic Information System (GIS). *Journal of Information Technology and Computer Science*, **10**(10), 1–4. https://www.researchgate.net/publication/354704876

[8] Mengist, W., Soromessa, T., & Feyisa, G. L. (2022). Estimating the total ecosystem services value of Eastern Afromontane Biodiversity Hotspots in response to landscape dynamics. *Environmental and Sustainability Indicators*, **14**(February), 100178. https://doi.org/10.1016/j.indic.2022.100178

[9] Rinc´on-Ruiz, A., et al., 2019. Applying integrated valuation of ecosystem services in Latin America: insights from 21 case studies. *Ecosystem Services* ,**36**, 100901. https://www.sciencedirect.com/science/article/abs/pii/S2212041618302572

[10] Arowolo, A.O., Deng, X., Olatunji, O.A., Obayelu, A.E., 2018. Assessing changes in the value of ecosystem services in response to land-use/land-cover dynamics in Nigeria. *Science of the total Environment* 636, 597–609. https://www.sciencedirect.com/science/article/abs/pii/S0048969718314542

[11] Juniyanti, L., Prasetyo, L. B., Aprianto, D. P., Purnomo, H., & Kartodihardjo, H. (2020). Land-use/land cover change and its drivers in Bengkalis Island, Riau Province (1990–2019). Jurnal Pengelolaan Sumberdaya Alam dan Lingkungan, **10**(3), 419–435. https://doi.org/10.29244/jpsl.10.3.419-435

[12] Adhiatma, R., Widiatmaka, & Lubis, I. (2020). Change and prediction of land cover/use change in South Lampung Regency. *Jurnal Pengelolaan Sumberdaya Alam dan Lingkungan*, **10**(2), 234–246. https://doi.org/10.29244/jpsl.10.2.234-246

[13] Christian, Y., Asdak, C., & Kendarto, D. R. (2021). Land use change analysis in West Bandung Regency. Jurnal Teknotan, **15**(1), 15. https://doi.org/10.24198/jt.vol15n1.3

[14] Fadhil, M. Y., Hidayat, Y., & Baskoro, D. P. T. (2021). Identification of land use change and hydrological characteristics in the Upper Citarum Watershed. Jurnal Ilmu Pertanian Indonesia, **26**(2), 213–220. https://doi.org/10.18343/jipi.26.2.213

[15] Smit, I. P. J., Maze, K., & van Wilgen, B. W. (2024). Land cover change in and around South African protected areas. *Biological Conservation*, **300**(October), 110844. https://doi.org/10.1016/j.biocon.2024.110844

[16] Ignasius, S. A., & Sunimbar. (2021). Analysis of land use change in Kupang City, East Nusa Tenggara, in 2010–2018 (Case study in Kelapa Lima, Oebobo, and Kota Lama sub-districts). Jurnal Geoedusains, **2**(1), 36–52. https://doi.org/10.30872/geoedusains.v2i1.564

[17] Muttaqin, T., & S.Hut, M. S. (2017). Erosion Rate Against Land Use Changes in Protected Forest Areas in the Agricultural Area of Sumber Brantas Village, Bumiaji District, Batu City. *Jurnal Daun*, **4**(2), 119–125.

[18] Sasmitha, R., & Nurlina. (2010). Land cover change analysis in Riam Kanan Reservoir and surrounding areas using Geographic Information System (GIS) and Landsat imagery data. *Jurnal Fisika FLUX*, *8*(2), 135–142. https://ppjp.ulm.ac.id/journal/index.php/f/article/view/3117

[19] Amalia, D., Desfandi, M., & Ruliani, R. (2022). Identification of Changes in Forest Land to Agricultural Land in Paya Dedep Village, Jagong Jeget District, Central Aceh Regency. *Journal of Geosphere Education*, **7**(2), 168–179. https://doi.org/10.24815/jpg.v7i2.23722

[20] Muttaqin, S., & Aini, Q. (2011). "Analysis of Changes in Forest and Plantation Land Cover in Jambi Province for the Period 2000 - 2008." Journal of Information Systems 4(2):1-8. *Journal of Information Systems*, **4**(2), 1–8. https://journal.uinjkt.ac.id/index.php/sisteminformasi/article/view/137/2380

[21] Ramadhan, R., Widiatmaka, W., & Sudadi, U. (2016). Land Use Change and Spatial Utilization in Landslide Vulnerable Regions of Banjarnegara Regency, Central Java. *Journal of Natural Resources and Environmental Management*, **6**(2), 159–167. https://doi.org/10.19081/jpsl.2016.6.2.159

[22] Widayani, P., Salsabila, H. N., & Andriantari, A. (2023). Mapping of population mobility in the suburbs of Bandung City. *Indonesian Geography Magazine*, **37**(2), 104–113. https://doi.org/10.22146/mgi.70636

[23] Mubarokah, A., & Hendrakusumah, E. (2022). The Effect of Conversion of Plantation Land Functions on Environmental Ecosystems. *Journal of Regional and Urban Planning Research*, **2**(1), 1–14. https://doi.org/10.29313/jrpwk.v2i1.754

[24] Handayani, H., A. Yusra, A. H., & Fitrianti, W. (2022). Impact of Forest Conversion to Agricultural Land on Village Development Case Study in Pinang Luar Protected Forest-Kubu Raya Regency. *Plantation and Tropical Land*, **12**(2), 48–58. https://doi.org/10.26418/plt.v12i2.60027

[25] Khan, M. F., Islam, M. K., & Chowdhury, M. A. (2024). Spatio-temporal assessment and prediction of wetlands: Examining the changes in ecosystem service value of RAJUK DAP area using Artificial Neural Network and Geospatial Techniques. *Heliyon*, **10**(14), e34327. https://doi.org/10.1016/j.heliyon.2024.e34327

[26] Deventer, H. V., Adams, J. B., Durand, J. F., Grobler, R., Grundling, P. L., Janse van Rensburg, S., Jewitt, D., Kelbe, B., MacKay, C. F., Naidoo, L., Nel, J. L., Pretorius, L., Riddin, T., & Van Niekerk, L. (2021). Conservation conundrum – Red listing of subtropical-temperate coastal forested wetlands of South Africa. *Ecological Indicators*, **130**, 108077. https://doi.org/10.1016/j.ecolind.2021.108077

[27] Aryani, U., Sulistyarini, S., & Christanto, L. M. H. (2022). Changes in Agricultural Land Use to Non-Agricultural Land in Sengah Temila District in 2009-2019. *Equatorial Journal of Education and Learning* (JPPK), **11**(7), 750–761. https://doi.org/10.26418/jppk.v11i7.56639

[28] Monsaputra. (2023). Analysis of changes in agricultural land use to housing in Padang Panjang City. *Tunas Agraria*, **6**(1), 1–11. https://doi.org/10.31292/jta.v6i1.200

[29] Dharmayanthi, E., Zulkarnaini, & Sujianto. (2018). Impact of Conversion of Rice Field Land to Oil Palm Plantation. *Social Environmental Dynamics*, **5**(1), 34–38. https://www.researchgate.net/publication/335020789

[30] Rahim, A., Kurniawan, E., MK, P., & Putriani. (2024). The Impact of Agricultural Land Conversion in the Latimojong Tourism Object Area on Farmers' Income. *Galung Tropika Journal*, **13**(1), 98–106. https://doi.org/10.31850/jgt.v13i1.1147

[31] Rezki, N. N., Aso, L., & Syahrun, S. (2020). Social and Cultural Changes in the Community After Conversion of Agricultural Land to Mining Land. Etnoreflika: *Journal of Social and Cultural*, **9**(1), 50–61. https://doi.org/10.33772/etnoreflika.v9i1.724

[32] Anh, H. H., Beaulieu, A., Da Hanh, T. M., & Tru, L. C. (2024). Economic effects of a controlled mangrove-to-pond coverage ratio policy on mangrove conservation and shrimp farming: A case study in Vietnam using instrumental regression analysis. *Trees, Forests and People*, **16**(29), 100579. https://doi.org/10.1016/j.tfp.2024.100579

[33] Saoum, M. R., & Sarkar, S. K. (2024). Monitoring mangrove forest change and its impacts on the environment.

Ecological Indicators, **159**(January), 111666. https://doi.org/10.1016/j.ecolind.2024.111666

[34] Rachman, T., Umar, H., & Bahtiar, I. H. (2022). The Impact of Coastline Changes on Coastal Land Utilization in Tamanlanrea District, Makassar City. *Journal of Marine Science and Technology Innovation*, **3**(1), 7–14. https://journal.unhas.ac.id/index.php/zonalaut

[35] Suwargana, N. (2008). Analysis of Mangrove Forest Changes Using Remote Sensing Data at Bahagia Beach, Muara Gembong, Bekasi. *Journal of Remote Sensing*, **5**, 64–74. http://jurnal.lapan.go.id/index.php/jurnal_inderaja/article/view/1170

[36] Yuliasamaya, Darmawan, A., & Hilmanto, R. (2014). Changes in Mangrove Forest Cover on the Coast of East Lampung Regency. Sylva Lestari Journal, 2(3), 111–124. https://doi.org/10.23960/jsl32111-124

[37] Van Deventer, H., Adams, J. B., Durand, J. F., Grobler, R., Grundling, P. L., Janse van Rensburg, S., Jewitt, D., Kelbe, B., MacKay, C. F., Naidoo, L., Nel, J. L., Pretorius, L., Riddin, T., & Van Niekerk, L. (2021). Conservation conundrum: red listing of subtropical-temperate coastal forested wetlands of South Africa. *Ecological Indicators*, **130**, 108077. https://doi.org/10.1016/j.ecolind.2021.108077

[38] Jiang, L., Wang, H., Wang, S., & Zhang, W. (2023). Process analysis and mitigation strategies for wetland degradation caused by increasing agricultural water demand: an ecology-economy nexus perspective. *Ecological Processes*, **12**(1), 1–13. https://doi.org/10.1186/s13717-023-00452-x

[39] Fan, X., Yu, H., Tiando, D. S., Rong, Y., Luo, W., Eme, C., Ou, S., Li, J., & Liang, Z. (2021). Impacts of human activities on ecosystem service value in arid and semi-arid ecological regions of China. *International Journal of Environmental Research and Public Health*, **18**(11121), 1–15. https://doi.org/10.3390/ijerph182111121

[40] Belay, T., Melese, T., & Senamaw, A. (2022). Impacts of land use and land cover change on ecosystem service values in the Afro-alpine area of Guna Mountain, Northwest Ethiopia. *Heliyon*, **8**(12), e12246. https://doi.org/10.1016/j.heliyon.2022.e12246

[41] Mathewos, Y., Abate, B., Dadi, M., & Mathewos, M. (2024). The Nexus between spatiotemporal land use/land cover dynamics and ecosystem service values in the wave river catchment, Omo Gibe River. *Environmental Challenges*, **17**(November), 101053. https://doi.org/10.1016/j.envc.2024.101053

[42] Shiferaw, H., Alamirew, T., Kassawmar, T., & Zeleke, G. (2021). Evaluating ecosystem services values due to land use transformation in the Gojeb watershed, Southwest Ethiopia. *Environmental Systems Research*, **10**(1). https://doi.org/10.1186/s40068-021-00227-3

[43] Berihun, M. L., Tsunekawa, A., Haregeweyn, N., Tsubo, M., & Fenta, A. A. (2021). Changes in ecosystem service values are strongly influenced by human activities in contrasting agroecological environments. *Ecological Processes*, **10**(1), 1–18. https://doi.org/10.1186/s13717-021-00325-1

[44] Hasanah, F., Setiawan, I., Noor, T. I., & Yudha, E. P. (2021). Mapping the distribution of paddy field conversion levels in Serang Regency. Jurnal Agrica, **14**(2), 171–182. https://doi.org/10.31289/agrica.v14i2.5039

[45] Latifah, S., & Handoyo, P. (2014). The impact of agricultural land conversion into elite housing. Jurnal Paradigma, **2**(3), 1–10. https://www.neliti.com/publications/250149

[46] Sorker, R. (2023). Variasi nilai jasa ekosistem sebagai respon terhadap perubahan penggunaan lahan di Distrik Dhaka dan Gazipur, Bangladesh. *Indonesian Journal of Ecology and Environment*, *5*(1), 10–24. https://doi.org/10.20886/ijee.2023.5.1.10-24

IOP Conf. Series: Earth and Environmental Science **1525** (2025) 012031 doi:10.1088/1755-1315/1525/1/012031

Determination of Landslide Prone Areas Based on Spatial Analysis Using Storie Index Method in Masama District, Banggai Regency, Indonesia

Andri Moh. Wahyu Laode[1], Muh. Altin Massinai[1*], Dewi Ainun Jariah[1]

[1]Department of Geophysics, Hasanuddin University, Makassar, Indonesia

*altin@science.unhas.ac.id

Abstract. This study examines the landslide potential in Masama Sub-district, Banggai Regency, an area frequently affected by landslides. The research aims to map landslide-prone areas using the Storie Index method, which considers parameters such as slope, rainfall, soil type, land use, and elevation. The results identify five categories of landslide vulnerability: very low, low, moderate, high, and very high, covering areas ranging from 263.2771 to 13,274.41 hectares. Overall, Masama Sub-district has a high landslide potential, with high and very high vulnerability concentrated in Minangandala, Simpangan, and Tompotika Makmur villages. Meanwhile, areas with very low to moderate vulnerability include 12 other villages, such as Tangeban, Cemerlang, and Purwo Agung. This research is expected to assist the Banggai Regency government, particularly Masama Sub-district, in disaster mitigation-based spatial planning.

1. Introduction

Masama District is one of the districts in Banggai Regency that has a high level of vulnerability to natural disasters, one of which is landslides [1]. Landslides is caused by a variety of conditioning parameters, these include the aspect and angle of the slope; steeper slopes and specific orientations tend to be more unstable [2][3]. Elevation influences potential for material movement and has an impact on gravitational forces. Because they saturate the soil and cause slope failures, hydrogeographical factors like rainfall patterns and drainage systems have a major impact on the occurrence of landslides [4]. The intricate dynamics of landslides are influenced by a variety of factors, including soil composition, geological and lithological features, human activity, and proximity to faults [5]. Effective mitigation techniques are required to reduce these devastating effects [6].

One of the effective steps to mitigate landslides is by mapping areas that have the potential to experience landslides. In mapping landslide prone areas, a method that can be used is the storie index method [7]. The Index Storie method is a landslide potential assessment by using four or more parameters to assess the potential for landslides. These parameters are processed using a specific equation to determine the landslide potential [8]. Given this context, the study aims to analyze the risk of landslide in Masama District as part of a disaster mitigation effort.

2. Research Method

2.1 Study Area

The study area is located in Masama District (Figure 1), Banggai Regency, Central Sulawesi Province, with an area of 11,616.89 hectares. This district is located in a subtropical region with

1,000–1,700 mm of annual rainfall and flat to extremely steep mountainous terrain with upland forest cover and predominant soil types are associated with clay. The geological formations in the study area consist of lonsio formation, ultramafic complex and alluvium, where alluvium dominates most of the study area.

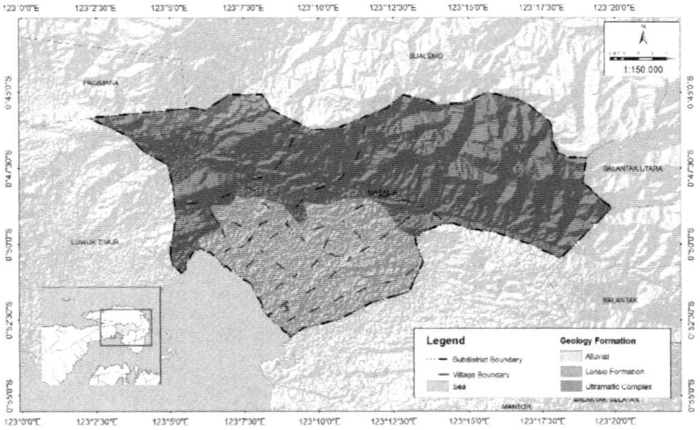

Figure 1. Study Area (Masama District, Banggai Regency, Indonesia)

Table 1. Score Values Based on Characteristics of Parameter of Landslide-Prone Determination.

No.	Parameter	Criteria	Score
1	Slope	- Steep, slope >75%	6
		- Very steep, slope 46-75%	5
		- Steep, slope 31-45%	4
		- Slightly steep, slope 16-30%	3
		- Sloping, slope 4-15%	2
		- Flat, slope 0-3%	1
2	Rainfall	- Rainfall > 3100 mm a year	6
		- Rainfall of 2800 - 3100 mm year	5
		- Rainfall of 2500 - 2800 mm a year	4
		- Rainfall of 2200 - 2500 mm year	3
		- Rainfall of 1400 - 2200 mm a year	2
		- Rainfall of <1400 mm year	1
3	Land Use	- Residential, Open Land	6
		- Grass, Shrubs	5
		- Rice field vegetation, swamp	4
		- Mixed gardens, yard crops	3
		- Plantation	2
		- Dense forest	1
4	Soil Type	- Clay	6
		- Dusty Clay	5
		- Clay, Clayey Loam, Sandy Loam	4
		- Sandy Clay, Dusty Clay Loam	3
		- Dusty Sand	2
		- Sand	1
5	Elevation	- 0 – 100 m	6
		- 100 – 200 m	5
		- 200 – 300 m	4
		- 300 – 400 m	3
		- 400 – 500 m	2
		- >500m	1

[11][9][8][12]

2.2 Data Analysis

The data used in this study are 2024 land use map, soil type map, geological map, rainfall map and slope map. The determination of landslide prone areas is done using the storie index method. The Storie index method was developed by William Storie in 1917 and has been widely used in various countries, including Indonesia for landslide potential [9]. The method is based on the principle that the susceptibility of a location to landslides is determined by a combination of several parameters [10]. The result of each parameter of the map created is then subjected to scoring or weighting process and overlapping function. The Storie Index method is done by applying the formula [11]:

$$L = A \text{ X } \frac{B}{10} \text{ X } \frac{C}{10} \text{ X } \frac{D}{10} \text{ X } \frac{E}{10} \qquad [1]$$

where, L is landslide potential, A is slope, B is rainfall, C is land use, D is soil type, and E is elevation. Utilizing GIS tools and the Storie Index method, landslide-prone areas were identified and a total score was obtained (see Table 1). Landslide-prone levels were divided into five classes or levels using the natural break method: Very High, High, Medium, Low, and Very Low.

3. Results and Discussion

3.1 Parameters of Landslide Potential

The first parameter, slope, is a fundamental factor influencing landslide susceptibility, as steeper slopes are more susceptible to gravitational forces that can trigger the downslope movement of soil, rock, or debris. The results obtained show 6 different classifications according to Table 1. The steep areas are dominant in the north of the study area and the sloping areas are dominant in the south of the study area (coastal), as shown in Figure 2.

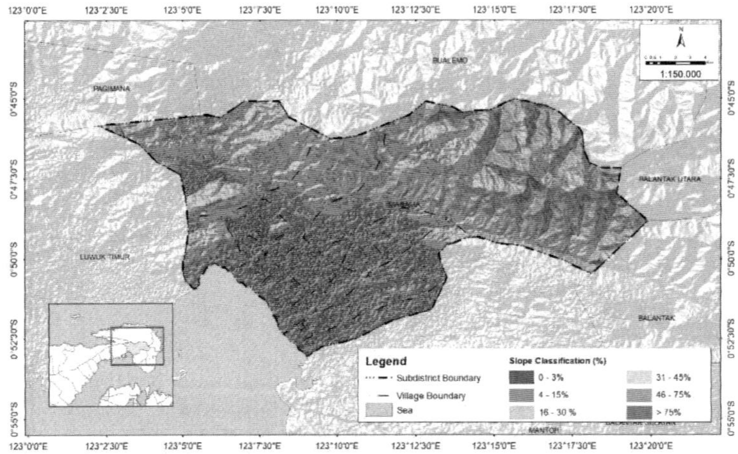

Figure 2. Slope Map (Masama District, Banggai Regency, Indonesia)

The next parameter is rainfall, rainfall increases pore water pressure within the slope, reducing frictional resistance and making the slope more susceptible to movement. The data analysis results identified two different classifications, as shown in Table 1 and Figure 3.

The land use parameter, which includes urban development, agriculture, forestry, and other purposes, significantly affects slope stability and landslide potential. The classification results (from Table 1) show six different categories: residential and open land, grass and shrubs, rice field vegetation and swamp mixed gardens and yard crops, plantation, and dense forest, as seen in Figure 4.

The soil parameter also plays a role, as different soil types have varying levels of susceptibility to landslides due to factors such as cohesion, permeability, and shear strength. From the results obtained, the overall study area has one common soil type, which is clay, as seen in Figure 5.

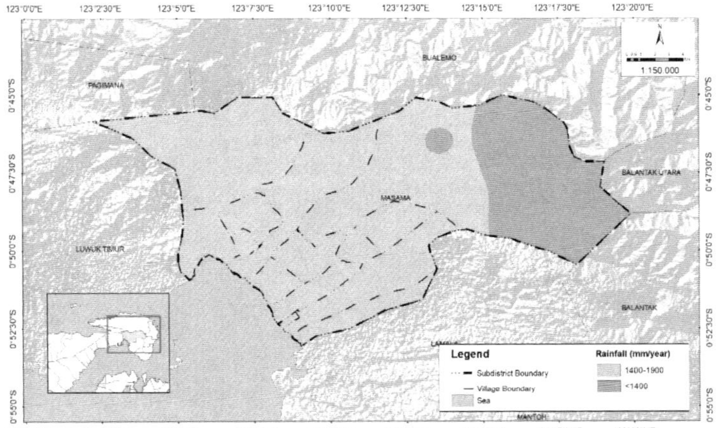

Figure 3. Rainfall Map (Masama District, Banggai Regency, Indonesia)

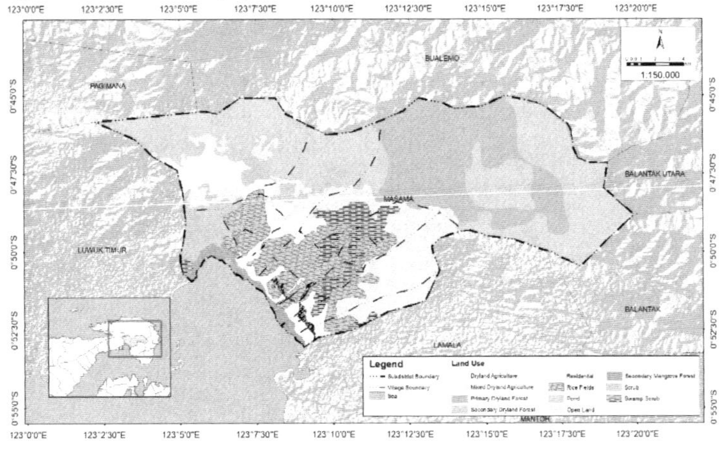

Figure 4. Land Use Map (Masama District, Banggai Regency, Indonesia)

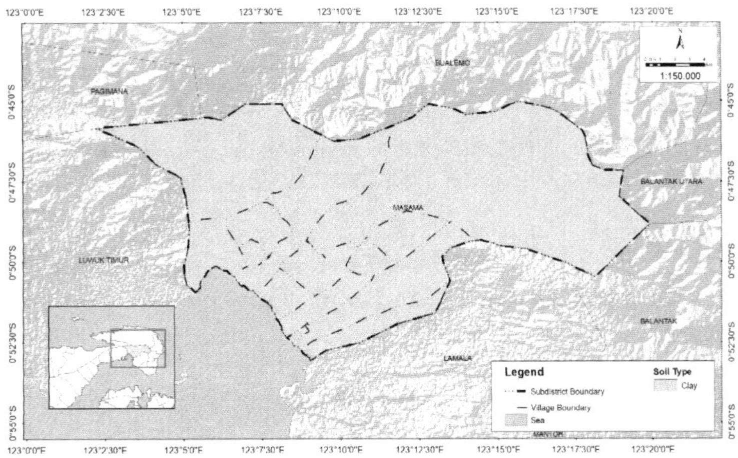

Figure 5. Soil Type Map (Masama District, Banggai Regency, Indonesia)

Moreover, elevation parameter may contribute to increased gravitational potential, making slopes more prone to downslope movement [13] . The results of the study area elevation classification obtained 6 classifications, as seen on Table 1 and Figure 6.

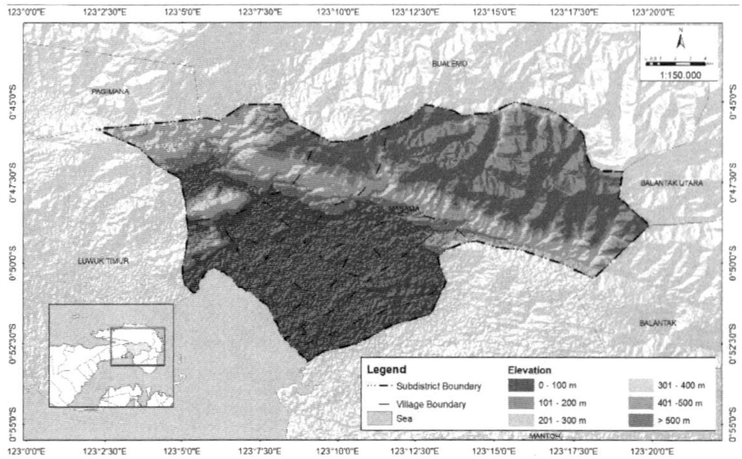

Figure 6. Elevation Map (Masama District, Banggai Regency, Indonesia)

3.2 Landslide Potential Analysis

Landslides are defined as the movement of a mass of rock or soil due to the downward pulling force of gravity. Landslides occur when the thrust on a slope is greater than the weight of the material that holds it. Based on five parameters of landslide potential using the Storie Index method, a landslide potential map of the study area was obtained (Figure 7).

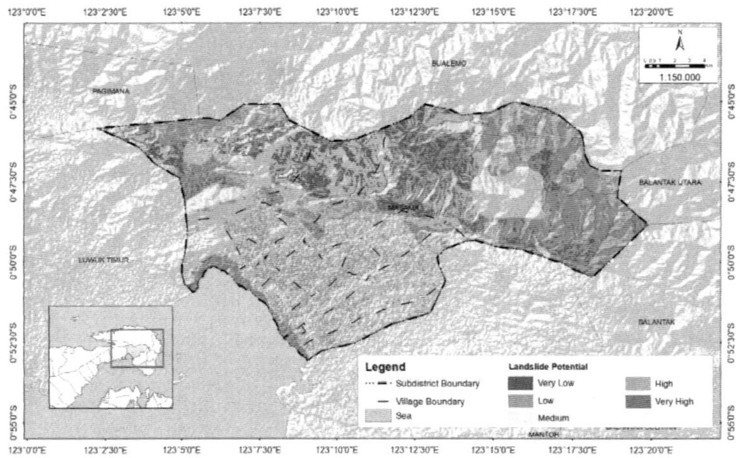

Figure 7. Landslide Potential Map (Masama District, Banggai Regency, Indonesia)

The landslide potential map is classified into five categories: Very Low, Low, Medium, High, and Very High. The area with Very Low potential covers 263.28 ha (1.01% of the total), located on the coast with flat slopes, dominated by mangrove forests, at an elevation of 0–100 m. The Low potential area spans 1,270.41 ha (4.83%), also located along the coast with flat to gentle slopes, primarily residential, at an elevation of 0–100 m. The Medium potential area is the largest, covering 13,274.41 ha (50.45%), with varied slopes (flat, gentle, slightly steep, and steep), dominated by rice fields, residential areas, and open land, at an elevation of 0–200 m. The High potential area spans 8,649.89 ha (32.87%), with moderate to very steep slopes, dominated by open land, residential areas, shrubs, and rice fields, at an elevation of 0–400 m. The Very High potential area covers 2,856.14 ha (10.84%), with very steep slopes, dominated by dryland forests, vacant land, and shrubs, at an elevation of 400–1,300 m. Based on village boundaries, villages with High to Very High potential include Minangandala, Simpangan, and Tompotika Makmur, while villages with Very Low to Medium potential include Tangeban, Cemerlang, Serese, Taugi, Duata Karya, Kospa Duata Karya, Ranga-Ranga, Padangon, Kembang Merta, Eteng, and Purwo Agung.

4. Conclusion

The results obtained show that the study area, Masama sub-district is an area with high landslide potential. This is characterized by an area with high category of 8649.885 ha or about 32.87% of the study area, and an area with very high category has an area of 2856.139 ha or 10.84% of the study area. Villages included in this classification include Minangandala village, Simpangan village, and Tompotika Makmur village. Meanwhile, the very low, low and medium classification areas have an area of 263.2771 ha, 1270.408 ha, 13274.41 ha respectively. There is a need for further handling and study related to landslide potential in the study area, especially the three villages with high landslide potential.

References

[1]. Syakir M, Ariani CD, Stani AH, Agus A. Increasing Community Awareness and Preparedness by Forming the DESTANA (Disaster Resilient Village) Group in Sinorang Village , South Batui District , Banggai Regency , Central Sulawesi. Pros Simp Nas Kelaut Dan Perikan. 2023;10:63–7.

[2]. Huang F, Tao S, Li D, Lian Z, Catani F, Huang J, et al. Landslide Susceptibility Prediction Considering Neighborhood Characteristics of Landslide Spatial Datasets and Hydrological Slope Units Using Remote Sensing and GIS Technologies. Remote Sens. 2022;14(18).

[3]. Aisyah SN, Supriatna S, Rustanto A. A spatial study of landscape patterns in landslide-prone areas at cikakak sub-district, west java. IOP Conf Ser Earth Environ Sci. 2022;1089[1].

[4]. Loche M, Alvioli M, Marchesini I, Bakka H, Lombardo L. Landslide susceptibility maps of Italy: Lesson learnt from dealing with multiple landslide types and the uneven spatial distribution of the national inventory. Earth-Science Rev [Internet]. 2022;232(February):104125. Available from: https://doi.org/10.1016/j.earscirev.2022.104125

[5]. Vojteková J, Vojtek M. Assessment of landslide susceptibility at a local spatial scale applying the multi-criteria analysis and GIS: a case study from Slovakia. Geomatics, Nat Hazards Risk [Internet]. 2020;11[1]:131–48. Available from: https://doi.org/10.1080/19475705.2020.1713233

[6]. Fan X, Dufresne A, Siva Subramanian S, Strom A, Hermanns R, Tacconi Stefanelli C, et al. The formation and impact of landslide dams – State of the art. Earth-Science Rev [Internet]. 2020;203(February):103116. Available from: https://doi.org/10.1016/j.earscirev.2020.103116

[7]. Abdul Muin, Heinrich Rakuasa. Spatial Analysis of Landslide Potential Using Modification of the Storie In-dex Method in the Wae Batu Gajah Watershed, Ambon City, Indonesia. Int J Sci Multidiscip Res. 2023;1[3]:107–16.

[8]. Ristya Y, Supriatna, Sobirin. Spatial pattern of landslide potensial area by SMORPH, INDEX STORIE and SINMAP method in Pelabuhanratu and surrounding area, Indonesia. IOP Conf Ser Earth Environ Sci. 2019;338[1].

[9]. Thoha AS, Sundari D, Patana P, Sulistiyono N. Spatial distribution of landslide vulnerability level in Dairi District, North Sumatera Province, Indonesia. J Phys Conf Ser. 2020;1542[1].

[10]. Utomo AH, Husin. Spatial analysis and mapping of landslide prone areas in Kemuning lor village, sub-district Arjasa, district of Jember. IOP Conf Ser Earth Environ Sci. 2021;672[1].

[11]. Arifin S, Carolina T, C W. Implementasi Penginderaan Jauh dan SIG untuk Inventarisasi Daerah Rawan Bencana Longsor. J Penginderaan Jauh dan Pengolah data Citra Digit [Internet]. 2006;3:77–86. Available from: http://www.jurnal.lapan.go.id/index.php/jurnal_inderaja

[12]. Ramdhoni F, Damayanti A, Indra TL. Smorph application for landslide identification in Kebumen Regency Smorph application for landslide identification in Kebumen Regency. IOP Conf Ser Earth Environ Sci. 2020;451:1–9.

[13]. Saleem N, Huq E, Yaw N, Twumasi D, Javed A. Parameters Derived from and / or Used with Digital Elevation Models (DEMs) for Landslide Susceptibility Mapping and Landslide Risk Assessment : A Review. Int J Geo-Information. 2019;8(545):1–25.

Remote Sensing for Shifting Cultivation Land Evaluation: An Idea of Maintaining Local Wisdom in West Papua

Kati Syamsudin Kadang Tola[1*], Sumbangan Baja[2], Risma Neswati[2] and Asmita Ahmad[2]

[1] Post Graduate Student of Agriculture Science, Hasanuddin University, Makassar, Indonesia
[2] Department of Soil Science, Hasanuddin University, Makassar, Indonesia

*E-mail: k.tola@unipa.ac.id

Abstract. This research emerged from an idea concern for the activities shifting cultivation (SC) of local communities in West Papua and was enriched by a review of articles. SC is currently experiencing rapid change due to population growth, climate anomalies, focus on market-oriented and industrial agriculture, and plantation land development. SC remains an important component for ethnic minority groups in the tropics. The research methods used were Unmanned Aerial Vehicle (UAV)-based photogrammetry methods and spatial analysis using GIS. The research stages include (1) Aerial Photography of SC Land, beginning with a site survey. The preparation phase includes selecting flight locations, planning flight paths, configuring flight altitude and image overlap, conducting aerial photography, and processing drone imagery. (2) Spatial Analysis of SC Land, which involves visual image interpretation, ground checking for validation, digitization, and calculating the land area of each SC field. The research results were obtained: 1). The average annual rainfall during the period 1992 to 2022 was 2804.4 mm; 2) During the period 2005 to 2019, the average area of dryland rice harvest was 251.7 hectares, production was 597.4 tons and productivity was 2.3 tons/ha, which continued to decrease from 2016; 3) An identification of the SC distribution in 2023, the smallest SC area opened was 0.01 hectares and the largest was 4.9 hectares. The average SC area of the locals was 0.6 hectares. In 2024, the smallest area of SC opened was 0.01 hectares and the largest was 11.2 hectares. The average SC area of the locals was 0.7 hectares. SC was found with a short fallow period and extended from fairly flat areas to hills and sloping areas. Land evaluation can be done by identifying sites, evaluation soil characteristics and determining land suitability for food production in SC using remote sensing technology. This activity is expected to maintain SC land sustainably.

1. Introduction

Shifting cultivation (SC) is one of the conventional agricultural practices for food production whose sustainability continues to be discussed worldwide [1]. This activity is still an important part for ethnic minority groups in tropical areas [2], [3]. More land is needed worldwide to overcome the lack of agricultural production and ensure the population's food consumption [4]. Remote sensing is one way to identify land cover in SC. When this activity is carried out, the results can be used to reconstruct the history of SC land use, although this method is not the only one [5] [6]. Monitoring SC is challenging because it is highly dynamic, small-scale and results in complex post-disturbance landscapes. The challenges of remote sensing in shifting cultivation lie in the high spatial heterogeneity, the time complexity, the small scale, the high dynamics, the presence of diverse SC and fallow landscapes and the emergence of complex post-disturbance landscapes [7], [8]. To date, no articles have been found that provide an assessment of SC in West Papua from the perspective of a land-based approach using remote sensing.

In West Papua, SC is one of the traditional farming patterns preserved to this day and can be found in all areas of West Papua managed by local communities. From the perspective of capital for SC, farmers are still highly dependent on nature to SC activities [9] and cosmopolitan attitudes accept social change due to social interaction with immigrant communities [10]. Local communities on SC areas as the first crop are planting upland rice, or locally referred to as dryland rice. In addition, after the harvest, the cultivation of tubers and other seasonal crops continues. 160 farming families in this area grow dryland rice [11] . They are the largest in the Manokwari Regency.

SC in Warmare District is practiced by the Arfak tribe, one of the major tribes in the Manokwari Regency. The location of SC naturally has sensitive soil conditions and is favored by the influence of anthropogenic factors [12]. Therefore, this condition needs to be addressed carefully so that the community does not lose land for livelihood activities. Based on this, SC research on local communities was carried out using a remote sensing land evaluation approach. From these activities emerges an idea of how the management of SC by local communities in West Papua can be sustained as local wisdom.

2. Research method

2.1 Study area

The research site is the SC land of local communities in villages active in SC including Gueintuy, Tanah Merah, Njuar, Imboisrati, Srandabey, Warmare Satu, Bahamyenti, Warmare, and Umcen, Warmare District, Manokwari Regency, West Papua Province.

2.2 Material, method, and analysis

The materials used in this study consist of primary and secondary data sources. The primary data were obtained from drone-based remote sensing conducted between 2023 to 2024. The secondary data include climate records from 1992 to 2022, as well as data on harvested area, production, and productivity of dryland rice in Warmare District, Manokwari Regency, from 2005 to 2019. The equipment used includes Mavic 3M and Mavic 2 Pro drones, along with a GPS device. The remote sensing method applied is Unmanned Aerial Vehicle (UAV) based photogrammetry, combined with spatial analysis using GIS. Research Procedures: (1) Aerial Photography of SC Land. The initial stage involved a field survey to understand the spatial distribution of dryland rice fields based on accessibility. The preparation phase included selecting flight locations, planning flight paths, configuring flight altitude, and setting image overlap (80% overlap, 70% sidelap). In March 2023, the Mavic 2 Pro drone was used at a flight altitude of 250 meters, completing 18 flights and capturing a total of 2,572 images, resulting in an orthophoto coverage of 1,913 hectares. In March 2024, the Mavic 3M drone was used at a flight altitude of 300 meters, completing 13 flights and capturing a total of 3,681 images, resulting in an orthophoto coverage of 2,507 hectares. The processing of drone imagery was conducted using Agisoft Metashape software. (2) Spatial Analysis of SC Land. Spatial analysis was carried out using ArcGIS Pro 2.8.0 software. Image interpretation was conducted through visual analysis to delineate the boundaries of dryland rice fields in SC areas. A ground check was performed to validate the interpretation results. Subsequently, the SC land boundaries were digitized, and the total land area of each SC field was calculated to compare the changes observed in SC imagery from 2023 to 2024.

3. Result and Discussion

3.1 Climate

The climate in an area provides information about long-term atmospheric conditions. According to the Koppen climate classification system, the area of Warmare District belongs to the Af climate, namely the tropical rainforest climate. The Warmare district lies at an average altitude of 241.7 m above sea level, with an average annual temperature of 25.6 ºC. The average annual rainfall during the period 1992 to 2022 was 2804.4 mm [13]. Monthly rainfall with wet months >200 mm from December to June, without dry months <100 mm. The highest monthly rainfall was 340.5 mm in March and the lowest was 156.6 mm in September.

3.2 Harvest area, production and productivity of dryland rice

Local farmers widely carry out dryland rice cultivation and its activities are carried out using a SC system. During the period 2005 to 2019, dryland rice had an average harvest area of 251.7 hectares, production of 597.4 tons and productivity of 2.3 tons/ha, which continued to decrease from 2016 [14], [15], [16]. In SC areas, the community does not use fertilizers and pesticides, soil fertility comes entirely from nature. Local wisdom possessed by the community realizes that the land is fertile again with the fallow period and the spread of pests and diseases can be stopped by moving fields and not settling or not managing the land intensively. The area dryland rice productivity in Wamare District is shown in Figure 1.

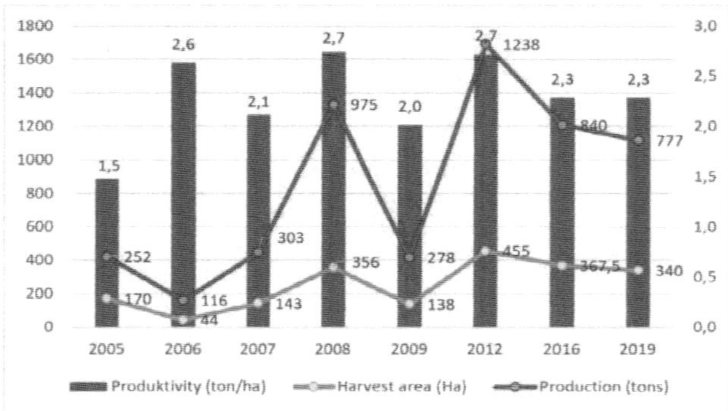

Figure 1. Harvest area, production and productivity of dryland rice in Warmare District from 2005 to 2019 (data processed)

3.3 Identification of SC distribution in Manokwari Regency, West Papua

SC sites include plots in villages inhabited by the indigenous Arfak population, actively engaged in SC activities, and located in areas with fairly flat to hilly relief. The identification of the SC distribution in 2023 and 2024 can be seen in Figure 2.

Figure 2 shows several villages that were detected carrying out SC activities. The distribution of SC is seen in the area used for SC in 2023, which is marked with a red polygon, and the area for SC in 2024, which is marked with yellow. quite extensive and spread across various locations, especially around Kampung Tanah Merah, Warmare Satu, Guentui, Warmare, Sraindabey and Njuar. Several areas indicating overlap, and adjacent areas between SC in 2023 and 2024 indicate a fallow period cycle. Remote sensing of SC locations was carried out when local people were harvesting dryland rice, which was ± 8 months after land clearance. In 2023, the smallest SC area opened was 0.01 hectares and the largest was 4.9 hectares in Njuar village. The average SC area of the locals was 0.6 hectares. In 2024, the smallest area of SC opened was 0.01 hectares and the largest was 11.2 hectares in Njuar village. The average SC area of the locals was 0.7 hectares. SC activities take place around the village where the locals live. The SC sites in 2023 and 2024 are not far from each other and will be randomly distributed according to the local community's ownership rights to SC land. Based on a study by [17], the SC activities of the Arfak tribe from one land clearance to the next are getting closer and the fallow period is becoming shorter, namely 2-4 years, where people return to their previous land more quickly, and development of opening an oil palm plantation [18].

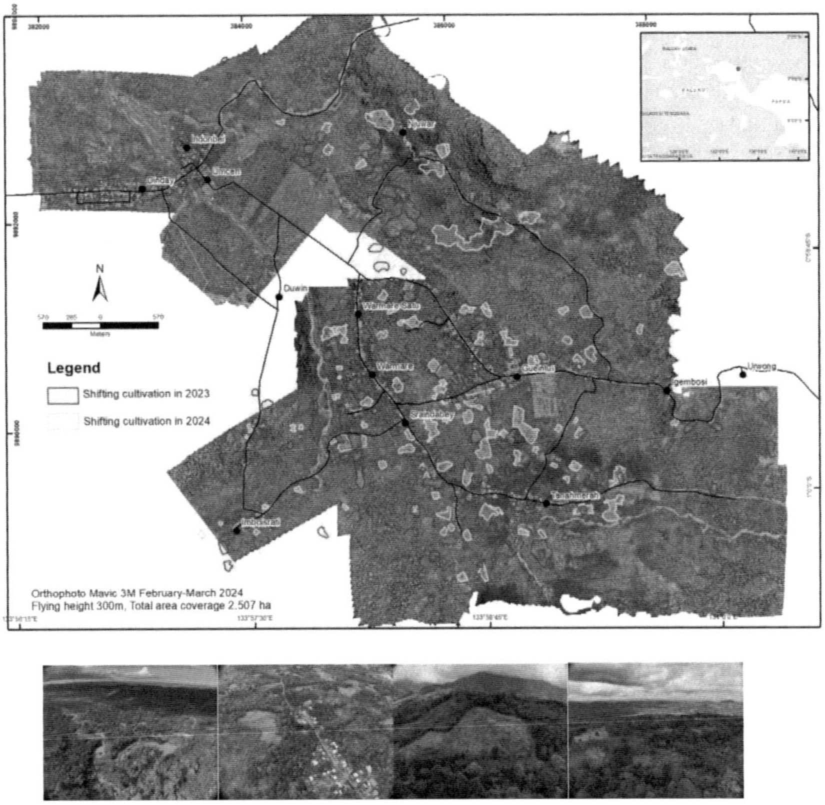

Figure 2. Identification of the distribution of shifting cultivation practices of local communities

The climate conditions of Warmare, which significant have wet months throughout the year, have a significant impact on the SC pattern. High rainfall supports land regeneration after the fallow period, but also has the potential to accelerate soil degradation due to nutrient leaching. In addition, with shorter fallow periods, land tends to be reused more quickly. Drone data shows that SC locations in Warmare are getting closer together and tend to experience an acceleration in the fallow cycle. This is in line with agricultural data showing a decline in harvested area and dryland rice yields since 2016, although land productivity remains fairly stable. This suggests that the decrease in production is more due to the shrinking area of planted land and the alleged decline in soil fertility. This condition indicates the need for evaluation and planning of land use so that the SC system remains sustainable and dryland rice productivity can increase.

Based on the location of the SC opening by the local community, which is a hilly area with a maximum area of 4.9 -11.0 hectares in Njuar Village, this condition will bring many advantages and disadvantages in terms of environmental damage. Several studies have found that land clearing increases rates of deforestation and land degradation [19], [20]. The fragmentation of SC areas impacts bird habitats. Planting on slopes leads to land degradation due to severe soil erosion [21]. Cutting down forest vegetation during land clearing increases runoff rate and soil erosion, resulting in nutrient losses [22]. It takes a long time for the ecological conditions and formation of secondary forests to resemble primary forests due to SC [23]. However, with the local knowledge of the community, the relocation of farming activities to hillside areas while clearing their land leaves large tree stumps that act as an obstacle to the rate of soil erosion. SC at the research location found that the land management was wise, namely not clearing land in landslide-prone areas, except for what happened in Njuar and Tanah Merah villages. SC was found to cause landslides during heavy rains. Several SC activities from land clearing, planting, harvesting, to landslides can be seen in Figure 3.

Figure 3. (a1 to a3) Land clearing leaving tree stumps and the process of planting dryland rice; (b1 to b3) SC dryland rice harvest by local communities; (c1 to c3) Landslides on SC on hills

Figure 3 shows SC activities (a1 to a3) including tree felling, cleaning the land floor, separating large and small woods, and continuing with burning. These activities can be found from August to September. The selection of new SC locations is agreed upon together within the extended family. Planting is carried out by family members and helping each other within the extended family. Some people plant in October and will harvest in March. There are no agronomic actions such as soil cultivation, fertilization, and pesticide application. Maintenance of dryland rice is only in the form of cleaning weeds until harvest time. Dryland rice harvesting activities (b1 to b3) are carried out by helping each other between farmer families. When harvesting, the best dryland rice seeds will be selected to be planted in the following planting season. Furthermore, some of the harvest is consumed by the family and some is sold to the inter-district market or to the district capital. Figure 3 also shows (c1 to c3) SC locations in small to hilly areas and large areas. This condition causes landslides when the rainfall intensity is high.

4. Conclusion

The local Arfak community carries out SC in addition to being a source of food as well as a livelihood for the community. The productivity of dryland rice is greatly influenced by land characteristics such as regional climate conditions that have high rainfall, and is also thought to be due to the increasingly short fallow period factor. It is necessary to carry out a package of research instruments that will protect the local wisdom of SC, such as research on mapping the distribution of SC based on changes in land use/land cover based on remote sensing, compiling the suitability of dryland rice land for shifting cultivation based on specific locations, and compiling effective SC management direction zoning. This will be useful for local governments in determining regulations and policies related to SC management as sustainable local wisdom in managing land ecologically.

References

[1] A. J. Nath, D. Reang, and G. W. Sileshi, "The Shifting Cultivation Juggernaut: An Attribution Problem," *Global Challenges*, vol. 6, no. 8, Aug. 2022, doi: 10.1002/gch2.202200051.

[2] A. Heinimann *et al.*, "A global view of shifting cultivation: Recent, current, and future extent," *PLoS One*, vol. 12, no. 9, p. e0184479, Sep. 2017, doi: 10.1371/journal.pone.0184479.

[3] P. Li and A. J. Nath, "The history and revival of swidden agriculture research in the tropics," Dec. 01, 2024, *BioMed Central Ltd.* doi: 10.1186/s43170-024-00298-z.

[4] B. Lanz, S. Dietz, and T. Swanson, "Global Economic Growth and Agricultural Land Conversion under Uncertain Productivity Improvements in Agriculture," *Am J Agric Econ*, vol. 100, no. 2, pp. 545-569, Mar. 2018, doi: 10.1093/ajae/aax078.

[5] L. P. Dutrieux, C. C. Jakovac, S. H. Latifah, and L. Kooistra, "Reconstructing land use history from Landsat time-series: Case study of a swidden agriculture system in Brazil," *International Journal of Applied Earth Observation and Geoinformation*, vol. 47, pp. 112-124, 2016, doi: 10.1016/j.jag.2015.11.018.

[6] P. Li, Z. Feng, C. Xiao, K. Boudmyxay, and Y. Liu, "Detecting and mapping annual newly-burned plots (NBP) of swiddening using historical Landsat data in Montane Mainland Southeast Asia (MMSEA) during 1988-2016," *Journal of Geographical Sciences*, vol. 28, no. 9, pp. 1307-1328, Sep. 2018, doi: 10.1007/s11442-018-1527-4.

[7] N. Jiang, P. Li, and Z. Feng, "Remote sensing of swidden agriculture in the tropics: A review," Aug. 01, 2022, *Elsevier B.V.* doi: 10.1016/j.jag.2022.102876.

[8] S. Chen, P. Olofsson, T. Saphangthong, and C. E. Woodcock, "Monitoring shifting cultivation in Laos with Landsat time series," *Remote Sens Environ*, vol. 288, Apr. 2023, doi: 10.1016/j.rse.2023.113507.

[9] Y. Holle and S. H. Sadiyah, "Modal Sosial Suku Arfak dalam Usahatani Padi Ladang di Kampung Guentuy Kabupaten Manokwari Papua Barat," *Jurnal Ilmiah Membangun Desa dan Pertanian*, vol. 7, no. 5, pp. 167-176, Sep. 2022, doi: 10.37149/jimdp.v7i5.343.

[10] T. Nofianti, "Perubahan Sosial Komunitas Suku Arfak Kabupaten Pegunungan Arfak Papua Barat," *Agribusiness Journal*, vol. 13, no. 2, pp. 95-107, Jan. 2020, doi: 10.15408/aj.v13i2.13956.

[11] BPS Manokwari, "Hasil Pencacahan Lengkap Sensus Pertanian 2023 - Tahap I Kabupaten Manokwari," BPS Kabupaten Manokwari. https://manokwarikab.bps.go.id.

[12] Taberima S *et al.*, *Kesesuaian Lahan dan Komoditas Unggulan Kabupaten Manokwari.* Indonesia: Fakultas Pertanian Unversitas Papua. ISBN: 978-623-95419-0-3. e-ISBN: 978-623-95419-1-0 xxi, 300. https://www.researchgate.net/publication/346953193 (eBook), 2021.

[13] climatecharts, "Temperature and Precipitation Warmare, West Papua, Indonesia. Years 1992-2022," https://climatecharts.net.

[14] "distrik-warmare-dalam-angka-2012".

[15] "kecamatan-warmare-dalam-angka-2020".

[16] "Kabupaten Manokwari dalam Angka 2016." [Online]. Available: www.panoramio.com

[17] Y. Ataribaba, I. Setiawan, and T. I. Noor, "Pola Pergeseran Nilai Kearifan Lokal Sistem Ladang Berpindah Pada Masyarakat Arfak," *Mimbar Agribisnis: Jurnal Pemikiran Masyarakat Ilmiah Berwawasan Agribisnis*, vol. 6, no. 2, p. 812, Jul. 2020, doi: 10.25157/ma.v6i2.3570.

[18] T. Tapi, G. Kurnia, I. Setiawan, and A. Sumule, "Shifting Cultivation of Arfak Farmer in Somi and Indisey Villages The development of Agriculture Polytechnic Manokwari, West Papua," *MIMBAR*, vol. 36, no. 1, pp. 175-185, 2020, doi: 10.29313/mimbar.v36i1.

[19] P. G. Curtis, C. M. Slay, N. L. Harris, A. Tyukavina, and M. C. Hansen, "Classifying drivers of global forest loss." [Online]. Available: http://science.sciencemag.org/

[20] A. Hazarika *et al.*, "The transition from shifting cultivation to indigenous agroforestry as nature-based solution for land restoration in the Indian Eastern Himalayas," *Ecol Indic*, vol. 162, May 2024, doi: 10.1016/j.ecolind.2024.112031.

[21] C. Wang, G. Zhang, P. Zhu, S. Chen, and Y. Wan, "Spatial variation of soil functions affected by land use type and slope position in agricultural small watershed," *Catena (Amst)*, vol. 225, May 2023, doi: 10.1016/j.catena.2023.107029.

[22] D. Hattori *et al.*, "Degradation of soil nutrients and slow recovery of biomass following shifting cultivation in the heath forests of Sarawak, Malaysia," *For Ecol Manage*, vol. 432, pp. 467-477, Jan. 2019, doi: 10.1016/j.foreco.2018.09.051.
[23] A. Murdjoko, F. Q. Brearley, A. Ungirwalu, D. A. Djitmau, and N. M. H. Benu, "Secondary Succession after Slash-and-Burn Cultivation in Papuan Lowland Forest, Indonesia," *Forests*, vol. 13, no. 3, Mar. 2022, doi: 10.3390/f13030434.

IOP Conf. Series: Earth and Environmental Science **1525** (2025) 012033 doi:10.1088/1755-1315/1525/1/012033

Investigation of Geomorphology and Petrology of the Central Point of Indonesia, Umpungeng area, Soppeng, South Sulawesi Province: Implications of Geological Evolution for Geological and Cultural Site

Asti Gindasari[1]* and Murni Sulastri[2]

[1] Department of Geological Engineering, Hasanuddin Univcrsitu, Makassar, Indonesia
[2] Geological Technology Study Program, Bandung Mining Energy Polytechnic, Bandung, Indonesia

*E-mail: astigindasari@unhas.ac.id

Abstract. Umpungeng (Posi Na Tanae) represents the Central Geographical Point of Indonesia, and its translation from the Bugis language denotes the "navel of the earth." This location has become a significant site in Indonesia, in terms of geography and culture. Identified as the central point of the Nusantara archipelago, it is situated in Umpungeng area, Soppeng, in South Sulawesi, Indonesia. This research aims to explore the geomorphological and petrological features of this area, along with the implications for regional geological evolution and its importance to local cultural practices. The geomorphological study will focus on the various landforms, geomorphic activities, and their interplay with the area. The geomorphological examination, bolstered by field observations and enhanced with petrographic laboratory data, seeks to deepen the understanding of the geomorphological and petrological dimensions of the Umpungeng locale. The results of the study indicate that the Umpungeng area possesses geomorphology characterized by steep and gently sloping hills, corroborated by petrographic assessments of local lithologies primarily made up of volcanic breccia. The upper layers, situated just below the topsoil, are encased in tuff or Quaternary volcanic sediments, rendering the region susceptible to instability and landslides. This research is anticipated to enrich the existing knowledge regarding the geological history of South Sulawesi, particularly emphasizing the tectonic and volcanic processes that have influenced its landscape.

1. Introduction

Soppeng Regency is located on the southwest arm of South Sulawesi Island, characterized by depression, consisting of Neogene and Quaternary sediments and Late Cenozoic granitoid volcanic formations [1]. With all of that geological characterized Soppeng Regency have opportunity to develop their tourism sector which focus on geological spot destination. To attract more visitors, the regency has prioritized infrastructure development and launched the "Visit Soppeng" campaign [2].

IOP Conf. Series: Earth and Environmental Science **1525** (2025) 012033 doi:10.1088/1755-1315/1525/1/012033

Looking ahead, Soppeng Regency will be proposed to become one of the next leading Geopark areas, particularly the Walannae Geopark (Bone, Soppeng, and Wajo) in South Sulawesi 2025. Soppeng regency, located around the Walanae Valley, is known for its abundant population of endemic bats, and the cultural-historical sites scattered like Umpungeng have proven to be one of the most important Indonesian archaeological sites. Due to its strategic position along a major fauna migration route from Asia to the archipelago has made this region a silent witness to the journey of human and animal evolution for millions of years [3,4,5,6].

The unique blend of natural, cultural, and scientific value makes Soppeng a strong contender for Geopark designation in 2025. The discovery of artefacts and paleolithic sites scattered around Soppeng strengthens the suspicion of cultural interactions in various regions.

Previous studies have shown that Soppeng and Pangkajene Caldera lava domes are interesting morphological forms around the Walanae Depression, while The Pangkajene Caldera lava dome is composed of trachyte and andesite [7, 8], and the lava dome units in Soppeng share similarities in tendency in the Lower Miocene formation period and lithology. The Soppeng Volcanic Rocks, predominantly composed of volcanic breccia and lava flows, exhibit sandy to lapilli tuff intercalations. Additionally, some rock units consist of basaltic pyroxene-basaltic leucite. Other studies utilizing deep geochemical analysis indicates that the magma from the Soppeng volcano is characterized by a shoshonitic affinity, typical of high-potassium volcanic suites. [9, 10].

Soppeng Regency presents a unique and complex volcanic landscape. The existence of a high-viscosity lava dome around Bulu Dua, Marioriwawo District, accompanied by pyroclastic deposits around Umpungeng, Lalabata Village, shows a varied and dynamic history of volcanic eruptions. From effusive eruptions forming lava domes to explosive eruptions producing pyroclastic deposits, the landscape indicates complex volcanic activity and changes in the characteristics of volcanic eruptions in the past.

The geomorphology of Umpungeng is in the form of land and hills, most of this area is still a forest area that needs to be preserved. The livelihoods of the surrounding community are the production of honey, palm sugar, and clove plants or secondary crops. Umpungeng Village is a fascinating area. Its sustainability is well maintained because it is considered a sacred place for the arajang ceremony (is a traditional Buginese ceremony, specifically referring to a ritual cleansing of royal heirlooms, passed down through generations) and was previously used for the inauguration of kings/datus [11]. However, the integration of these two characteristics into implication of geological evolution remains rare, thus this study offers a new perspective in this fied.

The objective of this research is to investigate the geomorphological and petrological characteristics of the Umpungeng area in Soppeng Regency, South Sulawesi Province, which is recognized as the central point of Indonesia. Through a detailed analysis of landforms and rock types, this study aims to reconstruct the geological history of the region and understand the processes that have shaped its present landscape. The geological features of this area are particularly significant, as Umpungeng represents one of the distinct morphologies associated with the Soppeng volcano.

In addition, this research explores the relationship between geological evolution and cultural development in Umpungeng, a site of major archaeological importance in Indonesia. By linking geological formations with archaeological evidence, the study provides insight into early human migration and settlement patterns in the Indonesian archipelago. The findings also serve as scientific support for promoting Umpungeng as a potential geosite within the proposed Walannae

Geopark. A deeper understanding of its geological and cultural value is essential for academic research, heritage conservation, sustainable tourism, and regional development. Overall, this study supports the integration of geodiversity and cultural heritage as a foundation for future sustainable development.

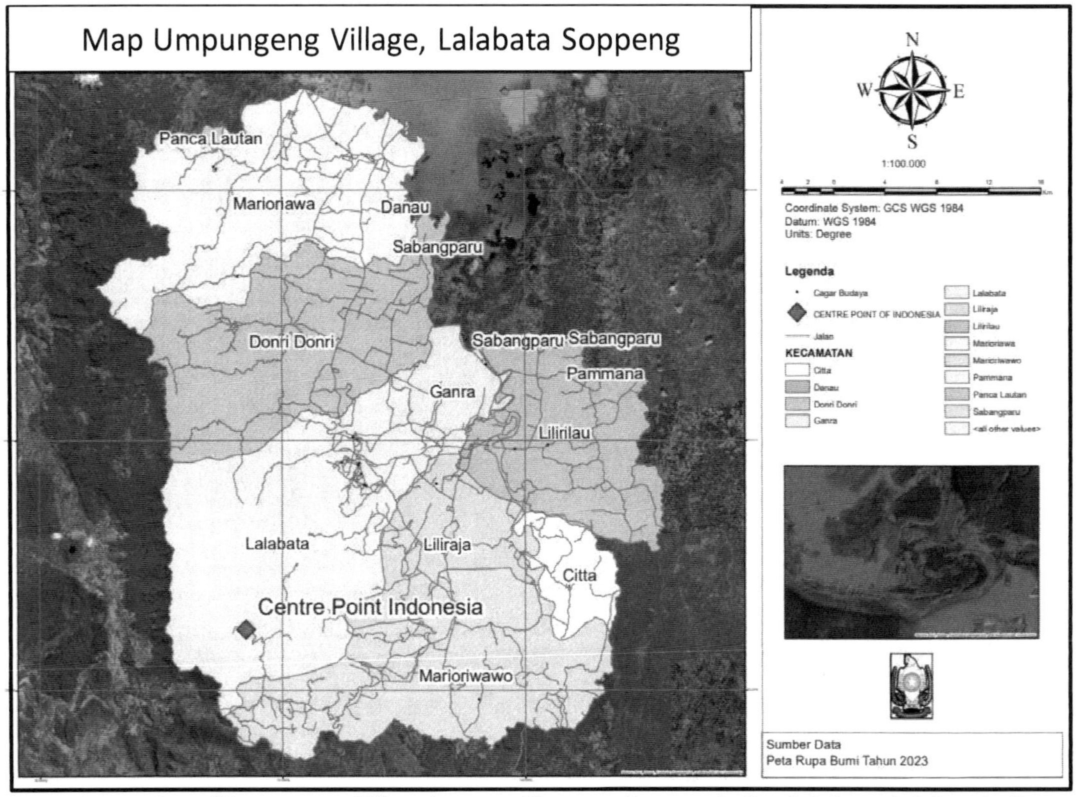

Figure 1. Location of Umpungeng Village in map of Soppeng Regency in South Sulawesi. Colour indicates the sub-districts in Soppeng Regency.

2. Samples and Methods

Samples for this study were collected by geological mapping around Umpungeng Village. Umpungeng Village is one of 10 villages/districts in the Lalabata Soppeng Regency. Geographically, Umpungeng Village is located between 119°48'31.4" East Longitude and -4°27'33.2" South Latitude and approximately 45 km or 1 hour 8 minutes from Soppeng City as shown in Figure 1. This area is intriguing as one of the iconic cultural sites in Soppeng Regency. Historically, it was considered the Center Point of Indonesia, predating the colonial era, and remains a sacred place for the local population.

This study employs a mixed-methods approach, combining deductive and inductive reasoning. Deductive reasoning was used to formulate hypotheses based on existing literature and established morphological concepts. Subsequently, inductive reasoning was applied to analyze field observations of morphological data [12] and lithological characteristics [13, 14], as well as petrographic data. This comprehensive analysis allowed for the identification of geomorphological units supported by mineral compositions and rock textures, leading to the formulation of conclusions. The flowchart of research is illustrated in Figure 2.

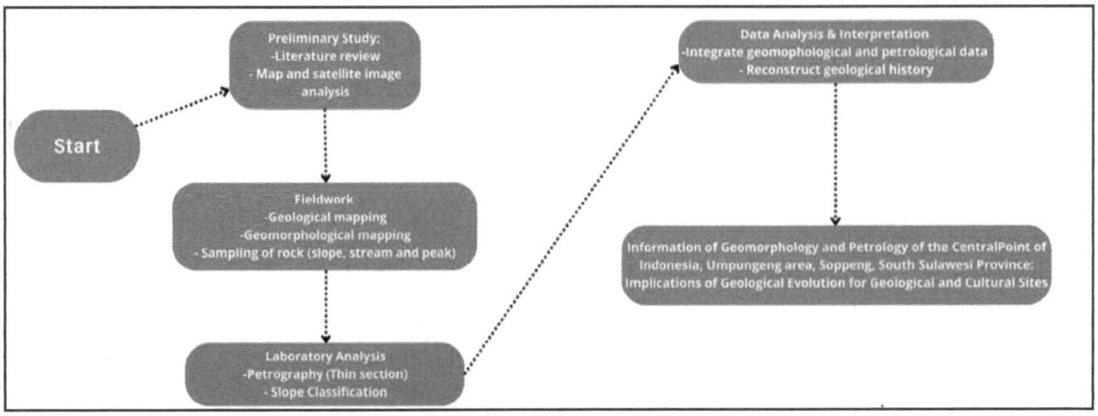

Figure 2. The flow chart of conducted research in this study.

3. Results and Discussion

3.1 Geomorphology of the Umpungeng area

Geomorphological processes are changes, both physical and chemical, that modify the Earth's surface [15]. They are divided into exogenous processes originating from outside the Earth (as destroyers) and endogenous processes originating from within the Earth's crust (as formers).

Endogenous is a process influenced by forces from within the Earth's crust, thereby changing the shape of the Earth's surface. Processes from within the Earth's crust include tectonics, intrusion, and volcanism. Exogenous is a process that is influenced by factors from outside the Earth's crust, such as climate (wind, rain, and temperature changes) and biological processes. Exogenous processes tend to gradually change the Earth's surface, namely the weathering of rocks into soil due to physical, chemical, and biological processes that end with erosion. This endogenous process is part of the morphotectonic study by discussing aspects of morphography (mountains, hills, and plains), morphometry (density of drainage patterns, contours, and slopes), morphogenetics (endogenous and exogenous factors), and the rocks that make it up.

Slope classification in the area is divided into slope categories (Table 1). The predominance of the Strongly Sloping and Moderately Steep classes in the Umpungeng area (Figure 3), (Figure 4), suggests that erosion processes have been significant in shaping the landscape. In the Umpungeng research area, the predominance of "Strongly Sloping" (15–30%) and "Moderately Steep" (30–70%) slope categories indicates a highly dynamic landscape shaped by both tectonic uplift and ongoing erosion. These steep slopes contribute to intense erosion processes, including surface runoff and mass movements such as landslides and rockfalls. As a result, they significantly affect soil development, vegetation distribution, land use potential, and the preservation of archaeological features. Furthermore, the steepness of the terrain suggests the presence of resistant volcanic rock types, such as breccia and dacite, which contribute to the rugged topography characteristic of the Umpungeng area.

Table 1. Slope Classification in Umpungeng.

Slope Class	Gradient (%)	Description	Implications
Flat	0–2%	Nearly level land	Minimal erosion; suitable for agriculture and construction.
Slightly Sloping	2–4%	Gently inclined terrain	Low erosion risk; ideal for farming and settlements.
Sloping	7–15%	Noticeable slope	Moderate erosion; can be cultivated with caution (e.g., terracing).
Strongly Sloping	15–30%	Moderately steep	Increased erosion risk; limited agricultural use without conservation.
Moderately Steep	30–70%	Steep hillside slopes	High erosion risk; unsuitable for most cultivation without strong soil management.
Steep	70–140%	Very steep land	Severe erosion potential; stable only if protected by vegetation or rock.
Very Steep	>140%	Cliff-like terrain	Extremely unstable; unsuitable for development, often exposed bedrock.

Figure 3. Some view in the Umpungeng Cultural Site

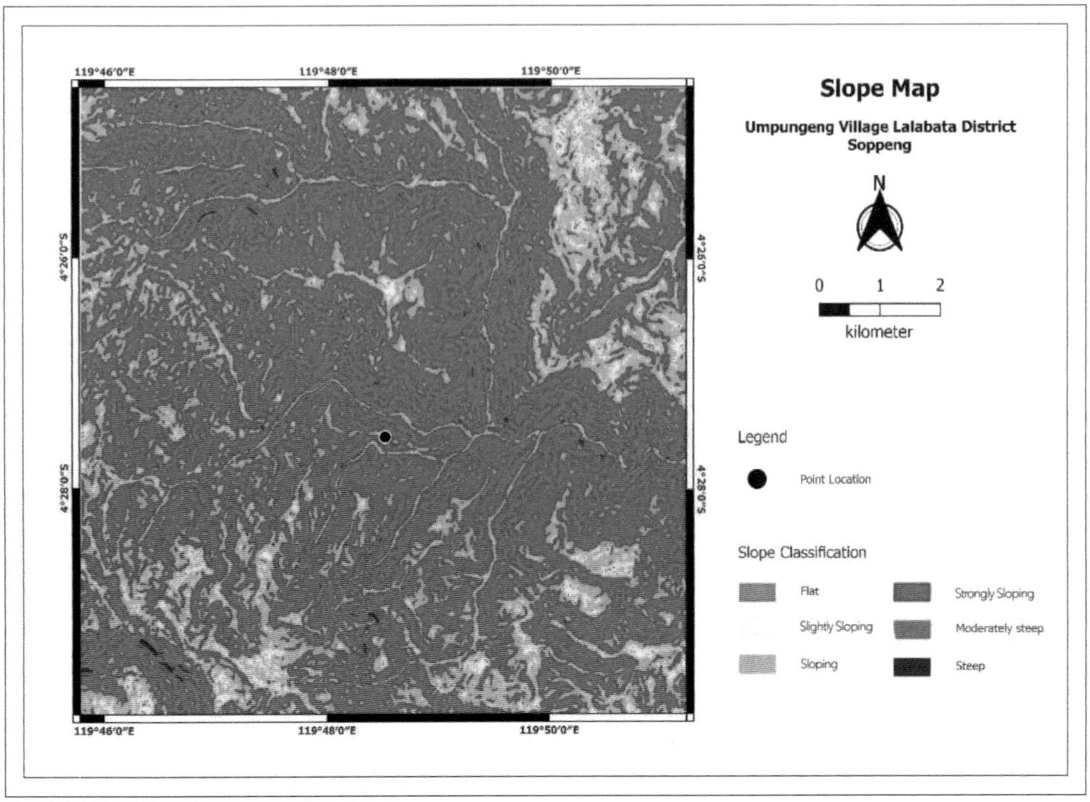

Figure 4. Slope Map in Umpungeng Village, Lalabata District. Soppeng

Figure 5. Aeral view and Morphology of Umpungeng Hills in Soppeng Volcanic Bulu Dua

3.2 Umpungeng area Units

Morphological (Fig. 4) and petrographic analysis strongly correlate landforms and rock types in the study area. The dominance of steep slopes (30-140%) indicates the presence of hard lithology, such as volcanic breccia and dacite (Fig.6-Fig. 10), which are widespread in the study area. Meanwhile, tuff is found dominantly at the peak, indicating the deposition of finer volcanic material in the area (Table 2).

Table 2. Mineral composition of rock samples.

Composition Minerals	Pyroclastic Breccia	Composition Minerals	Dacite	Composition Minerals	Victric tuff
Fragment		Groundmass		Groundmass	
Plagioclase	18%	Plagioclase	40%	Pyroxene	5%
Quartz	11%	Quartz	12%	Opaque	5%
Opaque Min	3%	Biotite	7%	Quartz	2%
Feldspar	5%	Opaque Min	2%	Matrix/Volcanic Glass	77%
Pyroxene	4%	Others/blank area	4%	Other/Blank area	12%
Rock Fragment/Lithic					
Matrix	45%				

3.2.1 Sample on Slope

Rock Type: Volcanic Breccia Rocks
Rock Name : - Piroclastic breccia
-Medium-verry fine breccia
Fragment: Dacite
Matrix : Vitric Tuff

Thin-section observations were carried out on rock samples of volcanic breccia at 10x Ocular magnification and 4x objective magnification. The samples has texture with size medium gravel grain (8 mm- 16 mm) to very gravelly smooth (2 mm- 4 mm), slightly sharp roundness-round, open packing, relationship between floating objects, type non-selective porosity in fabric (breccia), poorly sorted-very bad. The rock consists of quartz, plagioclase, feldspar, glass, pyroxene, sand, and opaque.

Figure 6. Thin section PPL (Plane Polarized Light) and XPL (Cross Polarized Light)

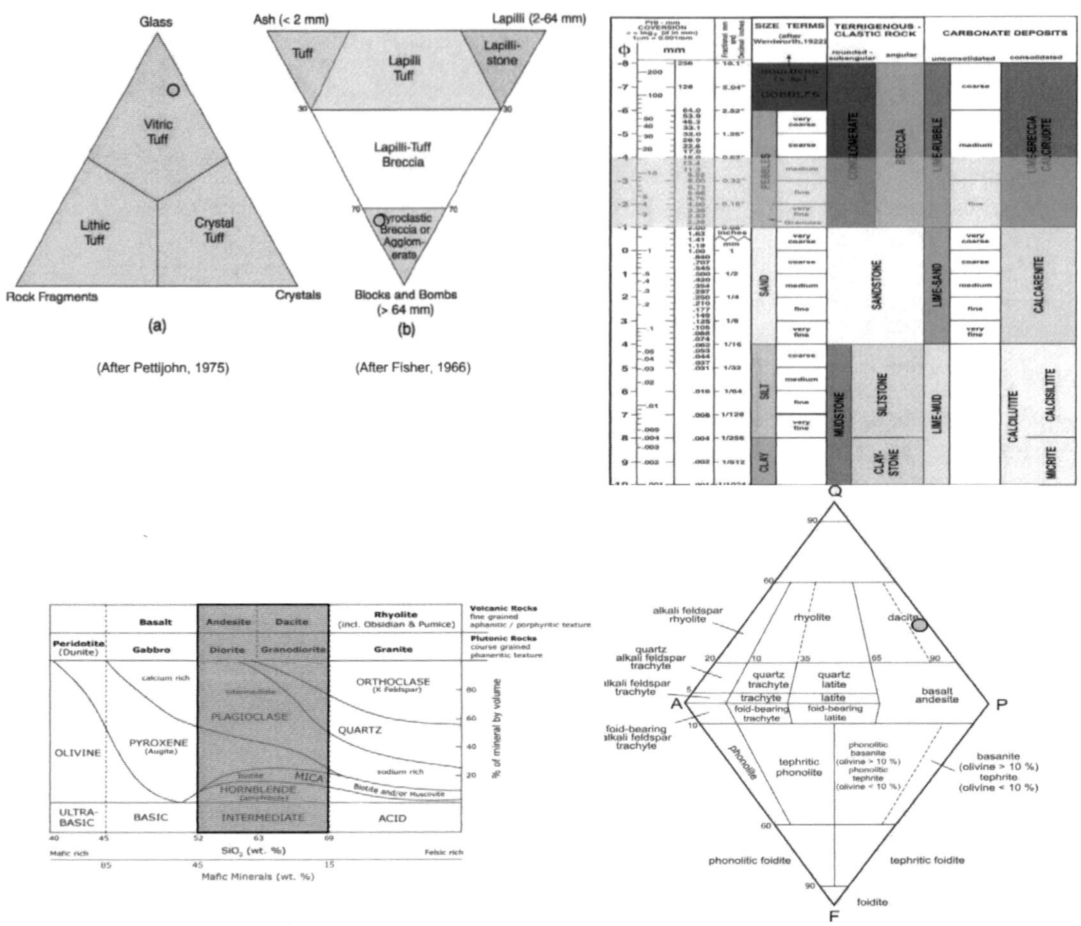

Figure 7. Result of Petrography analysis for the sample on slope [16, 17]

3.2.2 Sample on Stream

Rock Type : Igneous rock

Rock Name : Dacite

Thin section observations were made on igneous (Volcanic) rocks with 10x ocular magnification and 4x objective magnification. It has a general texture with a holocrystalline level of crystallinity, fine crystal size (<1mm), unbalanced granularity (porphyroophanitic), and euhedral-subhedral crystal shape. Special texture filter. The rock consists of plagioclase, opaque minerals, glass, biotite, feldspar, and quartz.

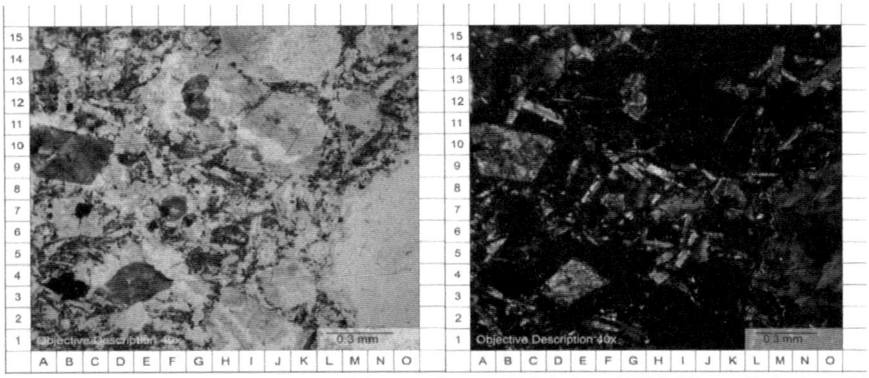

Figure 8. Thin section Dacite in PPL (Plane Polarized Light) and XPL (Cross Polarized Light)

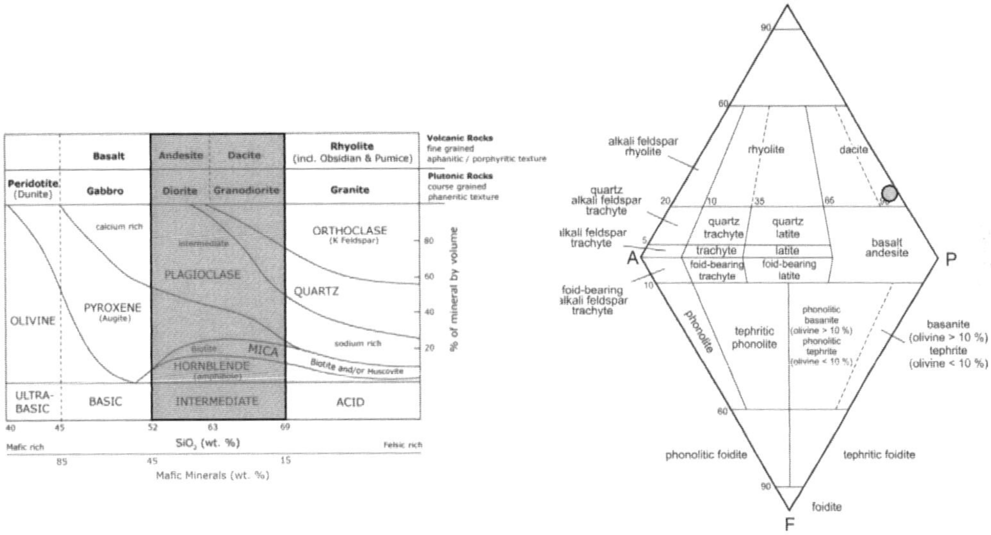

Figure 9. Result of Petrography analysis for the sample on stream [16, 17]

3.2.3 Sample on Peak/Top side

Rock Type : Pyroclastic

Rock Name : Vitric tuff

Tuff

Thin-section observations were carried out on pyroclastic rock samples with 10x ocular magnification and 4x objective magnification. It has a white color, medium sand grain size (0.25 mm- 0.5 mm) to very fine sand (0.0625 mm- 0.125 mm). Volcanic glass texture, poor sorting. The rock composition is glass, quartz, pyroxene, and opaque.

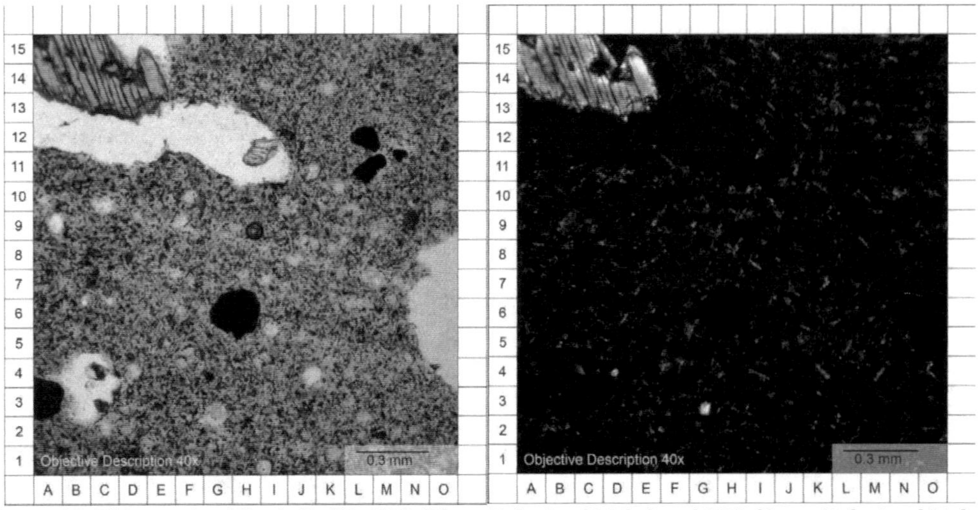

Figure 10. Thin section Victric Tuff in PPL (Plane Polarized Light) and XPL (Cross Polarized Light)

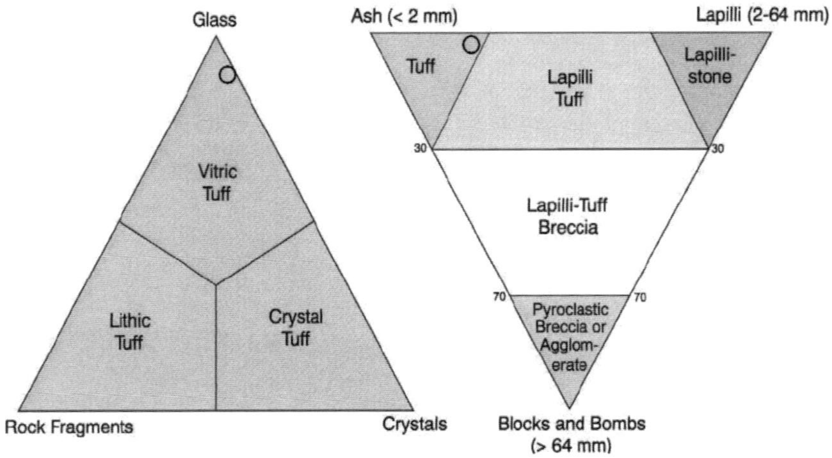

Figure 11. Result of Petrography analysis for the sample on peak/top side [16, 17]

4. Conclusion

In this study, we analyzed the Umpungeng geological site in Soppeng Regency, which is closely linked to complex geological features and a rich cultural history. Our analysis integrated geomorphological and petrological approaches to examine the correlation between landforms and rock types in the area. The geomorphological approach was used to classify slope gradients. Meanwhile, the petrological approach identifies characteristics and stages of rock weathering and erosion.

The results show that slope classification in the Umpungeng area is predominantly characterized by strongly sloping and moderately steep slopes, indicating that erosion processes have significantly shaped the topography. Geomorphological pattern also corresponds with hard lithologies, such as medium to very fine-grained volcanic breccia, dacite, and vitric tuff, which are widely distributed in the study area. Tuff deposits are found predominantly at the peak or summit areas, suggesting the accumulation of finer volcanic material. This distribution pattern reflects an interplay between volcanic activity and erosional processes, where past volcanic eruptions initially formed the landscape, and subsequent erosion gradually modified it, resulting in the present-day geomorphology.

References

[1] A. Maulana et al., "Origin and geodynamic setting of Late Cenozoic granitoids in Sulawesi, Indonesia," J. Asian Earth Sci., vol. 124, pp. 102–125, Jul. 2016, doi: 10.1016/J.JSEAES.2016.04.018.

[2] Peta Rupa Bumi Indonesia (Map of Indonesia). 2023 (in Indonesian)

[3] A. Hakim, 2018. "Sebaran Situs Paleolitik Di Tepi Aliran Sungai Walennae, Di Wilayah Bone Barat, Sulawesi Selatan," J. Walennae, vol. 16, no. 2, p. 85, doi: 10.24832/wln.v16i2.319.

[4] F. Umar, H. Winarso, and I. Kustiwan, , 2023. "'Bats are part of us': The role of bats in shaping community identity in Watansoppeng city, Indonesia," City Environ. Interact., vol. 21, no. July 2023, doi: 10.1016/j.2023.100139.

[5] Hasanuddin et al., 2016. Lembah Walennae : Lingkungan Purba dan Jejak Arkeologi Peradaban Soppeng. (in Indonesian)

[6] H. I. Fachroel Aziz et al, 2023. Atlas of Vertebrate Fossils Collection of Geological Museum. Badan Geology.

[7] Pettijohn, F.J., 1975. Sedimentary Rocks. 3rd Edition, Herper and Row, New York.

[8] Ria, Ulva., MS, Kaharuddin., A, Dadang, Nur, Irzal., 2013. Identification of Lava Domes in Pangkajene Caldera, South Sulawesi, Indonesia. Proceedings HAGI-IAGI joint Convention, Medan.

[9] Satya, Mutia Tri, Gatot Iwan Kurniawan, Muhammad Asdar, and Abdul Razak Munir. 2018. Swot Analysis of Marketing Tourism in Leading Tourism Object of Bandung and Soppeng. International Journal of Organizational Innovation 11 (2): 300–314.

[10] Sukamto, R. 1982. The geology of the Pangkajene and Watampone, Sulawesi, scale 1:250.000. Geological Research and Development Centre Bandung.

[11] Humairah, 2020. Umpungeng Community's Belief in Arajang (Case Study in Umpungeng Village, Soppeng Regency). Skripsi Thesis. UIN Alauddin. Makassar. (in Indonesian)

[12] Van Zuidam, R. A. (1983), Guide to Geomorphology Aerial Photographic Interpretation and Mapping, ITC, Enschede, The Netherlands.

[13] Kantu, Ribka R. 2022. Geokimia Batuan Kubah Lava Daerah Bulu Dua, Kecamatan Tanete Riaja Provinsi Sulawesi Selatan. Skripsi thesis, Universitas Hasanuddin. (in Indonesian)

[14] Fisher, R. V., 1966. Rocks composed of volcanic fragments and their classification. Earth Sci. Rev. https://doi.org/10.1016/0012- 8252(66)90010-9

[15] Tornbury, W. (1970). Principles of Geomorphology. New York: John Wiley and Sons Inc.

[16] Grabau, A.W. (1903) Paleozoic coral reefs. Geological Society of America Bulletin vol. 14, pp. 337 352.

[17] Grabau, A.W. (1904) On the classification of sedimentary rocks. American Geologist. vol. 33, pp. 228-247.

IOP Conf. Series: Earth and Environmental Science **1525** (2025) 012034 doi:10.1088/1755-1315/1525/1/012034

Analysis of Land Use and Land Cover Change (LULC) in Gowa Regency Using Multi-Temporal Satellite Imagery

Miftahul Jannah[1*], Samsu Arif[1]

[1] Department of Geophysics, Faculty of Mathemetics and Natural Sciences, Hasanuddin University, Makassar 90245, Indonesia

[*]E-mail: miftaalikwal@gmail.com

Abstract. This study investigates the dynamics of land use and land cover (LULC) change in Gowa Regency, South Sulawesi, using multitemporal Landsat imagery from 2001 (Landsat 7) and 2023 (Landsat 8). Four LULC classes; dense vegetation, sparse vegetation, built-up land, and water classified using a supervised Random Forest algorithm in Google Earth Engine, supported by NDVI and NDBI indices. The analysis employed the CROSTAB module in IDRISI for cross-classification of changes, revealing substantial shifts, particularly the conversion of vegetation to built-up land, signifying increased urban development. Accuracy validation using the Kappa index yielded a value of 0.9037, indicating a very good classification performance. These LULC transformations, primarily driven by urban expansion, population growth, and infrastructure development, have notable implications for ecological sustainability. The study underscores the importance of implementing sustainable land management policies to mitigate environmental impacts and support balanced regional development.

1. Introduction

The dynamics of land use and land cover (LULC) is one of the environmental changes that occur significantly, mainly due to the direct impact of various human activities [1]. Human-induced changes to LULC affect the physical, chemical conditions and distribution of living things on the Earth's surface [2]. Therefore, understanding the distribution and changes of LULC is a very important field of science [3].

Changes in LULC are influenced by various factors, such as urban growth or expansion, climate change, population increase, and government policies. Unplanned urbanisation and changes in LULC have a significant impact on the environment [4]. The identification of LULC is essential for evaluating environmental changes at global, regional and local levels. It becomes a fundamental element in sustainability research [5]. LULC analysis is an effective method to predict future urban development [6].

Gowa Regency is an area dominated by highlands, covering 72.26% of its area [7]. Rapid economic growth and uneven infrastructure development have led to significant LULC changes in the region. This study aims to analyse LULC changes in 2021 and 2023 in Gowa Regency. This analysis is expected to support environmental risk management strategies and support sustainable development in Gowa Regency.

IOP Conf. Series: Earth and Environmental Science **1525** (2025) 012034 doi:10.1088/1755-1315/1525/1/012034

2. Data and Method

2.1 Study Site

Administratively, Gowa Regency is located at the coordinates 119°21'54' East to 120°01'54' East and 5°5'20' LS to 5°34'10' LS. The administrative area of Gowa Regency is 1,883.33 km^2 (188,333 Ha) or equal to 3.01% of the total area of South Sulawesi Province and consists of 18 sub-districts.

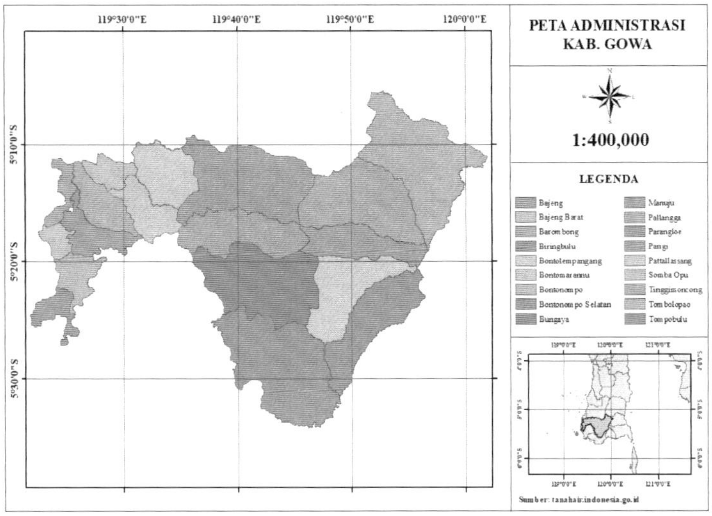

Figure 1. Administrative map of the research location.

2.2 Data Retrieval

In this study, 2001 Landsat 7 (ETM+) and 2023 Landsat 8 (OLI/TIRS) images were used, both of which have a spatial resolution of 30 meters. The image data was downloaded through Google Earth Engine (GEE). For land cover classification, 100 sample points each were taken for three classes: vegetation, built-up land and water. The sampling points were determined by utilizing band composite images as well as high-resolution satellite images available in GEE. The classification process was conducted using the Random Forest algorithm, resulting in land cover (LULC) maps for 2001 and 2023. Accuracy validation was conducted using the Kappa index, where a value of >70% indicates a good classification and is suitable for further analysis.

The classification data was converted to ASCII format and inputted into IDRISI software. The 2001 and 2023 LULC maps were then analyzed using the CROSSTAB method to see changes in land cover, both in terms of area and direction of transition. This information is an important basis in understanding the dynamics of land use and its impact on the environment in the study area.

3. Result and Discussion

The results of the analysis of LULC changes in Gowa Regency for the period 2001-2023 show a varied pattern of change as shown in 'Figure 2'.

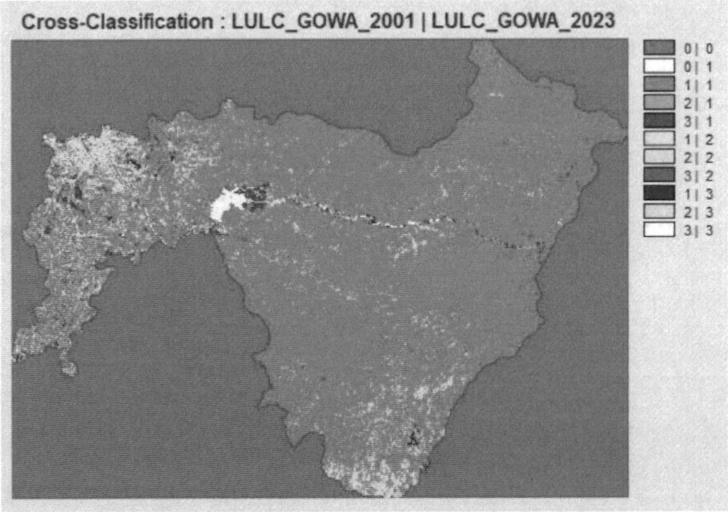

Figure 2. Cross classification of LULC in Gowa Regency 2001-2023. Code 1= Vegetation, 2= Built-up land, 3= Waters.

The results of the analysis of Land Use and Land Cover (LULC) change in Gowa Regency between 2001 and 2023 showed a significant change, especially in land conversion from vegetation to built-up land. The analysis was conducted using the CROSSTAB module of the IDRISI software, which produced 11 categories of LULC change, covering both land that changed between classes and land that remained in the same class. The most striking pattern of change was the conversion of vegetated land, such as forests or green areas, to built-up land (change code 1|2). This pattern indicates an increase in infrastructure and settlement development, especially in the central and eastern parts of the study area which have higher population density and proximity to the city activity center. Meanwhile, there are also areas that maintain their land cover types, such as vegetation that remains (1|1) and built-up land that does not change (2|2), reflecting the stability of land use in some areas.

In terms of area, vegetation land decreased from 15,551.37 hectares in 2001 to 14,791.41 hectares in 2023. In contrast, the area of built-up land increased significantly from 672.84 hectares to 1,517.22 hectares over the same period. In other words, over two decades, built-up land has more than doubled, largely replacing vegetated areas. There was also a small change from water to vegetation, most likely due to natural sedimentation or reclamation activities. Overall, however, the water area showed a decreasing trend, indicating the need for attention to water resources management in the region.

Furthermore, these changes not only show what is changing, but also encourage an understanding of why these changes are occurring and what their impacts are. Some of the main factors leading to these land use changes include population growth that drives the need for residential land, expansion of infrastructure and industrial activities, especially in peri-urban areas, as well as weak supervision of spatial planning implementation that causes land conversion to occur without adequate control. The impacts of these changes are quite evident, including decreased ecological functions such as loss of natural habitats and decreased biodiversity, increased local temperatures due to the urban heat island effect, and increased risk of disasters such as floods and landslides due to reduced water catchment areas and changes in land morphological characteristics.

The analysis of land cover change between 2001 and 2023 shows various categories of change with varying extent. This can be seen in the following table

Table 1. LULC in 2001 and 2023 in Gowa Regency.

Code	LULC	2001 (Ha)	2023(Ha)
1	Vegetation	15.55137	14.791,41
2	Built-up Land	672,84	1.517,22
3	Waters	309,78	226,71

4. Conclusion

The analysis revealed significant changes, particularly the conversion of vegetation to built-up land, reflecting increased urbanisation. While some areas maintain their land cover, others show active change dynamics. These changes have the potential to affect local ecosystems, so sustainable land management policies are needed to balance development with environmental conservation in Gowa Regency.

Acknowledgments

I would loke to thank my advisors Dr. Samsu Arif, M.Si. As well as my friends from the geophysics graduate program, Hasanuddin University.

References

[1] Risma, H Zubair & Paharuddin. (2019). Prediction of land use and land cover (LULC) changes using CA-Markov model in Mamuju Subdistrict. Journal of Physics: Conference Series. https//doi: 10.1088/17426596/1341/8/082033

[2] Mariane S. Reis, Maria Isabel, Escada & Luciano V. Dutra. (2018). Towards a Reproducible LULC Hierarchical Class Legend for Use in the Southwest of Pará State, Brazil: A Comparison with Remote Sensing Data-Driven Hierarchies. https//doi:10.3390/land7020065

[3] Md Shihab Uddin, Badal Mahalder, & Debabrata Mahalder. (2023). Assessment of Land Use Land Cover Changes and Future Predictions Using CA-ANN Simulation for Gazipur City Corporation, Bangladesh. https://doi.org/10.3390/su151612329

[4] Bubun Mahata, Siba Sankar Sahua, Archishman Sardab, Laxmikanta Ranaa & Mukul Maitya. (2024). Spatiotemporal dynamics of land use/land cover (LULC) changes and its impact on land surface temperature: A case study in New Town Kolkata, eastern India. http://www.keaipublishing.com/en/journals/regional-sustainability https://doi.org/10.3390/su151612329

[5] Md Mahadi Hasan Seyam, Md Rashedul Haque & Md Mostafizur Rahman. (2023). Identifying the land use land cover (LULC) changes using remote sensing and GIS approach: A case study at Bhaluka in Mymensingh, Bangladesh. https://www.sciencedirect.com/journal/case-studies-in-chemical-and-environmental-engineering

[6] Srishti Gaur, Ateeksha Mittala, Arnab Bandyopadhyay, Ian Holmanc & Rajendra Singha. (2020). Spatio-temporal analysis of land use and land cover change: a systematic model inter-comparison driven by integrated modelling techniques. International Journal Of Remote Sensing, VOL. 41, NO. 23, 9229–9255 https://doi.org/10.1080/01431161.2020.1815890

[7] DIKPLHD Kabupaten Gowa. (2019). Dokumen Informasi Kinerja Pengelolaan Lingkungan Hidup Daerah Kabupaten Gowa Tahun 2019. http://perpustakaan.menlhk.go.id/pustaka/images/docs/DIKPLHD-KABUPATEN-GOWA-2019.pdf

Mapping of Public and Private Green Space Distribution in Makassar City, South Sulawesi

Aswar Syafnur[1*], Syaeful Rahmat[2] and Mirnayani[3]

[1] Department of Geophysics, Hasanuddin University, Makassar 90245, Indonesia
[2] Study Program of Remote Sensing – Geographic Information System, Hasanuddin University, Makassar 90245, Indonesia
[3] Department of Geodetic Engineering, Gadjah Mada University, Yogyakarta 55281, Indonesia

*E-mail: aswar.syafnur@unhas.ac.id

Abstract. Green Open Space (GOS) is essential for creating sustainable, healthy urban environments by improving air quality, providing recreational areas, and supporting ecological balance. However, cities in Indonesia like Makassar, are facing a growing shortage of green spaces due to rapid urban expansion. While urban green space mapping has been studied globally, there is limited research specifically focused on the distribution of both public and private green spaces in Makassar. This lack of detailed data complicates efforts to plan and develop green spaces effectively, especially in line with national targets. This study aims to map and assess the availability of public and private green spaces in Makassar City using high-resolution satellite imagery and the Soil-Adjusted Vegetation Index (SAVI). The goal is to identify current green spaces and explore areas that could be developed to help reduce the city's green space deficit. Based on analysis in 2024, Makassar has only 12.39% GOS, well below the required 30%. Public green space makes up just 5.55%, and private space accounts for 6.85%. The city faces a shortfall of approximately 3,275.67 hectares, with districts like Biringkanaya and Tamalanrea experiencing the most significant gaps. The findings highlight the need for urgent action to expand GOS in Makassar. Potential solutions include converting agricultural land, developing roof gardens, and prioritizing green space in urban planning. These efforts are crucial to meeting regulatory targets and enhancing the overall sustainability and quality of life in the city.
Keyword: mapping, green open space, public, private

1. Introduction

Green Open Space (GOS) is essential aspect to achieve a healthy, comfortable, secure and sustainable environmental urban balance. Providing GOS means that maintain the land that can be used by vegetations to produce oxygen, absorb rain and become a water absorption area and organize fresh air circulation. The importance of GOS availability is seen in the United Nations making it one of the targets that need to be achieved by 2030 in Sustainable Development Goals (SDGs) point 11 by provision of green and public spaces [1]. In Indonesia, the significance of GOS has been recognized through national regulations, Law No. 26 of 2007 on Spatial Planning and Government, stipulate that urban areas should maintain at least 30% of total area as GOS [2].

IOP Conf. Series: Earth and Environmental Science **1525** (2025) 012035 doi:10.1088/1755-1315/1525/1/012035

Based on ownership, GOS is divided into public and private green space [2]. Public green space refers to open areas owned and managed by local governments for public use, while private green space refers to areas owned by institutions or individuals, with access limited to specific groups, such as gardens or private yards. Minimum proportion GOS is regulated by Government Regulation No. 21 of 2021 on the Implementation of Spatial Planning, consisting of 20% public green space and 10 % private green space [3].

However, despite these regulations, many urban areas, including Makassar, one of the biggest metropolitan areas in East Indonesia Region face significant deficits in GOS. According to Makassar Environmental Agency data, in 2023 Makassar was only 11.47% GOS [4]. The increasing demand for land for housing, infrastructure, and commercial development has further strained the availability of green spaces. In this context, understanding the spatial distribution of both public and private green spaces in Makassar is critical for urban planning and the sustainable development of the city.

Traditional methods for mapping GOS include manual field surveys, aerial photography, and remote sensing-based classification, each with its advantages and limitations. Satellite remote sensing-based GOS mapping is offering automation, extensive area coverage, and the ability to capture detailed spectral information. As advancements in high-resolution satellite imagery, this method has become increasingly popular for large-scale GOS monitoring, allowing for consistent temporal analysis and minimizing human error [5].

This study aims to map the spatial distribution of public and private GOS in Makassar City using advanced geospatial techniques. By leveraging high-resolution satellite imagery and the SAVI method, this research provides an accurate and detailed representation of existing green spaces and identifies potential areas for future development.

2. Material and Methods

2.1 Study area and data

The research was conducted in Makassar City (Fig.1), the capital city of South Sulawesi Province. This city covers a total area of 177,5 square kilometres and is located between 119°24'17'38" East Longitude and 5°08'06'19" South Latitude. It is divided into 15 districts and 153 subdistricts. The current estimated total population of the city is 1,474,393 people (6).

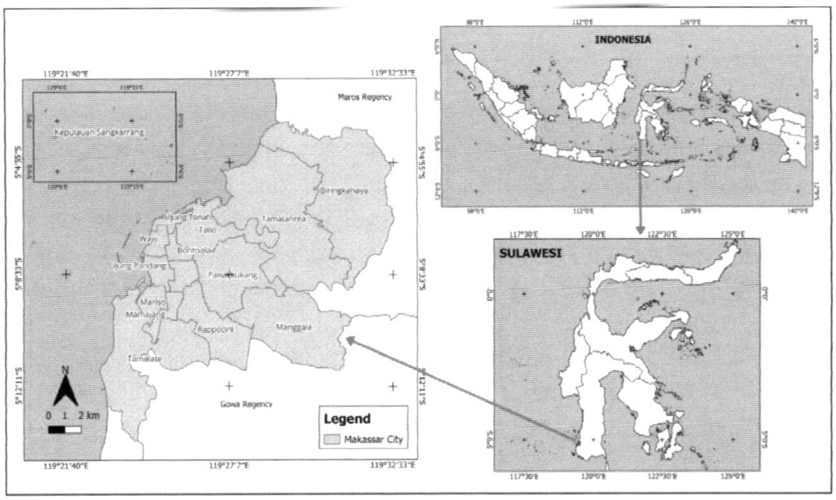

Figure 1. Geographical location of Makassar

This study utilized various data. Pleiades high resolution imagery with a multispectral resolution of 2 meters and a panchromatic resolution of 0.5 meters. The DEM data is taken from TERRASAR-X including a resolution of 5 meters by adding masspoint data from EGM2008 stereo-plotting. Furthermore, supplementary data included administrative boundaries and land cover maps from Makassar Spatial Planning (RTRW). Sentinel-2 MSI data was also used to improve the green space output. Specifically, Band 4 (Red) and Band 8 (NIR) from Sentinel-2 were utilized. The selection of Sentinel-2 imagery prioritizes the latest available data from 2024 with less than 10% cloud cover to ensure clarity and accuracy in the analysis.

2.2 Method

The method for mapping public and private green space in this research involves several stages. The process starts with imagery geometry correction through orthorectification to ensure spatial precision. This is succeeded by the on-screen digitization of the latest land cover maps using GIS Software to acquire comprehensive data. The next step entails identifying public and private green open spaces (Figure 2) based on field survey and Soil Adjustment Vegetation Index method to improve the precision of green area.

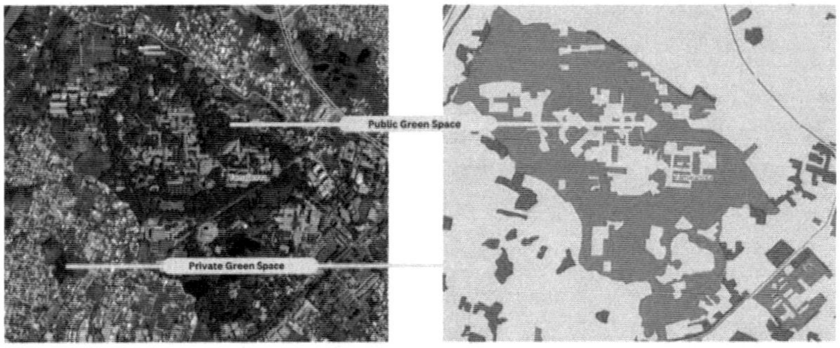

Figure 2. Ilustration of public and private green space classification

Utilizing remote-sensing derived spectral index using Soil Adjustment Vegetation Index (SAVI) for enchances identification of urban green space. SAVI is used to normalize the influence of soil background on various vegetation covers detected as green space. It was calculated using the formula 1. The L value used is 0.5 as a representation of the mid-value of vegetation cover [7].

$$SAVI = \frac{NIR-RED}{(NIR+RED+L)} \times (1 + L) \quad (1)$$

3. Result and discussion

Based on high resolution imagery interpretation, the results are presented in Figure 3. There are 23 land use classes, including: activity area, bare land, canal, drainage channel, dry land, fish pond, forest, golf course, GOS, lake, mining, mixed plantations, ocean, paddy field, paved surface, pond, reservoir, river, road, settlement, shrubs, swamp and waste disposal site. Approximately 9,985.95 hectares (56.23%) of Makassar City, are occupied by settlement and activity area. The large settlement areas, each over 1,000 hectares, are located in Biringkanaya (1,695.31 ha), Manggala (1,090.81 ha), and Tamalate (1,041.69 ha), while Tamalanrea recorded 953.21 ha. This enormous urban expansion substantially affects the availability of green open spaces. To rectify this, stringent restrictions are essential to guarantee that a minimum of 30% of residential zones is designated for green spaces, whether as thematic parks or private gardens.

Significant land uses in Makassar comprise paddy fields, fish ponds, and dry lands, covering 2,871.52 hectares (16.17%), 1,311.18 hectares (7.38%), and 1,333.27 hectares (7.50%), respectively. The regions, predominantly located in districts like Biringkanaya, Manggala, Panakkukang, Rappocini, Tallo, Tamalanrea, and Tamalate, are progressively being transformed for settlement or activity area, threatening their function in providing green spaces. It is essential for each zone to integrate adequate green spaces to reduce pollution and improve sustainability.

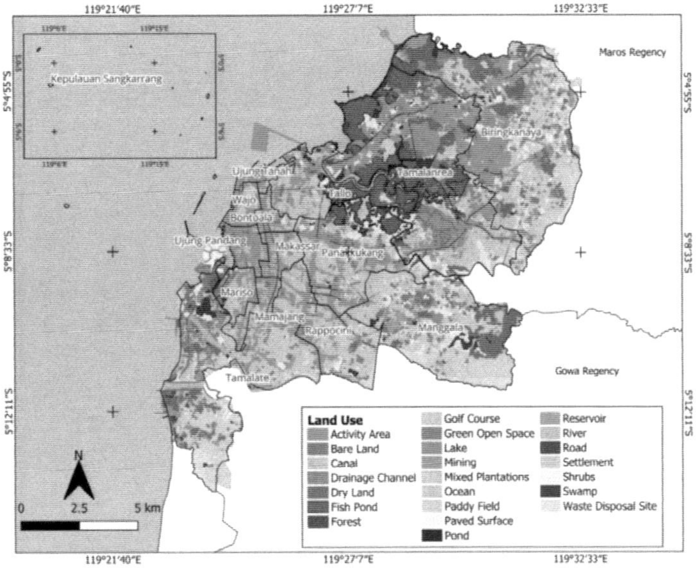

Figure 3. The spatial distribution of public and private green space in Makassar

The area of GOS of Makassar in 2024 is 2,200.40 ha or around 12.39% of the total area of Makassar City. The spatial distribution of public and private green open spaces of Makassar City in 2024 can be seen on the map in Figure 4. Private green open space was identified as 1,216.06 ha (6.85%) and public green open space as 984.34 ha (5.55%). The largest private green open

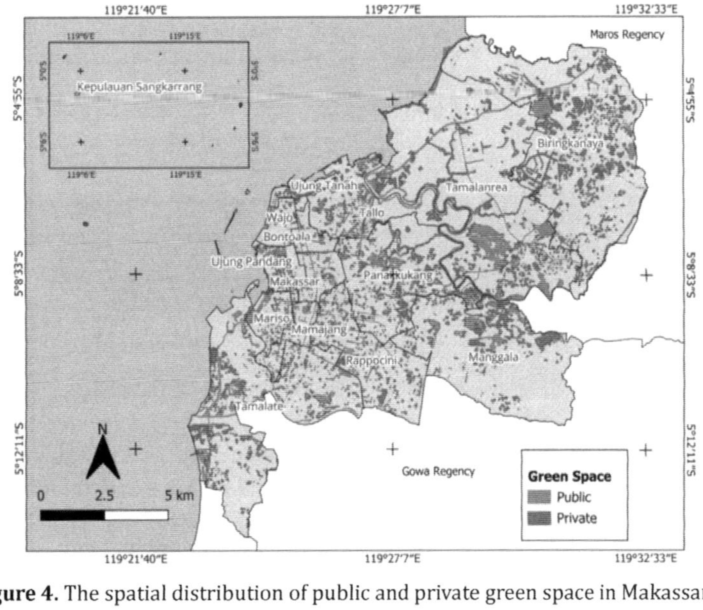

Figure 4. The spatial distribution of public and private green space in Makassar

space is in Biringkanaya District (373.18 ha), while the smallest is in Bontoala District (3.37 ha). Additionally, the largest public green open space is in Tamalanrea District (326.04 ha), while the smallest is in Kepulauan Sakarrang District (0 ha).

The largest proportion of green open space to the area of each district is in Mariso with an area of 54.47 ha (18.87%), while the smallest is in Wajo district with an area of 8.41 ha (4.12%) of the total area of the district. Furthermore, the highest public green open space is found in Mariso district with an area of 44.57 ha (15.44%) while the smallest is in Kepulauan Sangkarrang district (0%). For private green open space, the highest area is in Biringkanaya Sub-district 373.18 ha (10%) and Kepulauan Sangkarrang Sub-district 11 ha (12.77%) and while the smallest is in Bontoala Sub-district 3.37 ha (1.94%).

Table 1. Green Open Space in Makassar City in 2024

Districts	Public Green Space (ha)	Private Green Space (ha)	Total Area (ha)	Percentage (%)
Biringkanaya	139.80	373.18	512.98	2.89
Bontoala	9.35	3.37	12.72	0.07
Kepulauan Sangkarrang	0.00	11.00	11.00	0.06
Makassar	4.30	8.88	13.18	0.07
Mamajang	7.92	11.38	19.30	0.11
Manggala	101.55	150.53	252.08	1.42
Mariso	44.57	9.90	54.47	0.31
Panakukang	99.82	127.79	227.61	1.28
Rappocini	36.16	70.58	106.74	0.60
Tallo	49.30	67.00	116.30	0.65
Tamalanrea	326.04	202.95	528.99	2.98
Tamalate	132.00	149.24	281.24	1.58
Ujung Pandang	26.90	11.93	38.83	0.22
Ujung Tanah	3.43	13.15	16.58	0.09
Wajo	3.20	5.21	8.41	0.05
Total	984.34	719.23	1603.88	12.39

Source: Researchers analysis

According to the Ministry of Public Works regulation No. 5 of 2008, Makassar should possess 5,304.21 hectares of GOS, reaching 30% of the city's total area. The existing GOS encompasses just 2,028.53 hectares, representing 12.47%, resulting in a substantial deficit of 3,275.67 hectares (17.61%). Biringkanaya district exhibits the greatest demand for supplementary GOS, totalling roughly 705.3 hectares, whereas Kepulauan Sangkarrang has the least requirement. The

IOP Conf. Series: Earth and Environmental Science **1525** (2025) 012035 doi:10.1088/1755-1315/1525/1/012035

disparities are intricately linked to the dimensions of each district, with larger regions such as Tamalanrea, Tamalate, Manggala, and Panakkukang requiring more green space than their smaller counterparts. Based on area, the need of public and private green space in each district are presented in Figures 5 and 6.

Furthermore, no district in Makassar meets the 30% GOS criterion. Mariso possesses the

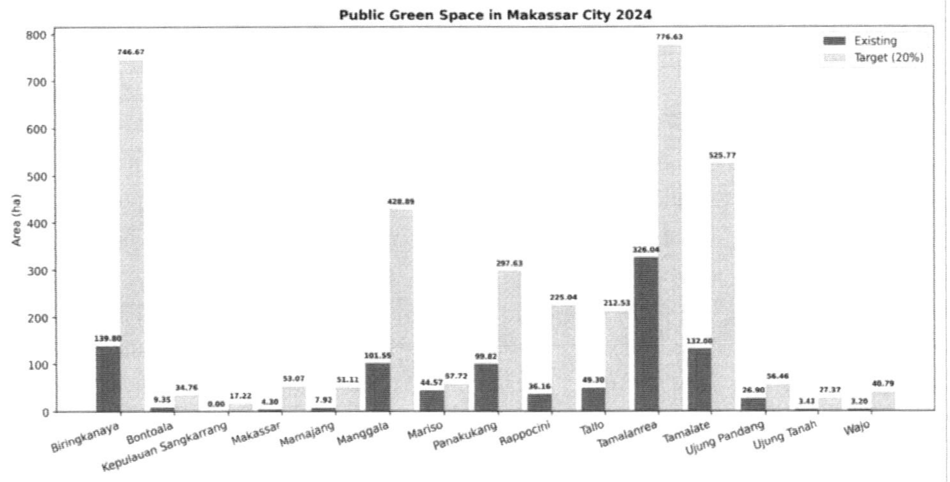

Figure 6. Public green space distribution of Makassar in 2024

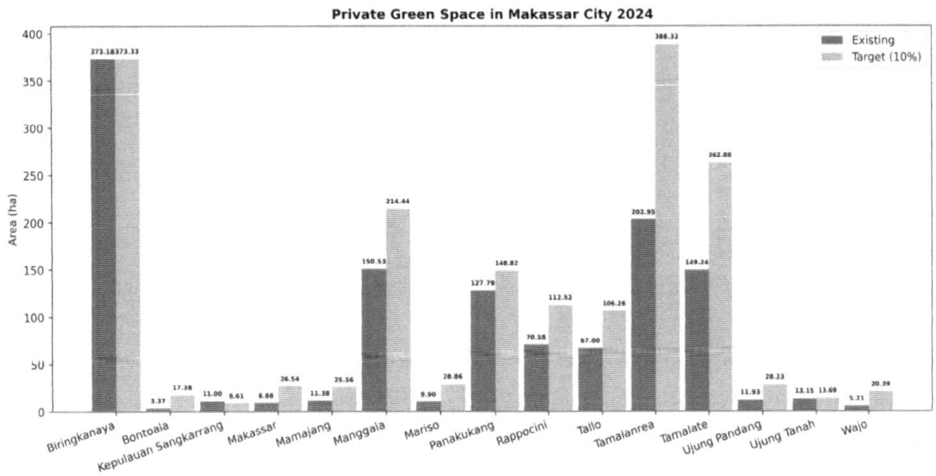

Figure 5. Private green space distribution of Makassar in 2024

highest percentage of green space at 19.31%, whereas Wajo has the lowest at merely 3.82%. Larger districts such as Tamalanrea, Biringkanaya, and Tamalate naturally show a higher demand for GOS, as their bigger areas require more green coverage. In contrast, smaller districts possess relatively minimal requirements. These variances underscore the necessity of customising green space development to the distinct conditions and needs of each district.

The potential for developing green open spaces in Makassar City can refer to the Regulation of the Minister of Home Affairs Number 1 of 2017 concerning the Arrangement of Green Open Spaces in Urban Areas. Where urban agricultural land and roof gardens are included in the types of green open spaces. Specifically in Makassar City, agricultural land has been identified in the

IOP Conf. Series: Earth and Environmental Science **1525** (2025) 012035 doi:10.1088/1755-1315/1525/1/012035

form of fields covering an area of 1333.27 ha and rice fields covering an area of 2871.52 ha. Shrubs are also potential land that can be developed for green open spaces. So that the total area of fields, bushes and paddy field that can be developed to meet the needs of 30% of green open space in Makassar City is 4375.14 ha.

Makassar City holds significant potential for developing green open spaces, incorporating diverse forms of green spaces, such as urban agricultural lands and innovative solutions like roof gardens, as integral components of sustainable urban planning. Leveraging this potential would not only help Makassar meet the regulatory requirement but also enhance the city's ecological balance and provide valuable public spaces for recreation and community activities.

4. Conclusion

This study finds that the critical shortage GOS in Makassar City, with only 12.39% of the city's area designated as GOS compared to the national target of 30%. Using combination of high-resolution satellite imagery, field survey and the SAVI method enabled a comprehensive analysis of GOS distribution, revealing significant disparities across districts and identifying potential areas for development, such as agricultural land and shrubs.

To bridge the GOS deficit of 3,275.67 hectares, strategic interventions and innovative solutions, including urban agricultural land use and shrubs, are essential. These efforts would not only meet regulatory requirements but also enhance the city's ecological balance, air quality, and community well-being. Policymakers and urban planners must prioritize GOS expansion and adopt sustainable land-use practices to ensure Makassar's environmental and urban resilience. Future research should focus on integrating community participation and advanced technologies for more effective GOS management and development.

References

[1] United Nations. (2015). *Transforming Our World: The 2030 Agenda for Sustainable Development.* http://www.un.org/

[2] Law No. 26 of 2007 on Spatial Planning. (2007). Republic of Indonesia. Access on https://peraturan.bpk.go.id/Details/161851/pp-no-21-tahun-2021

[3] Government Regulation No. 21 of 2021 on the Implementation of Spatial Planning. (2021). Republic of Indonesia. Access on https://peraturan.bpk.go.id/Details/161851/pp-no-21-tahun-2021

[4] Makassar Environmental Agency. (2023). Final Report on Green Open Space Identification Activities and Land Quality Index.

[5] Karan, S. K., Borchsenius, B. T., Debella-Gilo, M., & Rizzi, J. (2025). Mapping urban green structures using object-based analysis of satellite imagery: A review. *Ecological Indicators*, **170**, 113027.

[6] BPS – Statistics of Makassar Municipality. (2024). Makassar Municipality in Figure 2024.

[7] Sashikkumar, M. C., Selvam, S., Karthikeyan, N., Ramanamurthy, J., Venkatramanan, S., & Singaraja, C. (2017). Remote sensing for recognition and monitoring of vegetation affected by soil properties. *Journal of the Geological Society of India*, **90**, 609-615.

AUTHOR INDEX